Drinking Water Treatment 4

Drinking Water Treatment 4

Membranes Applied to Drinking Water and Desalination

Kader Gaid

ISTE

WILEY

First published 2023 in Great Britain and the United States by ISTE Ltd and John Wiley & Sons, Inc.

ISTE Ltd
27-37 St George's Road
London SW19 4EU
UK

www.iste.co.uk

John Wiley & Sons, Inc.
111 River Street
Hoboken, NJ 07030
USA

www.wiley.com

Library of Congress Control Number: 2022947257

British Library Cataloguing-in-Publication Data
A CIP record for this book is available from the British Library
ISBN 978-1-78630-786-6

Contents

Chapter 17. Microfiltration and Ultrafiltration 1
17.1. UF and MF: cut-off 2
17.2. UF and MF: materials 6
17.2.1. Cellulose acetate. 6
17.2.2. Polypropylene 7
17.2.3. Polyacrylonitrile. 7
17.2.4. Polyether sulfone/polysulfone. 7
17.2.5. Polyvinylidene fluoride 8
17.3. UF and MF: membrane types 8
17.4. UF and MF: implementation of membranes under pressure 11
17.4.1. Horizontal-vertical configuration. 13
17.4.2. Submerged membranes. 17
17.5. Filtration modes: frontal or tangential 19
17.5.1. Batch operation: filtration-backwash. 21
17.5.2. Filtration direction 21
17.6. Sizing parameters: membrane selection 22
17.7. Sizing parameters: horizontal or vertical configuration 25
17.8. Sizing parameters: flow 25
17.8.1. Instantaneous flow and net flow 26
17.8.2. Transmembrane pressure. 31
17.8.3. Resistance 33
17.8.4. Permeability 33
17.8.5. Principle of the calculation of the membrane surface
and water losses. 35
17.8.6. Pre-filters 37
17.9. Operating parameters 37
17.9.1. Evolution of the permeability 37
17.9.2. Clogging 38

17.9.3. Frequency and conditions of hydraulic and chemical backwashing. 46
17.9.4. Frequency and conditions of CIP 53
17.9.5. Membrane integrity 56
17.10. MF and UF's place in a treatment process 62
17.10.1. Turbidity and SS 63
17.10.2. TOC (and UV254) 63
17.10.3. Algae 63
17.10.4. Iron and manganese 63
17.11. Combination of coagulation and UF membranes 71
17.12. Combination of PAC and UF 75
17.13. Performance and guarantees 76
17.13.1. Turbidity 76
17.13.2. Supplier warranty on the life of the membranes 86
17.14. Advantages of MF and UF 87
17.15. Veolia's experience 87
17.16. Appendix: sheets 92
17.17. References 107

Chapter 18. Nanofiltration and Reverse Osmosis 111

18.1. Membranes 112
18.1.1. Materials 112
18.1.2. Membrane element configurations 115
18.2. Principles of operation and separation 119
18.2.1. Conceptual principle 119
18.2.2. Molecular weight cut-off 124
18.3. Treatment process including high-pressure membranes and parameters to be considered 127
18.3.1. Particulates and SS 127
18.3.2. Particle count 128
18.3.3. Conductivity 128
18.3.4. The SDI or MFI: clogging indices 128
18.3.5. The SDI 128
18.3.6. The MFI 130
18.3.7. Salts and metals 131
18.3.8. Biological clogging 132
18.3.9. Undesirable substances 133
18.3.10. Limit values of compounds at the inlet of high-pressure membranes 133
18.4. Sizing parameters 134
18.4.1. Temperature 134

18.4.2. Implementation configuration. 134
18.4.3. Calculation of the osmotic pressure 138
18.4.4. Mass flow diagram 139
18.4.5. Salt passage 140
18.4.6. Concentration factor 140
18.4.7. Hydraulic pressure loss. 140
18.4.8. Pressure tubes and number of modules per tube 141
18.5. Chemical conditioning of pre-treated water 143
18.5.1. Calculation of saturation indices and antiscalant dosage 143
18.5.2. Choice and implementation of the antiscalant 146
18.5.3. pH adjustment at the membrane inlet 147
18.5.4. Choice and application of the acid 148
18.5.5. Influence of sulfates 148
18.6. Design and implementation 148
18.6.1. Pre-treatment. 148
18.6.2. Treatment processes 149
18.6.3. Membrane station 152
18.6.4. Post-treatment 158
18.6.5. Cleaning units in place 159
18.7. Functional and operating parameters 162
18.7.1. Basic principles 162
18.7.2. Permeability (Lp) 166
18.7.3. Longitudinal pressure drop (ΔPfc) 166
18.7.4. Hydraulic resistance 167
18.7.5. Energy. 167
18.7.6. SDI 168
18.7.7. Chemical cleaning. 168
18.7.8. The fate of concentrates and used washing solutions 168
18.7.9. Methods for assessing the impact of concentrate discharges in the natural environment. 171
18.8. High-pressure membrane performance 174
18.8.1. Organic matter. 175
18.8.2. Pesticides, drug residues, endocrine disruptors and industrial residues 176
18.8.3. Various toxic and undesirable substances. 177
18.8.4. Salts 177
18.8.5. Micro-organisms 178
18.8.6. Overall performance 179
18.9. Lifetime warranties 179
18.10. Parameters affecting the performance of NF membranes 180
18.10.1. Taking clogging into account 181
18.11. Monitoring and control parameters: standardization of raw data . . . 182
18.12. Veolia's experience: examples of treatment processes 184

18.12.1. Surface water No. 1 184
18.12.2. Surface water No. 2 188
18.12.3. Groundwater No. 1 193
18.12.4. Groundwater No. 2 195
18.13. References 200

Chapter 19. Desalination by Reverse Osmosis 205

19.1. Characterization of the water to be treated. 205
19.1.1. Physical characteristics. 207
19.1.2. Chemical composition: ionic content 214
19.1.3. Chemical composition: organic substances 224
19.2. Fields of application 234
19.3. Operating principle of RO 234
19.4. The membranes used in desalination 237
19.5. Sizing parameters 238
19.5.1. Flow 239
19.5.2. Concentration polarization. 240
19.5.3. Conversion rate 240
19.5.4. Passage rate and rejection rate in salts 241
19.5.5. Influence of the temperature. 242
19.5.6. Determining the number of modules and pressure tubes 243
19.6. Implementation 244
19.6.1. Membranes 244
19.6.2. Pressure tube 245
19.6.3. Pass 246
19.7. Pre-treatment 246
19.7.1. Pre-treatment selection 246
19.7.2. Pre-treatment systems 250
19.8. Pre-chlorination 254
19.8.1. Pre-chlorination and development of micro- and macro-organisms 254
19.8.2. Implementation of chlorination 257
19.8.3. pH adjustment 263
19.8.4. Direct filtration 263
19.8.5. Chemical conditions of implementation. 273
19.8.6. Flotation 276
19.8.7. Settling 281
19.8.8. Membranes (UF and MF) 282
19.8.9. Conclusions on pre-treatment with UF membranes 291
19.9. Energy consumption. 292
19.9.1. Energy consumption without recovery 292

19.9.2. Energy consumption with recovery 293
19.9.3. Hydraulic exchanger systems 293
19.10. Operating parameters 305
19.10.1. Relationship between conductivity and salt concentration 305
19.10.2. Controlling RO membrane clogging 306
19.11. Performance of RO membranes used in desalination. 307
19.11.1. Boron removal 307
19.12. Post-treatment. 315
19.12.1. Indicators characterizing the aggressiveness or corrosiveness of the water. 317
19.12.2. Application to desalinated water 320
19.12.3. Treatments 322
19.13. Monitoring and control parameters 336
19.13.1. Standardization of raw data 336
19.13.2. Bromates 338
19.14. Veolia's new processes applied to seawater desalination 339
19.14.1. Flotation with the Spidflow® process 339
19.14.2. Spidflow® filter process applied to seawater desalination 344
19.14.3. BiopROtector. 349
19.14.4. Barrel (SIDEM Veolia). 353
19.14.5. Hiprode 355
19.15. Packaged solutions in desalination. 358
19.16. Veolia's experience (HP membranes). 360
19.17. References 374

Index 385

Summaries of other volumes 387

17

Microfiltration and Ultrafiltration

A membrane is a semi-permeable barrier that allows the selective separation of compounds that are present in water, under the effect of a pressure gradient. There are a large number of membranes on the market with different characteristics.

This chapter applies to ultrafiltration (UF) and microfiltration (MF) membranes that are used in drinking water, but can also be used for process water and tertiary wastewater filtration applications.

MF and UF are two techniques designed for solid/liquid separation through the removal of particles and macromolecules. MF is essentially aimed at the retention of particles between 0.1 and 5 µm in size, while UF is aimed at particles between 0.1 and 0.01 µm in size. They constitute barriers for suspended solids, colloids, bacteria and parasites. UF also eliminates viruses. They use working pressures of 0.2–1 bar for MF and in the range of 2–5 bar for UF.

There are different types of membranes, classified according to the type and size of the constituents they separate from the water. The best choice of membrane is not the most selective one, but the one that is sufficiently selective for the objectives set, and which offers the best operating and investment costs.

UF and MF membranes are porous membranes made from synthetic or ceramic polymers.

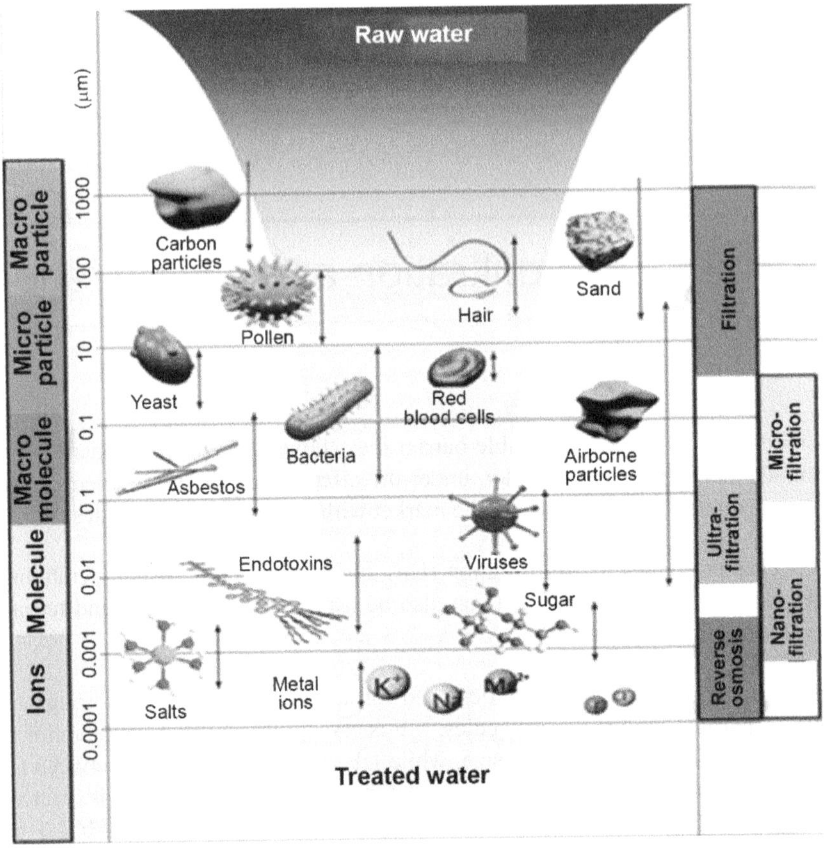

Figure 17.1. *Classification of different species according to their size and membrane cut-off. For a color version of this figure, see www.iste.co.uk/gaid/watertreatment4.zip*

17.1. UF and MF: cut-off

The cut-off is the molar mass of the smallest compound retained at 90% by the membrane. The lower the cut-off, the more the membrane is able to retain small molecules or colloids. Conversely, the higher the cut-off, the less the membrane is able to retain small molecules. In general, the cut-off is expressed in pore size (µm) for MF, while for UF, it is expressed in molecular weight in Dalton (or molecular weight cut-off [MWCO]), which is equivalent to 1 g per mole. The Dalton is equivalent to one-twelfth of the mass of a carbon atom 12, expresses the mass of a hydrogen atom and is equal to 1.66×10^{-27} kg. Thus, a protein with a molar mass equal to 75,000 $g{\cdot}mol^{-1}$ will have a mass equal to 75,000 Da or 75 kDa.

MF and UF are found in a fairly wide range of pore sizes: from 0.1 to 1 μm. Typically, MF membranes used in clean water (drinking water, process water) have a pore size between 0.05 and 0.2 μm, while UF membranes used for these applications have a pore size close to 0.01 to 0.03 μm or a MWCO of 80–200 kDa.

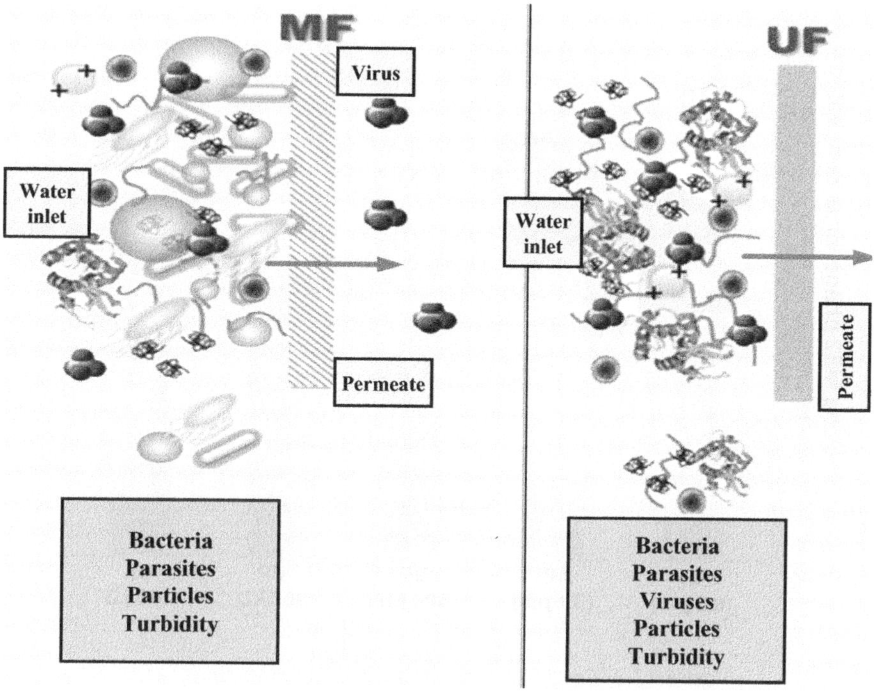

Figure 17.2. *Cut-off point for microorganisms with filtration, microfiltration and ultrafiltration. For a color version of this figure, see www.iste.co.uk/gaid/watertreatment4.zip*

MF and UF are often distinguished by the log virus removal. Using the MS2 phage as a reference, a membrane is accepted as UF when the log MS2 removal is at least 4 log (Figure 17.3).

UF has two characteristics that make it a particularly suitable unit operation for water treatment:

– On the one hand, the size of the pores allows the retention of bacteria and viruses while letting the dissolved salts pass through. This makes it possible to combine food safety and preservation of the mineral balance, both of which are essential for water intended for consumption.

– On the other hand, UF uses relatively low transmembrane pressures (TMPs) compared to nanofiltration or reverse osmosis. Consequently, this limits the operating costs associated with the energy consumption, related to the TMP.

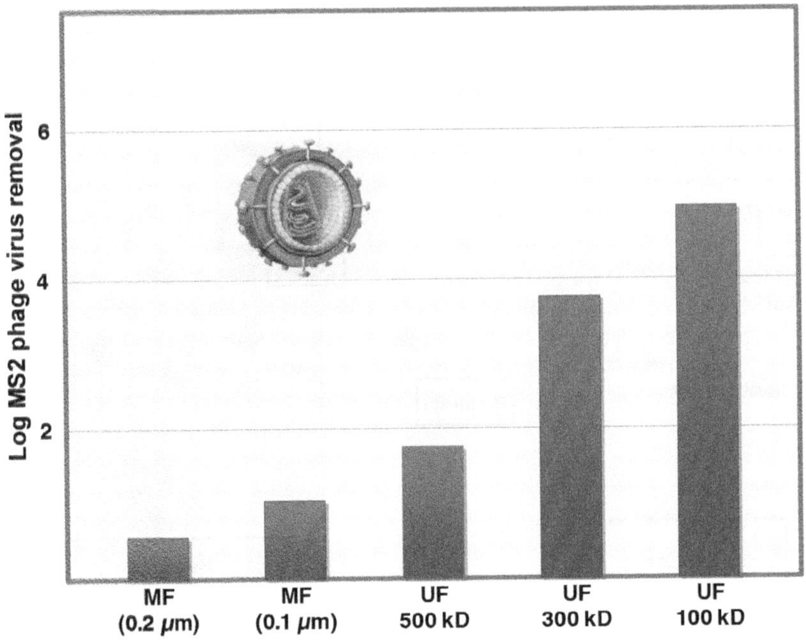

Figure 17.3. *Log MS2 removal according to the cut-off of MF and UF membranes. For a color version of this figure, see, www.iste.co.uk/gaid/watertreatment4.zip*

UF membranes with a low cut-off are mostly used in industrial applications (such as the food industry, ultra-pure water). However, there are some applications in drinking water where UF membranes close to nanofiltration are used to treat colored water, especially on units of small size.

Even though membranes are characterized with cut-off values (in μm or Da) in the data sheets given by the suppliers, a membrane does not have a uniform pore size, but rather a pore size distribution that is more or less tight. The pore size distribution is therefore as large as the advertised nominal pore size.

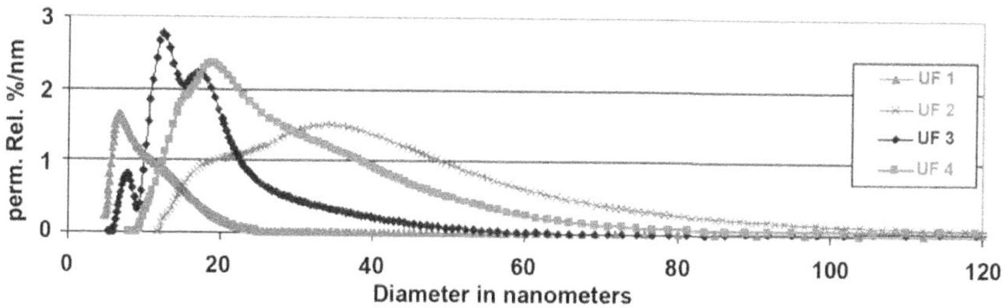

Figure 17.4. *Example of pore size distribution curves (source VERI Veolia). For a color version of this figure, see www.iste.co.uk/gaid/watertreatment4.zip*

The pore diameter can be obtained from the membrane cut-off (MWCO) and the following relationship:

$$dp = \frac{1.13\ \times k\ 10^{10}}{\pi\mu(\text{MWCO})^{-0.48}}$$

With:

– dp: pore diameter (m);

– k: Boltzmann constant ($1.38064852 \times 10^{-23}$ $m^2 \cdot kg \cdot s^{-2} \cdot K^{-1}$);

– μ: dynamic viscosity (1.31×10^{-3} $kg \cdot m^{-1} \cdot s^{-1}$);

– MWCO: membrane cut-off for a temperature of 25°C (298°K).

A membrane with a MWCO of 150,000 Da has a pore diameter of 11.5 nm. In contrast, an MF membrane with an MWCO of 1,000,000 Da has an estimated pore diameter of 100 nm. Veolia has a research and development center (VERI) which includes, among its many activities and applications, a technical expertise and assistance service specific to membranes. As such, the membranes on the market are listed, analyzed and tested for their physical, chemical and biological properties, as well as for their associated performance. Figure 17.4 is an example of the pore distribution curves on several membranes on the market. The definition of UF versus MF for a drinking water application is often taken in terms of virus removal.

Membranes	Operating modes	Types of separation	Cut-off characteristics (µm)	Permeate composition	Elements and compounds removed
Microfiltration	Hydrostatic pressure difference	Sieving	0.08–2	Dissolved substances	SS, turbidity, parasites, bacteria
Ultrafiltration	Hydrostatic pressure difference	Sieving	0.005–0.1	Dissolved substances (PM < 100 kDa)	SS, turbidity, parasitic macro-molecules, bacteria, viruses

Table 17.1. *General characteristics of microfiltration and ultrafiltration membranes*

17.2. UF and MF: materials

There is a wide range of membranes available today, depending on their material and intrinsic filtration capacity. MF and UF membranes can be made from a wide variety of materials that all have different properties. There are organic membranes and ceramic membranes.

Ceramic membranes are not currently used in drinking water applications because of their high cost. Therefore, only organic membranes are presented here. The membranes not only differ in the material used but also in their structure.

We distinguish isotropic membranes, called "symmetrical", whose structural properties are constant throughout their thickness, whether they are dense or porous. We also distinguish anisotropic membranes, called "asymmetrical", whose internal structure is different from one membrane to another.

Initially, the most used materials were cellulose acetate and polypropylene (PP). Nowadays, there are more and more polyether sulfone (PES)/polysulfone (PS) and polyvinylidene difluoride (PVDF) membranes. Here is a summary of the main properties of these materials.

17.2.1. *Cellulose acetate*

– It is naturally very hydrophilic.

– It is inexpensive raw material and relatively easy to manufacture membrane.

– It has low adsorption tendency (relatively stable permeability).

– It is relatively resistant to chlorine.

– It is sensitive to hydrolysis by acids and bases (pH between 4 and 8).

– It is sensitive to temperature (<35°C).

– It is subject to biological degradation.

17.2.2. *Polypropylene*

– It consisted of hydrophobic polymer, with relatively low permeability (limited ability to mix the polymer).

– It has good mechanical strength and flexibility of the fiber.

– It shows risk of membrane oxidation by chlorine.

– It has been widely used for the manufacture of MF membranes.

17.2.3. *Polyacrylonitrile*

– It is hydrophobic polymer by nature; it is modified to make it more hydrophilic.

– It has good chemical resistance (e.g. pH 1.5–11).

– It has moderate tolerance to chlorine (e.g. 100–200 ppm).

17.2.4. *Polyether sulfone/polysulfone*

– PES is a hydrophobic polymer by nature. It is modified with polymer blends to produce hydrophilic membranes.

– Membrane is quite easy to manufacture and different pore sizes can be easily obtained.

– It has good mechanical resistance.

– It has very good chemical resistance (e.g. pH 1–13).

– It has moderate tolerance to chlorine (e.g. 100–200 ppm per cleaning).

– It has good resistance to a wide range of temperatures (e.g. 1–125°C for the polymer, but may be limited to 40°C related to other module materials).

– It is commonly used for UF membranes with a tight pore size distribution.

17.2.5. *Polyvinylidene fluoride*

– PVDF is a low hydrophilic polymer.

– It has very good chemical resistance (e.g. pH 1.5–11).

– It has very high chlorine resistance (e.g. 500 or 1,000 ppm).

– It has good resistance to a wide range of temperatures (e.g. 1–40°C).

– It is very flexible material and mechanical resistance (ideal for applications with air scouring).

– It is mostly used for MF membranes, but also for some UF.

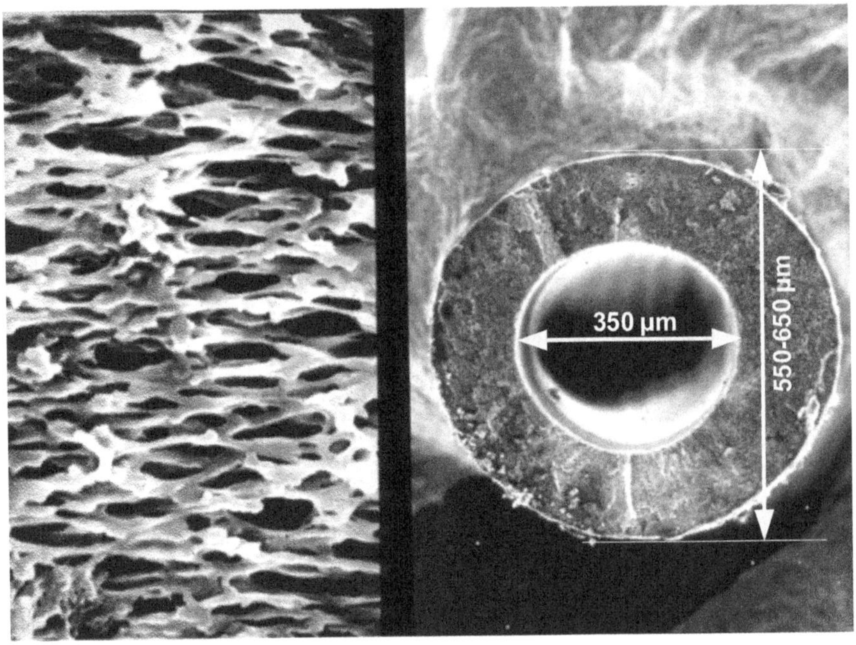

a) b)

Figure 17.5. *(a) Electron microscope view of a polypropylene MF membrane (0.2 µm) and (b) dimensions of the fibers. For a color version of this figure, see www.iste.co.uk/gaid/watertreatment4.zip*

17.3. UF and MF: membrane types

UF and MF membranes come in different forms. Their dimensions and implementation are not standardized. It can sometimes be difficult to change membranes if the latter has not been considered at the beginning of the project. However, some membrane manufacturers can offer replacement products with the same dimensions that use the same equipment, when replacing membranes in a plant.

The most common type of membrane in the MF/UF drinking water market is the hollow fiber, because it is compact and its implementation is well adapted to this type of application (cleaning, frontal or tangential modes possible, etc.). These membranes correspond to cylinders of variable length and of very small diameter. The internal diameter can vary between 0.3 and 1.5 mm.

There are two types of hollow fiber membranes: single bore and multi-bore.

Single bore hollow fibers	
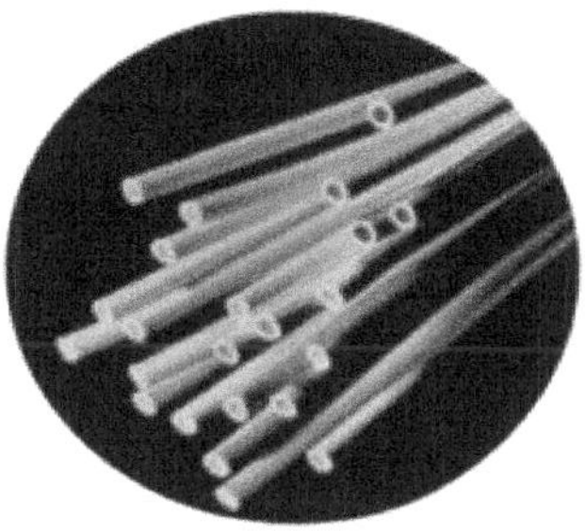	These are the ones proposed by most of the membrane manufacturers. They can be found in a pressurized or submerged configuration. They can sometimes have a support layer adapted for membranes that require a high mechanical resistance. The filtration can be done according to the type of modules: from the inside to the outside of the fibers or from the outside to the inside.
Multi-bore hollow fibers	
	In this configuration, each fiber has several internal channels, which provides a better mechanical resistance to the fiber. The material used is hydrophilized polyether sulfone. The filtration is only done from the inside of the fiber to the outside. This particular type of hollow fiber is proposed by very few membrane manufacturers.

Table 17.2. *Single bore and multi-bore hollow fiber membranes*

The hollow fibers are grouped into bundles of several thousand fibers, and each bundle is placed in a cylindrical module. The ends of the fibers are sealed on both sides in a resin, leaving access to the feed water through the interior of the fibers. The feed water enters from the inside of the fibers and can only pass through the tiny pores along the fiber, while the particles remain inside.

The filtered water (or permeate) is isolated from the feed water compartments by the resin that was used to seal the fibers. It is then collected in a central tube and discharged through the ends of the modules.

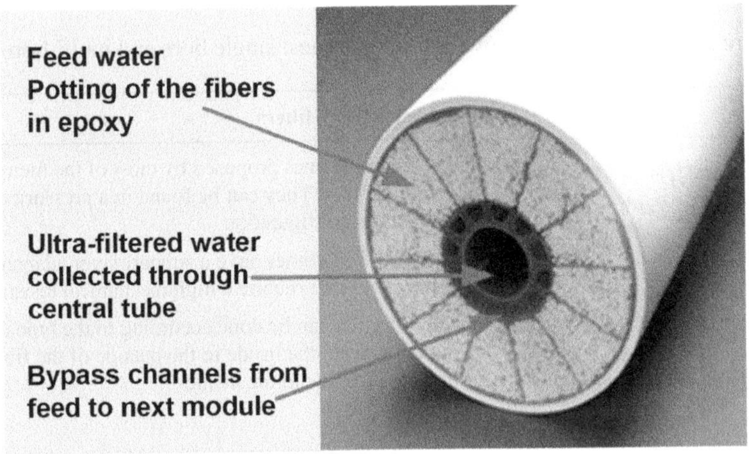

Figure 17.6. *Ultra-filtered feed water and water outlet points. For a color version of this figure, see www.iste.co.uk/gaid/watertreatment4.zip*

The hollow fibers are supported in cylindrical housing modules for pressurized membranes, or in more or less open modules for submerged membranes. The fibers are fixed by an epoxy potting at their extremity, which ensures the physical separation between the feed side and the permeate side of the membranes.

In some cases, pressurized modules contain a central collector that requires gaskets to seal between the feed and the permeate.

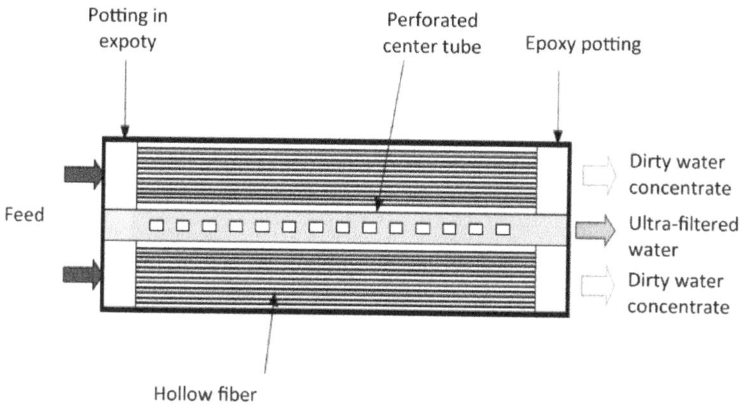

a)

b)

Figure 17.7. *Example of a hollow fiber module in an internal–external configuration with a central collector. For a color version of this figure, see www.iste.co.uk/gaid/watertreatment4.zip*

17.4. UF and MF: implementation of membranes under pressure

The membranes are inserted into housings or pressure tubes, which are installed on a skid and pressurized with a booster pump. The pressure on the feed side of the module or pressure tube provides the necessary pressure gradient for the water to pass through the membrane pores.

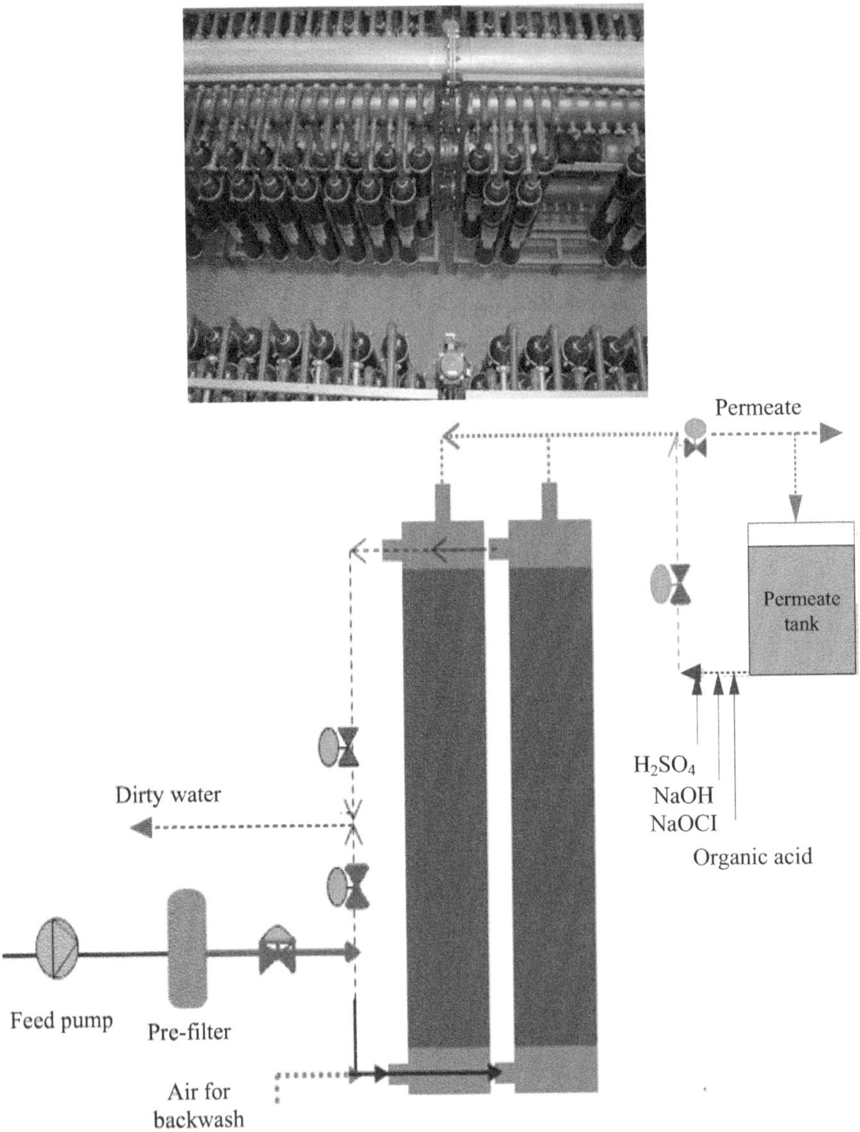

Figure 17.8. *Membranes in a pressurized configuration. For a color version of this figure, see www.iste.co.uk/gaid/watertreatment4.zip*

17.4.1. *Horizontal-vertical configuration*

Depending on the application, it is possible to install pressurized or submerged membranes in a horizontal or vertical configuration.

a) b)

Figure 17.9. *Vertical (a) and horizontal (b) configuration for pressurized membranes. For a color version of this figure, see www.iste.co.uk/gaid/watertreatment4.zip*

The MF-UF membranes under pressure are most often installed in a vertical configuration.

Some membrane suppliers offer a horizontal arrangement in pressure tubes with several modules per pressure tube. In some cases, this configuration provides a more compact solution in terms of floor space.

Inside the pressure tube, the membranes work in parallel. In the case of a four-element tube, the ultra-filtered water is fed and discharged on both sides. The modules have a bypass system surrounding the permeate collector, which supplies the central modules with raw water and discharges the dirty backwash water. Interconnecting parts between the modules ensure the hydraulic operation in the pressure tube.

Since the vertical configuration allows for tangential filtration and features air backwash conditions (backwash, forward rinse, air scour and chemical backwash), it is recommended for variations in raw water quality throughout the year, due to its more robust operation and cleaning facilities. In addition, a horizontal configuration is only possible for water qualities below 10 NTU, especially when there is no pre-treatment.

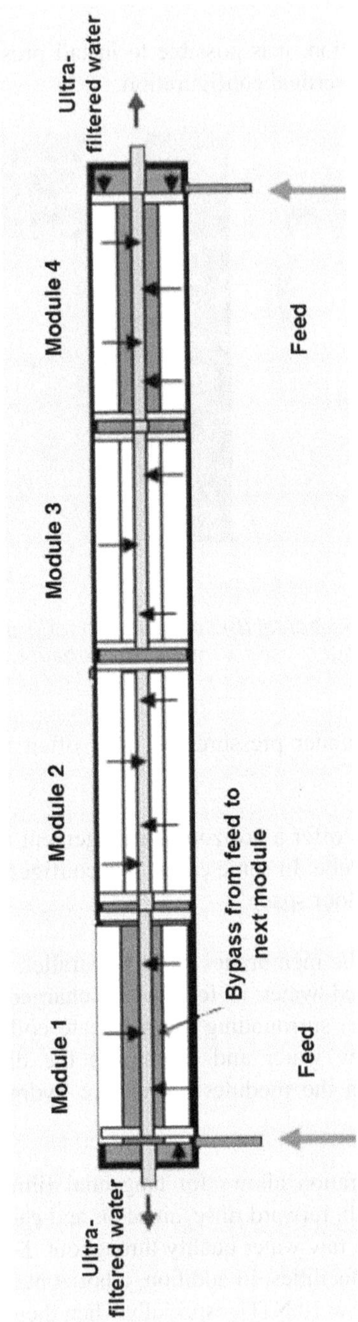

Figure 17.10. *Filtration mode in a horizontal configuration (Xiga for Cabot-Norit). For a color version of this figure, see www.iste.co.uk/gaid/watertreatment4.zip*

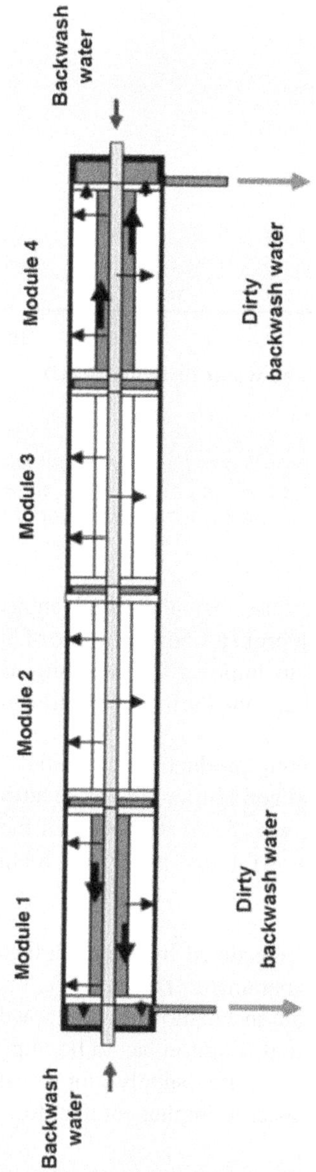

Figure 17.11. *Backwash filtration mode in a horizontal configuration (Xiga for Cabot-Norit). For a color version of this figure, see www.iste.co.uk/gaid/watertreatment4.zip*

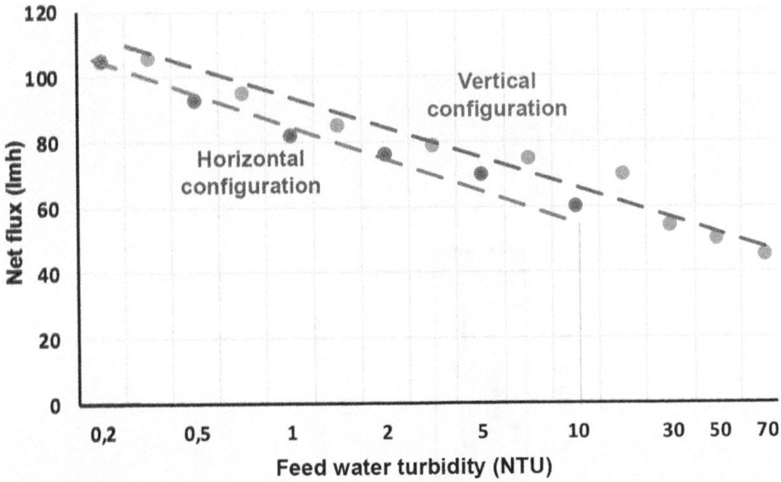

Figure 17.12. *Recommended net flows for both configurations depending on turbidity. For a color version of this figure, see www.iste.co.uk/gaid/watertreatment4.zip*

The horizontal configuration also has the disadvantage of making it more difficult to identify a module with broken fibers. The use of air is rather restricted to integrity testing and is not used to improve backwashing, due to the difficulty of evacuating air during re-filtration and the horizontal position of the modules.

In terms of footprint, for high production capacities and good feed water qualities (<1 NTU), the horizontal configuration is more advantageous due to larger standardized units. For example, with 55 m^2 Cabot-Norit membranes, the footprint of a 120-module vertical unit is 41.4 m^2, while the footprint of a 192-module horizontal unit is 16.7 m^2.

The footprint is a parameter considered by the manufacturer, but the operating conditions are often the selective parameter. For example, membranes are chosen in a vertical configuration for small installations when powdered activated carbon (PAC) has to be added, if tangential filtration has to be implemented, if turbidity is variable (especially in direct raw water supply), or because the addition of a coagulant and integrity tests are easier to implement than for vertical modules.

The vertical configuration is therefore the most frequently used. It allows for more complex backwash sequences, if necessary with air and water phases, and thus for the treatment of more heavily loaded water. Furthermore, as the vertical modules are individually connected to the collector and the horizontal modules are loaded into pressure tubes, the allowable operating pressure for horizontal configurations is higher.

17.4.2. *Submerged membranes*

Submerged MF membranes are available in a vertical configuration with, for example, vertical modules for Evoqua's CMF-S® (formerly known as Memcor) and Zenon modules.

Figure 17.13. *Configuration of submerged membranes. For a color version of this figure, see www.iste.co.uk/gaid/watertreatment4.zip*

The membranes are submerged in a tank (steel, HDPE, coated concrete, etc.) called a "cell", and the filtration is done by sucking water through the membrane thanks to a pump placed at the permeate outlet of the membrane cell.

The CMF-S® cell uses hollow fiber membranes. Feed water flows into a cell containing submerged membrane modules and is drawn through the membrane fibers in an external–internal filtration direction. The microfiltered water exits from the top of the module and is led to a central collector. CMF-S® systems can be equipped with both PP and PVDF membranes.

A backwash system uses a combination of air and water backwash to remove the build-up of solids that have been deposited on the outer surface of the membrane. Backwash frequencies range from 15 to 60 min.

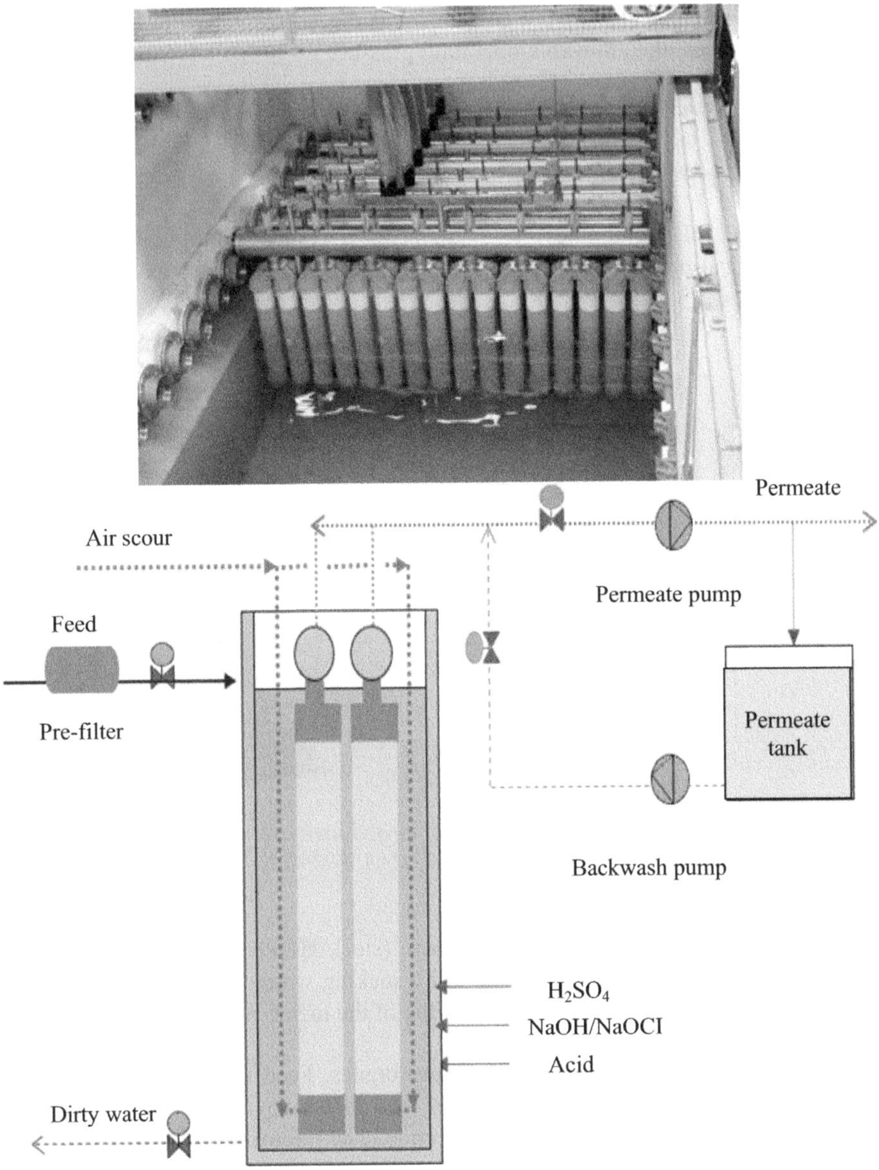

Figure 17.14. *Membranes in submerged configuration. For a color version of this figure, see www.iste.co.uk/gaid/watertreatment4.zip*

Chemical cleanings are programmed for the removal of organic matter, mineral elements and biofilm that have not been eliminated during the hydraulic and chemical backwashes.

The CMF-S® system is designed with an automated integrity test, known as the "pressure decay test" (PDT).

The washing efficiency is obtained using the following equation:

$$Washing\ efficiency = \left(1 - \frac{R3 - R1}{R2 - R1}\right) 100\%$$

With:

– R1: resistance calculated after a previous wash ($m^{-1} \times 10^{12}$);

– R2: resistance calculated before a wash, after an operating time corresponding to a volume of filtered water ($m^{-1} \times 10^{12}$);

– R3: resistance calculated after the wash ($m^{-1} \times 10^{12}$).

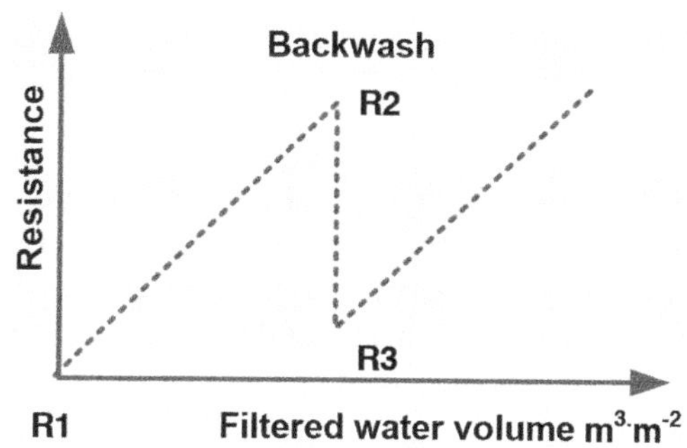

Figure 17.15. *Washing efficiency based on the resistance to fouling*

17.5. Filtration modes: frontal or tangential

UF can be conducted according to two flow modes, depending on the quality of the water to be treated and the objectives to be reached: tangential filtration or frontal filtration.

In tangential filtration, the fluid flows parallel to the membrane, resulting in a shear which limits particle accumulation. This mode of operation produces water continuously. In this implementation method, there is therefore inevitably an inlet for the feed and two outlets: one for the permeate and the other for a flow of water which does not pass through the membrane, which is known as concentrate or retentate.

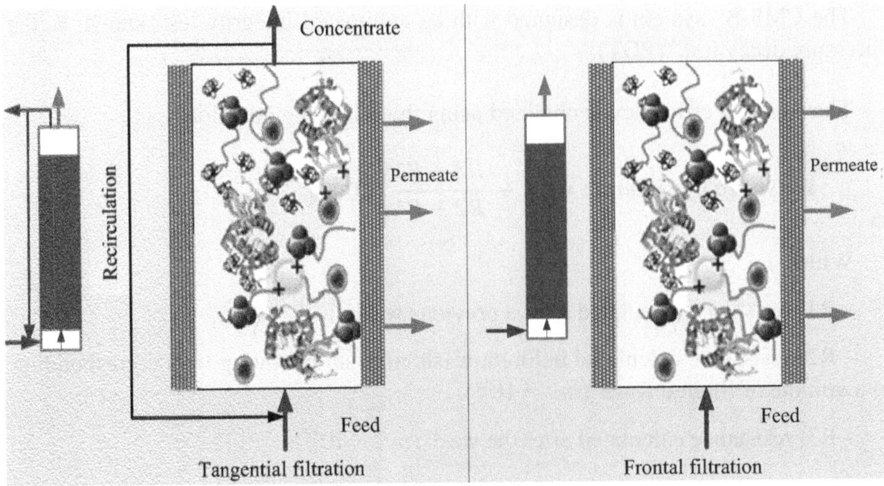

Figure 17.16. *Frontal and tangential filtration. For a color version of this figure, see www.iste.co.uk/gaid/watertreatment4.zip*

This filtration method limits clogging and deposit formation by creating shear forces. However, the recirculation of the flows imposes a higher number of members and an additional cost due to recirculation energy. Tangential filtration is used for applications with loaded water or when adding PAC. It ensures that the membrane is scavenged and that deposits on the membrane surface are, to a certain extent, avoided.

In frontal filtration, the feed flow is perpendicular to the membrane and the particles are directly filtered by the membrane. There are no recirculation pumps and the feed flow is equivalent to the product flow. As filtration proceeds, the particles accumulate and form a deposit that clogs the membrane. Frontal filtration is the most common type of filtration for hollow fiber membranes in relatively clear water applications (drinking water, process water, tertiary wastewater). It consumes less energy than tangential filtration, which requires pumping much larger volumes of water.

17.5.1. *Batch operation: filtration-backwash*

In order to remove the particles that accumulate on the membrane (especially in the frontal mode), backwashes are regularly performed. Backwashes can consist of different phases usually including a washing sequence in the opposite direction of the filtration using permeate. However, they also include phases of sweeping the fiber with water or air. The sequences depend on the filtration direction (internal–external or external–internal). These backwashes are done very frequently, but last for a very short time.

Frequencies can, for example, be from one backwash every 15 min to one backwash every 90 min, and their duration can be from 30 s to 3 min.

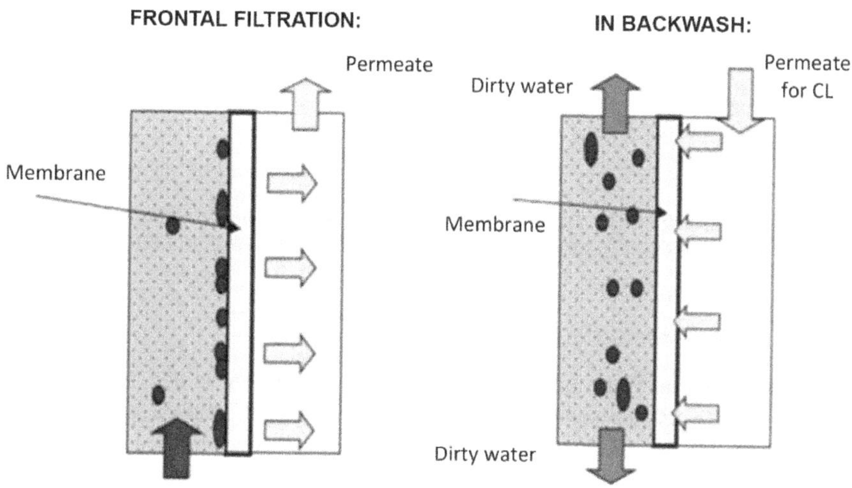

Figure 17.17. *Filtration-backwash diagram. For a color version of this figure, see www.iste.co.uk/gaid/watertreatment4.zip*

17.5.2. *Filtration direction*

The filtration direction can be internal–external; in other words, the feed water enters the interior of the fibers and exits through the external surface. The filtration direction can also be external–internal; in other words, the feed water enters through the outer surface and exits through the inside of the fibers.

17.5.2.1. *Internal–external filtration direction*

Internal–external filtration has hydraulic advantages. The backwashing is done from the outside to the inside at a flux often two to three times the filtration flux.

The velocities are high and allow very efficient permeate backwashes.

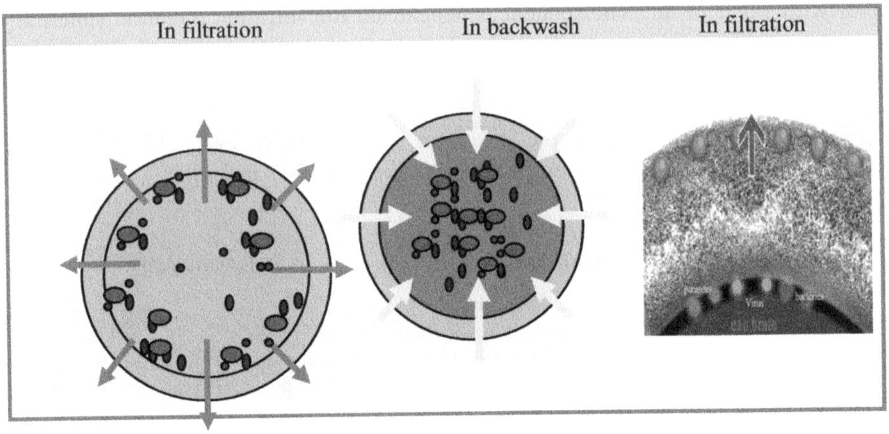

Figure 17.18. *Internal–external filtration. For a color version of this figure, see www.iste.co.uk/gaid/watertreatment4.zip*

17.5.2.2. *External–internal filtration direction*

External–internal filtration allows operation at higher SS values. It has a larger filtration surface for the same fiber diameter. However, the flows applied on this configuration are lower than on the internal–external configuration. An external sweep of the fiber with air is usually used in backwashing to optimize the backwash, and create sufficient turbulence to carry away the cake dropped by the water backwash.

17.6. Sizing parameters: membrane selection

The first phase of sizing is choosing the membrane type. Veolia is not a membrane manufacturer, and project managers select the most suitable membrane on a case-by-case basis, taking into account various considerations such as the quality of the water to be treated, the type of pre-treatment, the local context, the treatment objectives and associated guarantees, the approvals for each country and the economic costs (investment and operation).

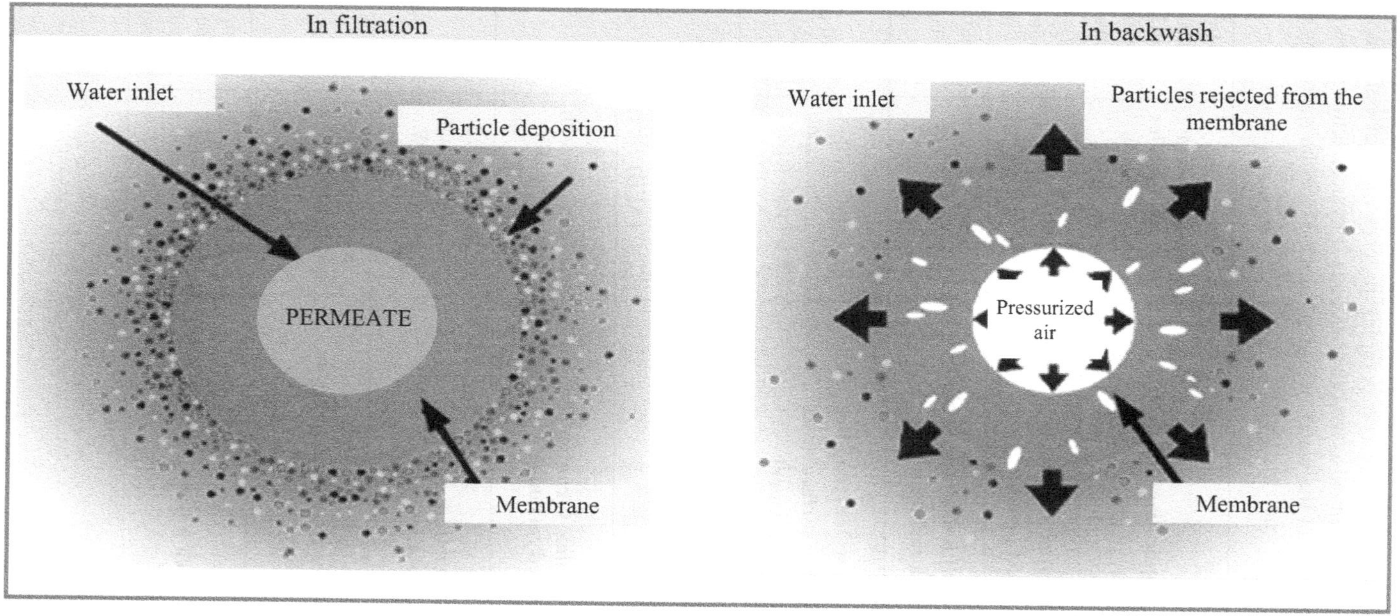

Figure 17.19. *External–internal filtration. For a color version of this figure, see www.iste.co.uk/gaid/watertreatment4.zip*

The choice of a membrane is a complex process that not only depends on the treatment objectives, but also on the various physicochemical and biological conditions that prevail in the water treatment process. Veolia's strategy is to choose the best membrane for each project.

Applications	**Vertical configuration (such as Aquaflex membrane or equivalent)**	**Horizontal configuration (such as Xiga membrane or equivalent)**
Refining (after sand filters, GAC, etc.)	Well-adapted configuration	Well-adapted configuration (large plants)
Groundwater with medium turbidity **Peak turbidity**	Turbidity < 5 NTU Turbidity < 100 NTU	Turbidity < 5 NTU Turbidity < 15 NTU
Pre-treated surface water (coag-floc-settling) **Good quality surface water (e.g. lake water)**	Recommended for small and medium-sized installations and for turbidities > 2 NTU	Well adapted for turbidities ≤ 2 NTU
Veolia process: Opaline S configuration after Multiflo Carb or Actiflo Carb	Recommended configuration with air during BW (AEB)	–
Veolia process: Opaline C configuration with powdered activated carbon (PAC) and UF	Recommended configuration with air during BW (AEB)	–
Backwash water from a primary UF installation	Recommended configuration with air during BW (AEB)	–
Backwash water from sand filters or activated carbon filters	Recommended configuration with air during BW (AEB)	–
Sea water	–	Example: Seaguard type module is installed in Xiga configuration

Table 17.3. *Some recommendations for the choice of a horizontal or vertical configuration of ultrafiltration modules*

For a construction project, the choice of membrane type is based on a number of parameters such as:

– the type of membrane: MF or UF;

– the type of membrane: hollow fiber, tubular, spiral etc. Usually for UF-MF in drinking water, hollow fibers are often required by the specifications;

– the type of filtration (for hollow fibers): internal–external or external–internal;

– if it is a pressurized or submerged system;

– the density of the fibers in a module (compactness of the system);

– other specificities: configuration with vertical or horizontal membranes, multi-bore fiber, etc.

17.7. Sizing parameters: horizontal or vertical configuration

The choice of a horizontal (e.g. horizontal with Cabot-Norit Xiga modules) or vertical (Cabot-Norit Aquaflex modules) configuration is made on a case-by-case basis. The Xiga concept installs up to four filtration elements in horizontally placed, overlapping tubes. This arrangement allows a certain compactness for large plants and is well suited for lightly loaded water, especially in refining.

The Aquaflex concept uses modules in housings, individually mounted in a vertical position. As the Aquaflex solution allows more flexibility in treatment and washing modes, it includes fiber sweeping phases and, if necessary, the use of air.

These two phases allow for more efficient cleanses, and therefore operation, on heavier or more difficult water.

Table 17.3 shows recommendations based on water type and turbidity for horizontal and vertical configurations.

17.8. Sizing parameters: flow

Darcy's law, applied to water treatment, is used to translate the proportional relationship between the permeate flux J and the TMP:

$$J = \frac{\text{TMP}}{\mu s\ Rm} = Lp - \text{TMP}$$

With:

– Rm: membrane resistance;

– Lp: permeability of the membrane;

– μs: viscosity at given water temperature.

In the case of water containing a solute retained by the membrane, such as salts or organic substances, there is added resistance to transfer via an osmotic pressure difference ΔΠ, opposite to the pressure gradient, due to the difference in concentration on either side of the membrane. Darcy's modified law then allows us to describe the flow according to the relationship:

$$J = \frac{\text{TMP} - \Delta\Pi}{\mu s \ \text{Rm}}$$

The permeate flux no longer changes linearly with the TMP. Although the difference in osmotic pressure is often neglected in MF and UF because of the large size of the compounds retained and the low osmotic pressure, the phenomenon of polarization concentration exists.

Organic substances and particles can clog the membrane in a more or less irreversible way, due to specific physical and/or chemical interactions between them and the membrane, resulting in additional resistance to transfer. These interactions are generally attributed to one or more of the following mechanisms: (a) deposition formation, (b) pore blockage and (c) adsorption. The relative preponderance of each of these phenomena may depend on the membrane (material, cut-off) as well as on the nature of the substances, the characteristics of the medium (pH, temperature, ionic strength, etc.) or the operating conditions:

$$J_{NET}(T^\circ) = \frac{Q_{net}(T^\circ) \times 10^3}{S_f}$$

17.8.1. *Instantaneous flow and net flow*

The permeate flux corresponds to the flux rate through the membrane per m^2. It is usually expressed in liters per m^2 per hour ($L \cdot m^{-2} \cdot h^{-1}$ or Lmh), but can also be expressed in meters per day ($m \cdot d^{-1}$). A distinction is made between net flux and instantaneous flux in filtration. For the nominal daily flow rate required, the instantaneous flux must be based on the minimum water temperature. The minimum and maximum temperature values, as well as the capacity of the plant according to the seasons if there are important variabilities (e.g. seaside resort with maximum production in summer alone), are taken into consideration during the design.

The flux rates and filter cycles (intervals between conventional backwashes) are strongly related to the presence of particles in the feed water of the UF. The higher the SS content of the water, the lower the flux rate and the higher the backwash frequency.

For the sizing of the membranes, instantaneous flux is preferred, which is the real flow passing through the membranes.

The instantaneous flux at a given temperature T (°C) is determined by the following formula:

$$J_{Inst}(\text{T °C}) = \frac{Q_{inst}(T\ °C)\times 10^3}{S_f}$$

With:

– J_{Inst}: instantaneous flux at a temperature T ($L\cdot m^{-2}\cdot h^{-1}$);

– Q_{Inst}: instantaneous filtration flow at a temperature T ($m^3\cdot h^{-1}$);

– S_f: total filtration surface (m^2).

The net flux at a given temperature (T) is determined by the following formula with:

$$J_{net}(T°) = \frac{Q_{net}(T°)\times 10^3}{S_f}$$

– J_{net}: net flux at a temperature T ($L\cdot m^{-2}\cdot h^{-1}$);

– Q_{net}: net filtration flow rate at a temperature T ($m^3\cdot h^{-1}$), daily flow rate produced (excluding UF washing volumes)/hours of operation of the membrane plant;

– S_f: total filtration surface (m^2).

As the temperature has an influence on the flow, it is necessary to relate the flow to a reference value related to the temperature. For example, when sizing for water at 12°C and water at 6°C, the flux and the production flow rate will be about 16% lower at 6°C than at 12°C for the same operating pressure. The instantaneous flux rate used for sizing is always reduced to a normalized flow rate at 20°C in order to compare performances under identical conditions:

$$J_{Inst}(20\ °C) = J_{inst} \times \frac{\mu(T\ °C)}{\mu(20\ °C)}$$

With:

– J_{inst}: instantaneous flux at T °C ($L{\cdot}m^{-2}{\cdot}h^{-1}$);

– J_{inst} (20°C): instantaneous flux at 20°C ($L{\cdot}m^{-2}{\cdot}h^{-1}$);

– μ (20°C): dynamic viscosity at 20°C (mPa·s = centipoises);

– μ (T °C): dynamic viscosity at temperature T°C (mPa·s = centipoises).

Or:

$$J_{inst}(20°C) = J_{inst}\cdot\ \mu\ (T\ °C).\exp\ (3.056\ (\frac{20-T}{105+T}))$$

T (°C)	2	3	4	5	6	7	8	9	10	11	12	13	14	15	16	17
Viscosity (mPa.s)	1.671	1.618	1.567	1.519	1.472	1.428	1.386	1.346	1.307	1.271	1.235	1.202	1.169	1.139	1.109	1.081
T (°C)	18	19	20	21	22	23	24	25	26	27	28	29	30	31	32	33
Viscosity (mPa·s)	1.053	1.027	1.002	0.978	0.955	0.933	0.911	0.890	0.871	0.851	0.833	0.815	0.798	0.781	0.765	0.749

Table 17.4. *Viscosity of conventional surface water according to temperature*

The flux values are therefore the maximum instantaneous values that can be taken as guide values for sizing. These flux are applicable at any temperature above 12°C.

A temperature correction of the instantaneous flux is made for any temperature below 12°C. For temperatures above 12°C, the instantaneous flux is not increased. For temperatures below 4°C, special care must be taken in calculating the flux due to the viscosity of the water.

Feed water of the UF	Maximum turbidity (*) NTU	Maximum instantaneous flux at 20°C ($L \cdot m^{-2} \cdot h^{-1}$)	Interval between min CLs	Coagulant (ppm Fe or Al)	Maximum frequency between CLC chlorine (**) h	Estimate of hydraulic efficiency (***) %
Refining (after a sand and/or GAC filter)	≤0.2	130–132	90–120	0	140–160	≥98
	> 0.2 ≤ 0.5	127	90	0	140–160	98.1
	> 0.5 ≤ 1.0	120–125	60	0	140–160	96.1
	> 1.0 ≤ 2.0	100–110	40	0	140–150	94
Groundwater (direct filtration with addition of coagulant)	≤0.2	125–132	90–120	0	140–160	98.5
	> 0.2 ≤ 0.5	127	90	0	140–160	97.5–98.1
	> 0.5 ≤ 1.0	120–125	60	0	140–160	96.7
	> 1.0 ≤ 2.0	100–110	40	0	140–150	94.6
	> 2.0 ≤ 4.0	85	30	1–2	90	92
	> 5.0 ≤ 10.0	80	20	2	48	87.6
	> 30.0 ≤ 70	<55	15	3–5	24	87.2
Pre-treated surface water (e.g. coagulation–flocculation--settling)	≤0.2	127	90–120	0	24	96.7–97.3
	> 0.2 ≤ 0.5	110	90	0	24	95.8
	> 0.5 ≤ 1.0	95–100	60	0	24	95.1
	> 1.0 ≤ 2.0	90	35	0–0.5	24	92.7–93.6
	> 2.0 ≤ 5.0	80	30	1–2	24	90.5–91.5
	> 5.0 ≤ 10.0	75	20	2	24	86.5

Feed water of the UF	Maximum turbidity (*) NTU	Maximum instantaneous flux at 20°C $L \cdot m^{-2} \cdot h^{-1}$	Interval between min BW	Coagulant (ppm Fe or Al)	Maximum frequency between CEB $_{CL2}$ (**) h	Estimate of hydraulic efficiency (***) %
Surface water	≤1.0	90–100	50–60	0	24	95.2
	>1.0 ≤ 2.0	90	40	0	24	93–93.9
	>2.0 ≤ 4.0	80	30	1–2	24	90.59–1.6
	>4.0 ≤ 8.0	70	20	2	24	85.5
	>8.0 ≤ 10	60	15	2–2.5	24	80

*Turbidity: For turbidities higher than 15 NTU, special measures must be taken with forward flush cycles or recirculation.

**Disinfection backwash: The number of disinfection backwashes may be higher if the temperature exceeds 20°C. In addition, the frequency of chlorine backwashes is strongly dependent on the biological clogging capacity of the water.

***Estimation of hydraulic efficiency: These values are given as an indication. They are useful to estimate a first approach.

Table 17.5. *Examples of flux sizing and chemical cleaning (CEB_{CL2}) frequency for pressurized membranes*

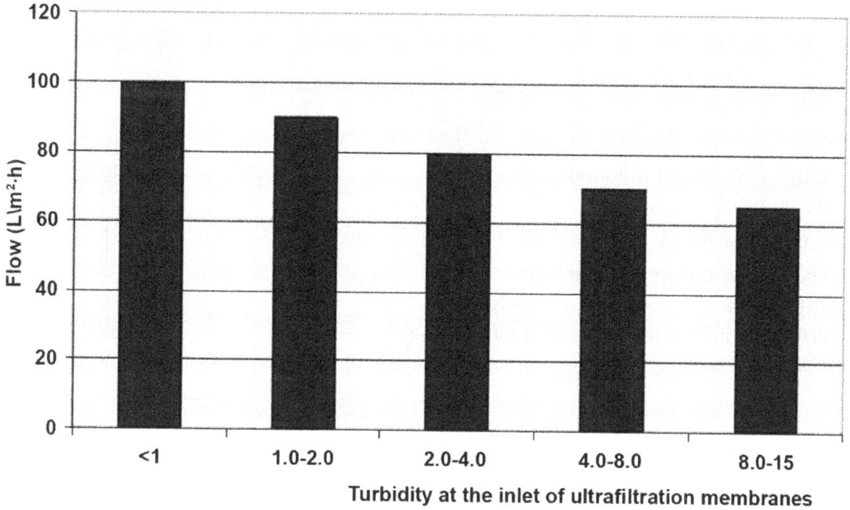

Figure 17.20. *Flux applicable on a surface water treated directly on a UF (with coagulant). For a color version of this figure, see www.iste.co.uk/gaid/watertreatment4.zip*

17.8.2. *Transmembrane pressure*

In drinking water plants, filtration operations are conducted by imposing either the pressure gradient or the permeate flow. The pressure loss across the membrane (TMP) is around 1 bar.

In general, a pressure of about 2 bars must be supplied to the feed water of the membranes. Since the membranes are thin (about 0.5 mm), surface filtration is used. The materials separated from the water accumulate on the surface of the membrane and at very little depth. The TMP is the pressure loss across the membranes, and it corresponds to the pressure difference between the feed side and the permeate side.

Depending on whether submerged or pressurized membranes are used, it is:

– positive in filtration and negative in backwash for pressurized membranes;

– negative in filtration and positive in backwash for submerged membranes.

17.8.2.1. *Application to pressurized membranes*

$$TMP = \frac{Pfeed + Pconc}{2} \text{ - P perm}$$

With:

- TMP: TMP during filtration (bar);
- Pfeed: feed pressure (bar);
- Pconc: concentrate pressure (bar);
- Pperm: permeate pressure (bar).

Note: The sensors must be altimetrically corrected to bring the measurement to the same level.

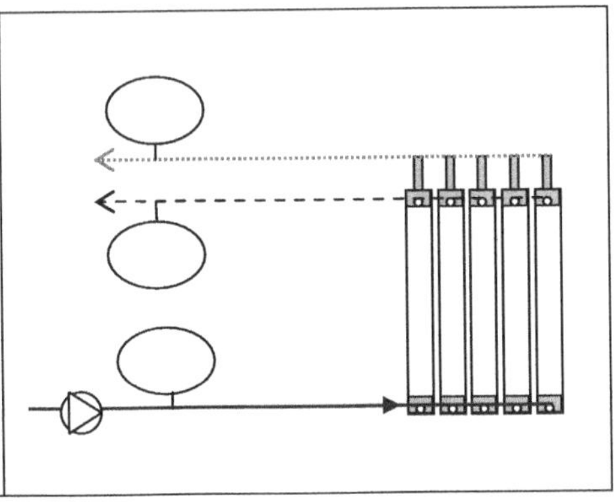

Figure 17.21. *TMP calculation for pressurized membranes*

17.8.2.2. *Application to submerged membranes*

$$TMP = Pperm - \rho gh + \Delta P_{reseaux}$$

With:

- TMP: TMP during filtration (Pa);
- Pperm: permeate pressure (Pa);
- density of the feed water (kg/m^3);
- g: acceleration of gravity (m·s^{-2});
- h: geometrical height between the pressure sensor and the feed water level in the cell (m);
- ΔP networks: pressure loss in the network (Pa).

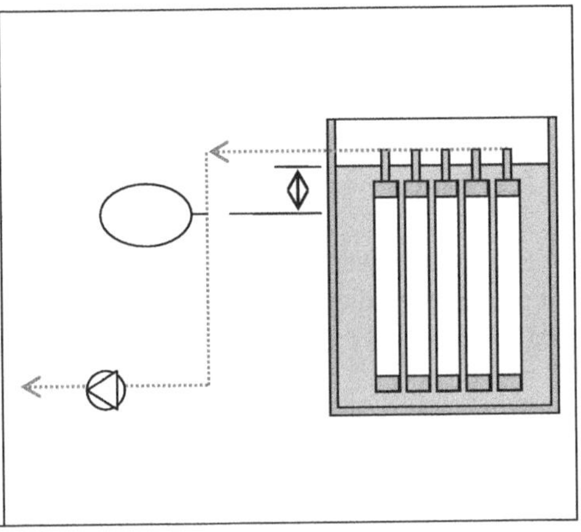

Figure 17.22. *TMP calculation for submerged membranes*

The TMP indicates the pressure loss across the membrane. It therefore varies with flow rate, temperature and the degree of clogging of the membrane. The clogging of a membrane cannot be tracked by TMP monitoring alone unless it is operating at a fixed flux and temperature. Nevertheless, it is an important parameter in both filtration and backwashing, since for each type of membrane there are TMP values that should not be exceeded, in order to avoid irreversible membrane clogging or mechanical damage.

17.8.3. *Resistance*

The filtration resistance is calculated from the instantaneous flux, the TMP and the corrected viscosity at the reference temperature (20°C).

Given that resistance is a direct measure of the state of membrane fouling, set points for this parameter are often used to determine the need for backwashing or a clean-in-place (CIP) cycle.

The resistance value is adjusted for viscosity at the feed temperature. As the feed temperature drops, its viscosity increases and the TMP required to maintain flow also increases:

$$R = \frac{3.6\,TMP \times \mu\,(20°C)}{\mu(T\,°C).\ J\ inst}$$

With:

– R: resistance ($m^{-1} \times 10^{12}$);

– TMP: transmembrane pressure (kPa);

– μ (T °C): viscosity of water at temperature T (mPa·s);

– J inst: instantaneous flux ($L \cdot m^2 \cdot h^{-1}$).

17.8.4. *Permeability*

The water permeability of a membrane is defined by the flow rate of water passing through it or permeate flow rate Qp per unit area S by the action of an average transmembrane pressure ΔTPM of 1 Pa. Permeability (Lp20) is an indicator of the long-term stability of filtration performance, despite variations in flow or temperature. It corresponds to the filtration capacity per m^2 and per bar. It should be set at 20°C, so as to monitor performance and clogging without the impact of operating temperature and, therefore, viscosity.

A high initial permeability is a sign of an energy-efficient membrane (low pressure needed to produce the desired permeate). However, the stability of the permeability is also important, since it is this stability that will allow the membrane system to operate with the least amount of chemical cleaning.

Parameters	Definitions	Relations
Transmembrane pressure (TMP)	The average of the feed pressure and the concentrate pressure minus the permeate pressure	$TMP = \frac{Pfeed + Pconc}{2} - P\,perm$
Permeate flux	The permeate flow rate divided by the membrane surface	$J_{NET}(\mathrm{T\ °C}) = \frac{Q_{net}(T\ °\mathrm{C}) \times 10^3}{S_f}$
Effectiveness or rejection rate	The ratio between the concentration of the removed substances and the inlet concentration	$\%\, Rej = \frac{Cin - Cout}{Cin}$
Permeability	The ratio between flux and TMP	$L_{\mathrm{p\ (20\ °C)}} = \frac{J_{inst}(20\ °C)}{TMP}$
Resistance	The ratio between TMP, viscosity and instantaneous flow	$R = \frac{3.6 \times TMP \times \mu\,(20\ °C)}{\mu(T\ °C) \times J\,inst}$

Table 17.6. *Main parameters characterizing the operation of the membrane plant*

Permeability is calculated as follows (for any type of membrane):

$$L_{\mathrm{p\ (20°C)}} = \frac{J_{inst}(20°\mathrm{C})}{\mathrm{TMP}}$$

With:

– L_p (20°C): permeability at 20°C ($L \cdot h^{-1} \cdot m^{-2} \cdot bar^{-1}$);

– J_{inst} (20°C): instantaneous flux at 20°C ($L \cdot m^{-2} \cdot h^{-1}$);

– TMP: transmembrane pressure in bar.

L_p (20°C) = L_p (T°C)/correction factor (20°C):

$$Correction\ factor\ (20°C) = Exp(6.433 - \frac{1.885}{(273 + T\ °C)})$$

In these operations, the filtration is carried out in tangential mode in order to limit the accumulation of material on the surface (clogging), which is the main phenomenon causing the separation. This operating mode allows continuous operation with a constant permeate flux, but requires additional energy consumption to ensure the flow and overcome the pressure losses. It is also possible to operate using frontal filtration, which simplifies the operation and reduces the energy consumption of the process, but requires sequential backwashing (circulation of the fluid in the opposite direction of permeation) in order to eliminate the clogging.

17.8.5. *Principle of the calculation of the membrane surface and water losses*

MF or UF are processes that operate intermittently with frequent skids or cell which are stopped for washing, chemical cleaning and integrity testing. This creates production and downtime for each skid or cell.

When the specifications require a constant flow rate for the membrane plant, the membranes operate with a variable flux; an N flux when all the skids are in filtration and an N–1 flux when one of them is stopped in production. The sizing flux of the membrane plant is thus based on the N–1 flux. This is the same for gravity filtration.

When the plant is designed for a nominal daily flow rate that takes into account the skid shutdowns, the skids are sized to produce the required daily flow rate (including water losses due to washing), while considering the shutdown times. A buffer tank is used to smooth the flow between the upstream sector and the needs downstream of the membrane plant.

The membrane surface is determined from the duration of the non-production phases (or downtime) of each skid, the water losses related to the different washes of the MF/UF and the non-production periods.

The number of daily production hours of the membrane plant is defined, as well as its operating mode (fixed or variable flux). For each skid, the number of hours related to backwash, chemical cleaning (CEB chlorine, acid or base) and integrity test cycles is determined because the skid is stopped during these maintenance periods. As each skid undergoes the same downtime, this implies that the non-production time of each skid is equivalent to the total non-production time of the plant in 1 day.

With few chemical cleanings in place, downtime may not be accounted for.

During the backwash sequences, part of the permeate and, in some cases, part of the feed water is consumed and discharged to the sewer. The calculation of the daily volume of permeate used for the washes gives the water losses of the membrane plant. This is therefore scaled while considering the volume consumed. For some membranes, a volume of feed water is also consumed during the wash cycles (sweeping of the fibers, emptying of cells, etc.) and must therefore be considered.

For each type of sequence (backwashes, CEB, integrity test and CIP), the following calculations are performed:

– number of sequences per day (per skid or cell and total);

– duration of each sequence;

– volume of permeate consumed by each sequence;

– volume of feed water consumed by each sequence.

We then deduce the following for each skid:

– the duration of non-production of the plant, which corresponds to the sum of the number of sequences multiplied by the duration of the sequence;

– the total volume of permeate consumed by each skid or cell, which corresponds to the sum of the number of sequences multiplied by the volume of permeate of the sequence;

– the total volume of feed water consumed by each skid, which is the sum of the numbers of sequences multiplied by the volume of feed water of the sequence;

– the volumes consumed will be multiplied by the total number of skids or cells to obtain the global water losses.

This gives the total daily production volume, taking into account the internal permeate requirements of the plant. The total real production time of the plant considers the production decrease when a skid or a cell is not in production.

This information makes it possible to calculate the total membrane surface that is necessary to produce the required permeate flow, from a flux defined at the beginning (depending on the water quality to be treated).

17.8.6. *Pre-filters*

The use of prefilters is intended to protect the membranes from particles that could either obstruct the internal channel of the fibers (case of internal–external membranes), or damage the fibers by particles that are too large in size. The mesh size of the pre-filters depends on the membrane chosen, but also on the application. It is generally 120–140 μm. In the case of a coupling an Actiflo-UF, the mesh size of the pre-filter must be lowered to 60 μm in order to trap the micro-sand in the settled water.

17.9. Operating parameters

The main operating parameters are as follows:

– permeability and flux;

– the interval between physical washes and the mode of implementation;

– intervals between CEB and their modes of implementation;

– intervals between CIP and their modes of implementation.

Depending on the membrane used, the operating parameters depend on the quality of the water to be treated and the pre-treatment upstream of the membrane plant.

When possible, pilot tests are recommended so as to optimize the washing frequencies under real operating conditions. Indeed, basic parameters such as turbidity are most often considered. However, other parameters affect the behavior of the membrane in terms of clogging and permeability, such as organic matter, color, high concentration of microorganisms and algae. The pre-treatment operation also has an impact on the membranes, such as excess reagents (aluminum, iron, polymer promoting clogging and affecting permeability).

17.9.1. *Evolution of the permeability*

The evolution of the permeability on each skid is one of the main operating parameters. It is regularly monitored by the operator for quick intervention in case of a sudden drop in this parameter. The loss in permeability of the membranes depends on the quality of the feed water, its clogging substance content (organic matter) and its turbidity. It is always corrected to a reference value (20°C) in order to assess its development under reference conditions. Figure 17.23 shows the evolution of permeability in a plant treating groundwater with a karstic tendency. The UF skids are fed directly with a micro-coagulation when turbidity increases. The applied flux is 70 $L \cdot m^{-2} \cdot h^{-1}$.

This evolution over more than 7 years of operation shows that the permeability drops by about 50 $L \cdot m^{-2} \cdot h^{-1} \cdot bar^{-1}$ per year of operation. This is a value that is found on other plants operating on water of equivalent quality, or on surface water that has undergone advanced treatment of turbidity and organic matter.

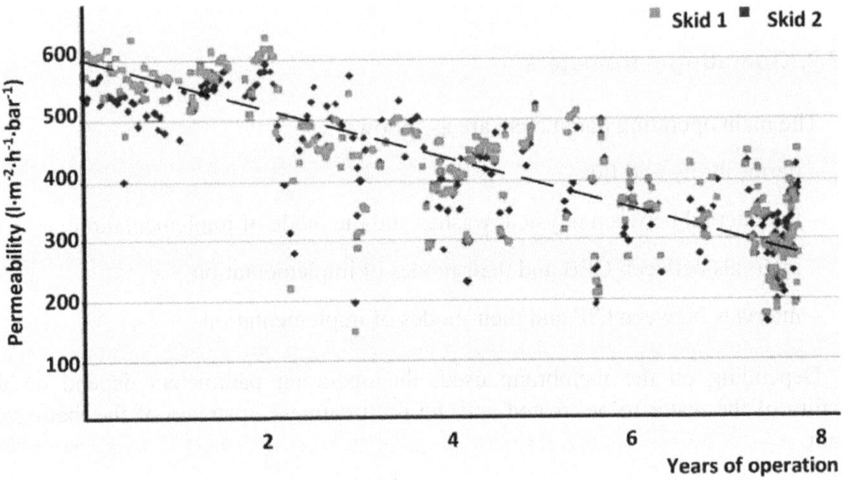

Figure 17.23. *Evolution of the permeability on an ultrafiltration fed from a groundwater with variable turbidity. For a color version of this figure, see www.iste.co.uk/gaid/watertreatment4.zip*

17.9.2. *Clogging*

If the equilibrium state is not reached (frontal filtration or tangential filtration in the case where the diffusion is weak compared to the convective flow), the concentration at the membrane boundary Cm increases until it reaches the solubility limit of the macro-molecules, thus leading to the formation of a gel or the aggregation of colloids, which inevitably leads to the formation of a deposit.

This deposit forms a porous layer, generating an additional resistance of varying magnitude according to its structure. The expression used for the flux includes, in addition to the osmotic back pressure and the membrane resistance, a deposition resistance noted as Rd:

$$J = \frac{\text{TMP} - \Delta\Pi}{\mu s\ (\text{Rm} + \text{Rd})}$$

The deposition resistance itself is a function of the deposited mass, M, per membrane area, where α is the specific deposition resistance which is often related to particle characteristics:

$$\text{Rd} = \alpha\ \text{M}$$

Lastly, we obtain the evolution of the TMP as a function of time:

$$\frac{\text{TMP}}{TMPo} = 1 + \frac{\alpha\ \text{Cin}}{\text{Rm}} Jt$$

With:

– t: operating time;

– Cin: concentration of the water to be treated.

In the case of membrane filtration by MF and UF, the osmotic pressure ΔΠ can be neglected. Clogging due to an accumulation and interactions of deposits with the membrane can be explained by several mechanisms: adsorption, concentration polarization, deposit formation and pore blockage.

17.9.2.1. *Concentration polarization*

During filtration, an accumulation of solute in the vicinity of the membrane leads to a concentration gradient. This leads to a back-scattering flow in opposition to the convective flow, according to Fick's law of diffusion. When the concentration of the accumulated material reaches saturation, a deposit/gel or precipitation of solutes is formed.

17.9.2.2. *The formation of a deposit*

A deposit is formed when particles, larger than the pore size, accumulate on the surface of the membrane. The deposit formed can be more or less porous, according to the size of the particles, and more or less compressible. The resistance of the deposit is then proportional to the material accumulated on the surface of the membrane.

17.9.2.3. *Pore blockage*

Pore blockage is a steric property: particles with a diameter close to or smaller than the size of the pores come to be included in the latter and block them, thus reducing the surface available for the flow. This mechanism requires backwashes at a pressure significantly higher than the filtration pressure. This type of clogging can lead to irreversible clogging.

17.9.2.4. *Adsorption*

Adsorption results from physicochemical Van der Waals type of interactions or from electrostatic forces between the membrane and the substances present in the water. These compounds adsorb on the surface of the membrane as well as inside the pores, resulting in the formation of layers leading to a decrease in the cross-sectional area of the pores and, consequently, to an increase in hydraulic resistance. In many cases, clogging by adsorption leads to irreversible clogging.

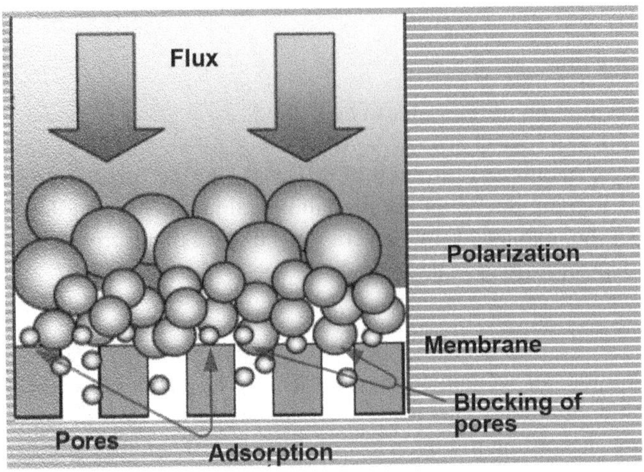

Figure 17.24. *Mechanisms of transfer of substances on a membrane. For a color version of this figure, see www.iste.co.uk/gaid/watertreatment4.zip*

The hydraulic resistance of the deposit (Rd) is proportional to the amount of material deposited per unit of membrane surface (qm) and the specific resistance of the deposit (Rs), which depends on the nature of the deposit:

$$Rd = qm\ Rs$$

The clogging of membranes is the handicap of the membrane system and results from the accumulation of matter on the surface of the membranes. The operation of a membrane filtration process is conditioned by the clogging, which is the phenomenon that occurs during the filtration.

When a system is operated at constant TMP, clogging results in a decrease in permeation flow over time. In constant flow operations, such as water treatment operations, the direct consequence is a pressure increase.

The concept of critical flow is defined as the phase transition between reversible clogging (concentration polarization) and irreversible clogging (deposit formation). The method of determining this transition is based on the use of pressure and flow stages, as well as the critical filtered volume defined as the filtered volume at which clogging becomes irreversible. It appears that charged colloidal particles do not adsorb to the membrane surface irreversibly, but do after a filtration time or critical time. The critical filtered volume is determined as the intersection between the tangents of the resistance versus net filtered volume curves.

Clogging is caused by an accumulation of material at the interface between the membrane and the water. This results in an increase of the TMP during constant flow filtration. Whatever the mechanism involved, clogging can be either reversible or irreversible. Reversible clogging occurs when the clogging deposit is removed after hydraulic or chemical cleaning. This results in a very good recovery of the permeability and TMP, which return to their initial values when filtration is resumed. Irreversible clogging occurs if a hydraulic wash, CEB or CIP does not allow recovery of the permeability. Consequently, when filtration is resumed after the washing procedure, the TMP will be higher than its initial value.

Clogging (Figure 17.25) can be of the following nature:

– particulate (deposition of particles on the surface of the membrane);

– biological (deposition of a biofilm on the surface of the membrane);

– organic (dissolved organic carbon, color adsorbed on the membrane);

– inorganic (calcium carbonate, iron, aluminum, manganese, etc.).

The accumulation of material leads to frequent hydraulic and chemical cleaning, which affects the life of the membranes, water losses and operating costs.

The physicochemical properties of the membrane determine whether they are hydrophobic or hydrophilic. Hydrophilic membranes are characterized by the presence of active groups that form hydrogen bonds with water, while hydrophobic membranes have little attraction to water molecules. This tends to promote clogging with hydrophobic particles and substances. Other parameters such as pore size, pore distribution and pore geometry influence the clogging mechanisms through pore blocking and accumulation of surface material. The same is true for particles and substances present in the water, whose physical and chemical properties may or may not influence clogging.

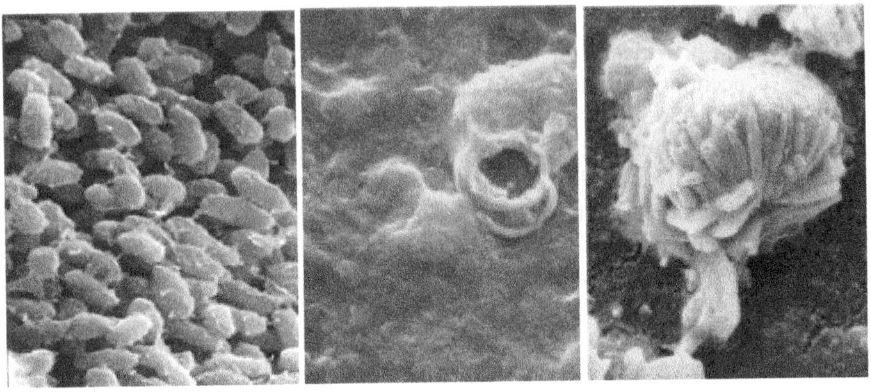

a) b) c)

Figure 17.25. *Different types of clogging: a) biofilter; b) organic clogging; c) mineral clogging*

Surface filtration favors the accumulation of materials (SS and high molecular weight molecules) on the surface of the membrane. Thus, as the water is filtered, there is an increase in the pressure required to keep the same flow rate through the membranes.

Many parameters make the behavior of a UF or MF membrane more complex:

– the formation of a cake on the surface of the membranes, which promotes depth filtration behavior;

– given that the surface of the membranes are electrically charged, electrical repulsions are involved in the separation mechanisms;

– molecules with a molecular weight lower than the membrane's cutting power, such as certain organic matter, can sometimes adsorb to the membrane surface.

	Internal–external pressure membranes	External–internal pressure membranes	External–internal submerged membranes
Turbidity and SS	Best adapted for low turbidities. If turbidity is high (>15 NTU), shorter cycle times, lower flux, adapted cleaning sequences, or a tangential mode are required so as not to accumulate too much material in the fibers.	Withstands high turbidity peaks better. Gives better performance in case of high SS or turbidity than internal–external type membranes.	Withstands high turbidity peaks better. Gives better performance in case of high SS or turbidity than internal–external type membranes.
Organic matter	Risk of adsorption on the membrane, creating a decrease in permeability. Possibility of using a low dosage of coagulant to limit clogging.		
Metals (iron, manganese, aluminum)	Risk of precipitation during chemical backwashing with chlorine or in the backwash tank if the water is chlorinated. These precipitates can clog the outer surface of the fiber and even damage it. Plan specific cleanings and avoid using chlorinated water for cleanings.		
Bacteriology	Risk of biofilm, provide CEB with high chlorine dosages or even soda.		
Algae	Algae pose clogging problems, especially for this type of membrane. Provide micro-coagulation if short and punctual episodes.		Algae clogging problems may be more manageable on these membranes thanks to a coagulation stage.

	Internal–external pressure membranes	External–internal pressure membranes	External–internal submerged membranes
Hardness (Ca) and alkalinity	Preferably, feed the membranes with slightly aggressive or calco-carbonic equilibrium water. Provide acid for CEB to avoid the risk of carbonate precipitation. Beware of washing with sulfuric acid if there is a lot of $CaCO_3$: risk of calcium sulfate precipitation.		
Powdered activated carbon (PAC)	PAC alone does not pose a clogging issue as long as its evacuation is ensured by frequent backwashing or forward flushing, and a tangential operation is used (dose 5–10 ppm).	PAC alone does not pose a clogging issue as long as its evacuation is ensured by frequent backwashing or purging and a tangential operation is used (dose 5–10 ppm).	PAC alone does not pose a clogging issue, provided that it is evacuated with relatively frequent backwashing (dose 5–10 ppm).
	On the other hand, be warned that when PAC is combined with a polymer, it becomes sticky, even at very low doses (carbon leakage). Provide micro-coagulation, and preferably daily CEB soda and acid.		
Polymers	No cationic polymers or emulsion polymers in the upstream treatment file. Be careful with the doses of anionic polymer upstream of the membranes. Provide a micro-coagulation upstream, so as to limit the risks of clogging, as well as a high dose of CEB chlorine for PVDF membranes and CEB soda (then CEB acid) for the others.		
Oils and grease	High sensitivity of organic membranes to clogging by effluents containing oil and grease: – must be in emulsion form, tends to increase chemical cleaning and impact production; – in dissolved form: oils will pass through the MF/UF membranes; – oil and grease separated from the aqueous phase: irreversible clogging of the membrane (forbidden by some suppliers).		

Table 17.7. *Comparison of configurations according to the type of clogging*

17.9.2.5. *Feed fouling index*

The feed fouling index (FFI) is a parameter used to characterize the nature of the fouling of feed water, as well as to monitor the performance of an MF cell. This parameter is similar to other parameters such as the SDI (silt density index), mostly used for reverse osmosis and nanofiltration membranes. The FFI of a feed flow is a measure of the feed fouling rate normalized to a feed temperature of 20°C. The calculation is an estimate of the increase in the rate of resistance when a given volume of water is filtered per unit of membrane surface. Many variables affect the FFI such as solids (nature, size, compressibility and organic content), as well as their interactions with other parameters such as pH, temperature and use of coagulants.

The FFI is calculated using the initial resistance when the cell returns to filtration after a backwash, the final resistance calculated before the backwash begins, the total permeate flow rate produced between these intervals and the membrane surface:

$$\mathrm{FFI} = \frac{(R_2 - R_1) \times S}{V_{(R1 \to R2)}}$$

So:

$$\mathrm{FFI} = \frac{(R_2 - R_1) \times 1{,}000 \times 60}{J_{inst} \times t_R}$$

With:

- FFI: feed fouling index (in $m^{-2} \times 10^{12}$);
- R_1: initial resistance ($m^{-1} \times 10^{12}$);
- R_2: final resistance calculated after backwash ($m^{-1} \times 10^{12}$);
- S: filtration surface (m^2);
- $VR_1 \rightarrow R_2$: volume of filtered water between the interval R_1 and R_2 (m^3);
- J_{inst}: instantaneous flux ($L \cdot m^{-2} \cdot h^{-1}$);
- t_R: filtration time between R_1 and R_2 (min).

17.9.3. *Frequency and conditions of hydraulic and chemical backwashing*

The proper cleaning of a membrane is essential to its proper functioning. It includes:

– Physical cleaning to remove particles and cake from the surface of the membrane. Physical cleaning is based on mechanical treatment dislodges and removes foulants from the membrane surface. Physical methods include hydraulic cleaning, vibration and air scouring.

– Chemical cleaning with an oxidant to inactivate and remove microorganisms present on the surface. Chemical cleaning depends purely on chemical interactions to remove foulants from the membrane surface. Chemical agents react with deposits, scales, corrosion products and other foulants. The chemical should loosen and dissolve the foulant, keep the foulant in dispersion, avoid new fouling and not attack the membrane.

It is recommended to maintain flexibility on the frequency and duration of chemical backwashing in order to increase the frequency during certain periods of the year.

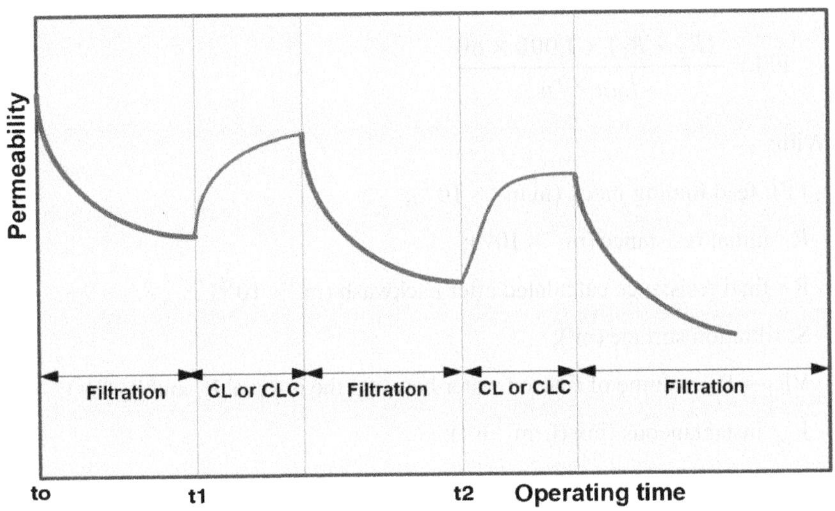

Figure 17.26. *Sequential diagram of permeability recovery after clogging. For a color version of this figure, see www.iste.co.uk/gaid/watertreatment4.zip*

17.9.3.1. *Hydraulic backwashes (air and water)*

To prevent this continuous increase in the TMP, the UF blocks will be exposed to a series of automatic operations, namely:

– backwashing (BW): periodic clearing of the membranes with only water.

Backwashing consists of reversing the direction of the flow and the pressure gradient in order to send the permeate through the membrane, and then extracting and evacuating the deposition and particles involved in pore blockage. In order to effectively prevent the continuous increase of the TMP, the physical cleaning of the membrane is performed regularly. The duration of the filtration cycles between each physical cleaning mainly depends on the turbidity of the feed water and the flow used. Typically, these cycles last between 15 and 90 min, depending on the amount of accumulated material. The physical washes implemented depend on the type of membranes and the following sequences. Once the block has been put back into filtration, this operation is a way of bringing down the trans-membrane pressure. Backwashing is carried out with ultrafiltered water (permeate).

17.9.3.1.1. Backwash phases in countercurrent filtration

This is the basic sequence of a hydraulic wash. A volume of permeate is flushed in the opposite direction to the normal filtration direction, so as to loosen the cake that has formed on the fiber. Most UF/MF (hollow fiber) membrane systems operate with this type of backwash phase, either in submerged or pressurized membranes. With pressurized membranes, and depending on the filtration direction of the membrane, backwashing is done from the outside to the inside, at flows in the range of two to three times the filtration flow for membranes with internal–external filtration flow.

When the filtration flow is external–internal, backwashing is done from the inside to the outside, often including an air cleaning phase outside the fiber.

If the modules are in a vertical configuration, the dirty water is evacuated from the top and/or from the bottom, depending on the type of membrane and the direction of the feed water (top or bottom).

With submerged membranes, backwashing is done from the inside to the outside of the fiber because the filtration flux is always external–internal.

17.9.3.1.2. Sweep phase of the fiber

Primarily used for pressurized membranes, the concentrate outlet is opened while the module is being fed, creating a tangential sweeping of the fiber (Figure 17.27). This allows some of the material in the module that is not compacted into a cake to be removed, ensuring a more efficient wash. It can also be used during the cycle to deconcentrate the material that accumulates in the modules during a filtration cycle (in the case of water loaded with SS, for example).

17.9.3.1.3. Air cleaning phase

Most membrane systems with external–internal filtration include air cleaning during the wash phase (Figure 17.28). Air is often injected between the fibers to loosen the cake on the fiber surface. Air-enhanced backwashing is simply pressurized air applied to the membrane and is included more particularly in the chemical-enhanced backwashing cycles. This air cleaning phase is sometimes used for external–internal pressure membranes. In some cases, internal–external filtration flow membranes are provided with air injected inside the fiber. The air injection is performed on each module or on the feed collector.

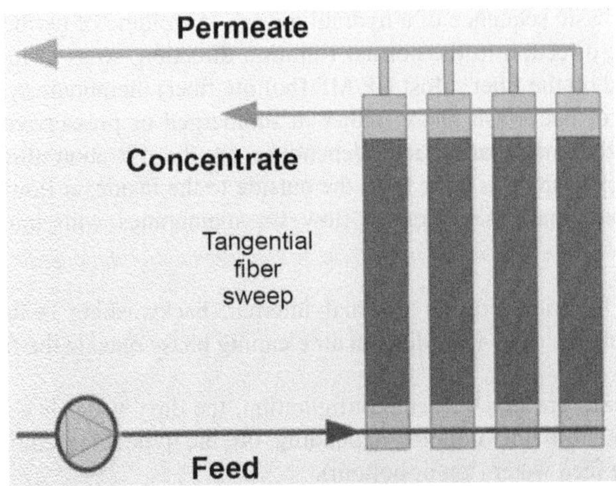

Figure 17.27. *Example of the sweep phase of the fiber. For a color version of this figure, see www.iste.co.uk/gaid/watertreatment4.zip*

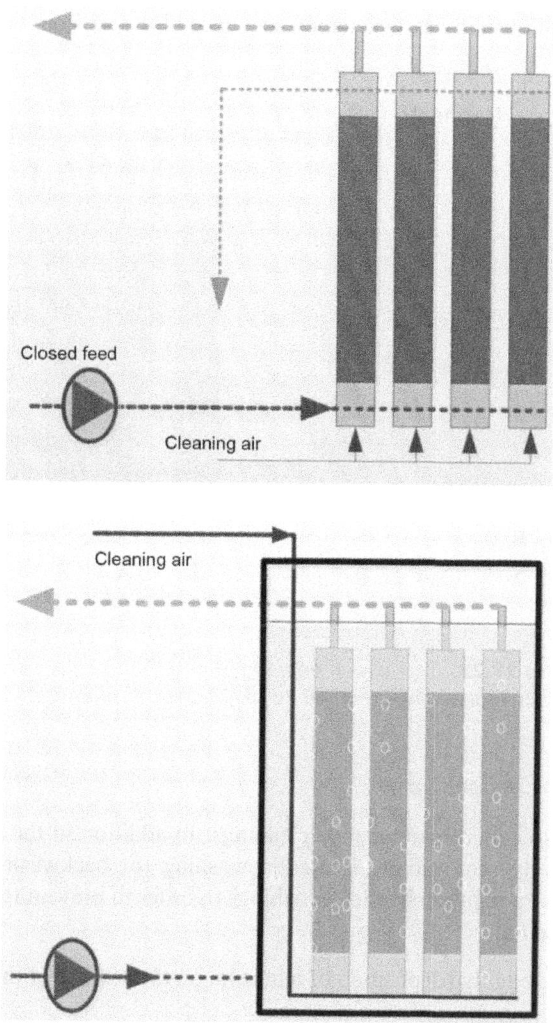

Figure 17.28. *Air cleaning phase with pressurized and submerged membranes. For a color version of this figure, see www.iste.co.uk/gaid/watertreatment4.zip*

17.9.3.1.4. Air and water backwash phase

The air phase can be combined in an air + water phase, which sometimes follows an air-only sequence (Figure 17.29). The (air + water) sequence is systematic for submerged membranes following the air-only phase. For pressurized membranes, the air phase is followed by an evacuation sequence of the air from the system.

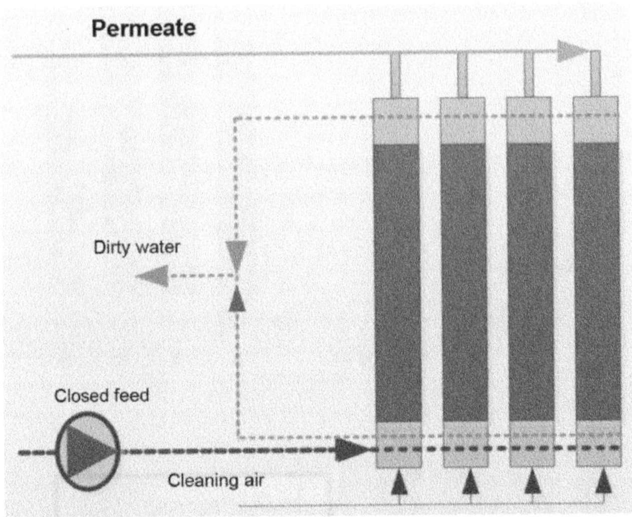

Figure 17.29. *Air + water washing phase. For a color version of this figure, see www.iste.co.uk/gaid/watertreatment4.zip*

17.9.3.2. *Chemical backwash*

These are also called maintenance cleanings. In addition to the partial clogging that is preventively removed by hydraulic washing (or backwashing), membrane installations require regular chemical washings in order to prevent the other types of clogging encountered:

– biological clogging (deposition of a biofilm on the surface of the membrane);

– organic clogging (dissolved organic carbon, color adsorbed onto the membrane);

– inorganic clogging (calcium carbonate, iron, aluminum, manganese, etc.).

Chemical washing consists of injecting a chemical solution that is adapted to the type of clogging during a contact time, depending on the type of clogging and the characteristics of the membrane. A rinsing phase finalizes the chemical washing in order to remove the deposits.

Chemically enhanced backwashing with chlorine (CEB_{Cl2}): membrane disinfection, removal of biological deposits. Chlorine backwashes maintain permeability stability by reducing the risk of biological clogging (Figure 17.26). Chemically enhanced backwashing is usually carried out using the same water circuit as simple backwashing, with the same pump, but with an in-line injection of reagents.

Chemically enhanced backwashing with acid or alkaline solution (CEB_{acid}; CEB_{base}): removal of inorganic or organic deposits. CEB is usually carried out using the same water circuit as simple backwashing, with the same pump, but with an in-line injection of reagents. Soda backwashes recover permeability losses due to organic clogging. Acid backwashes help to dissolve mineral clogging (iron, Mn, Al and carbonates, which are precipitated during other chemical backwashes).

Chemical washing typically includes:

– an injection phase of the washing solution, which is done with a flow rate and an injection temperature that is calculated to fill the skid with the chemical solution;

– a soaking phase, the duration of which depends on the reagent, the type of clogging and the water temperature;

– a rinsing phase, which allows the removal of deposits and ensures the global rinsing of the system so that when the filtration is restarted, the quality of the treated water is in conformity.

Aside from the periodical chemical cleaning, which is part of the automated process control of the installation, a more intensive chemical cleaning might be necessary. This known as "cleaning in place" (CIP), which can be automatic or semi-automatic. If CIP is not able to clean the membranes, membranes are replaced by new one.

The dirty water from chemical backwashes is neutralized before being evacuated. Generally, CIP is programmed every 3–6 months and therefore does not necessarily need to be fully automated.

This cleaning is very often carried out with the same reagents as the chemically enhanced backwashes, but the solution is first heated to be more efficient and is recirculated for some time inside and through the fibers.

The PES membrane has a relatively low tolerance to free chlorine at $pH < 10$, due to a hydrophilic additive (PVP) that is used on PES membranes, and tends to be removed from the membrane surface when NaOCl is applied. The typical maximum allowable instantaneous chlorine concentration is 500 ppm and the cumulative allowable chlorine concentration over the life of the membrane is 250,000 $ppm \cdot h^{-1}$. Above this threshold, the membrane warranty may be in question.

Frequency of CEB base	**Minimum values according to TOC of the feed water**
TOC < 1	Once per month
1 ≤ TOC < 2	Once per week
2 ≤ TOC < 5	3 times a week
TOC ≥ 5	Once a day
Frequency of CEB acid	**Minimum values according to the hardness of the feed water**
Fresh or moderately soft water (≤150 $mg \cdot L^{-1}$ $CaCO_3$)	Every 4–6 soda CEB, at least once every 2 weeks
Medium soft to hard water (150–400 $mg \cdot L^{-1}$ $CaCO_3$)	CEB soda, at least once a week
Hard to very hard water (≥400 $mg \cdot L^{-1}$ $CaCO_3$)	CEB soda, at least once a day

Table 17.8. *Minimum interval for chemical cleanings CEB (base and acid)*

By providing a CEB_{acid} after each CEB_{soda}, one can self-neutralize both solutions, only needing a small top-up of reagents if necessary.

If the water is hard, care must be taken not to cause precipitation on the membrane with CEB_{soda}. In high-risk cases (Ca hardness > 150 $mg \cdot L^{-1} CaCO_3$), a CEB_{acid} is systematically done after a CEB_{soda}, so as to dissolve any possible carbonate precipitate.

Figure 17.30. *MF membrane that is (a) clean, (b) clogged and (c) still clogged despite chemical cleaning*

17.9.4. *Frequency and conditions of CIP*

CIP is less frequent than chemical backwashing and requires more severe conditions with longer soaking and recirculation times (several hours), as well as the use of a heated solution. The frequency of these cleanings is, on average, two to four times a year. However, this frequency can be monthly if the quality of the feed water to the membranes requires it.

CIP is a corrective cleaning procedure to restore the permeability of a membrane. The part of the clogging that cannot be recovered by a backwash is qualified as totally irreversible. It is generally attributed to adsorption and/or precipitation phenomena.

When membrane's capacities drift to a limit defined by the manufacturer, a more thorough cleaning must be carried out in order to restore the membrane's permeability (this is complementary to hydraulic and chemical backwashing).

The choice of reagents used depends on:

– the type of clogging encountered, and therefore the water to be treated and the pre-treatment;

– the membrane used and its chemical tolerance (reagents to be validated by the membrane supplier);

– authorizations for use in a drinking water system according to the regulations of the country in question;

– discharge authorizations and environmental constraints.

The most commonly used reagents are:

– acids (sulfuric acid, hydrochloric acid, citric acid);

– caustic soda;

– sodium hypochlorite.

In agreement with the membrane supplier, other products are listed such as detergents, oxalic acid or Hydrex No. 4920, or products specifically developed for membrane cleaning (Veolia-Hydrex).

Cocktails of products, such as caustic soda + sodium hypochlorite, citric acid + sulfuric acid, detergent + caustic soda, are also often recommended.

Types of clogging	**Reagents**
Biofilms	High dose of chlorine when the membrane allows it caustic soda + chlorine
Iron	Citric acid Sulfuric, ascorbic or phosphoric acids Hydrex No. 4920
Manganese	Citric acid Sulfuric or phosphoric acid (sometimes with the addition of EDTA to complex manganese when authorized)
Carbonates	Sulfuric acid
Organics	Caustic soda Caustic soda + chlorine Caustic soda + detergent
Silica	Caustic soda Caustic soda + detergent
Algae	Caustic soda Caustic soda + chlorine
Polymers	High dosage of chlorine when the membrane allows it caustic soda + chlorine
Polymers, PAC, coagulants	Caustic soda + chlorine Reagents to be used individually or together according to the type of clogging observed Hydrex No. 4920, oxalic acid Sodium bisulfite

Table 17.9. *Reagents to be used according to the type of clogging observed*

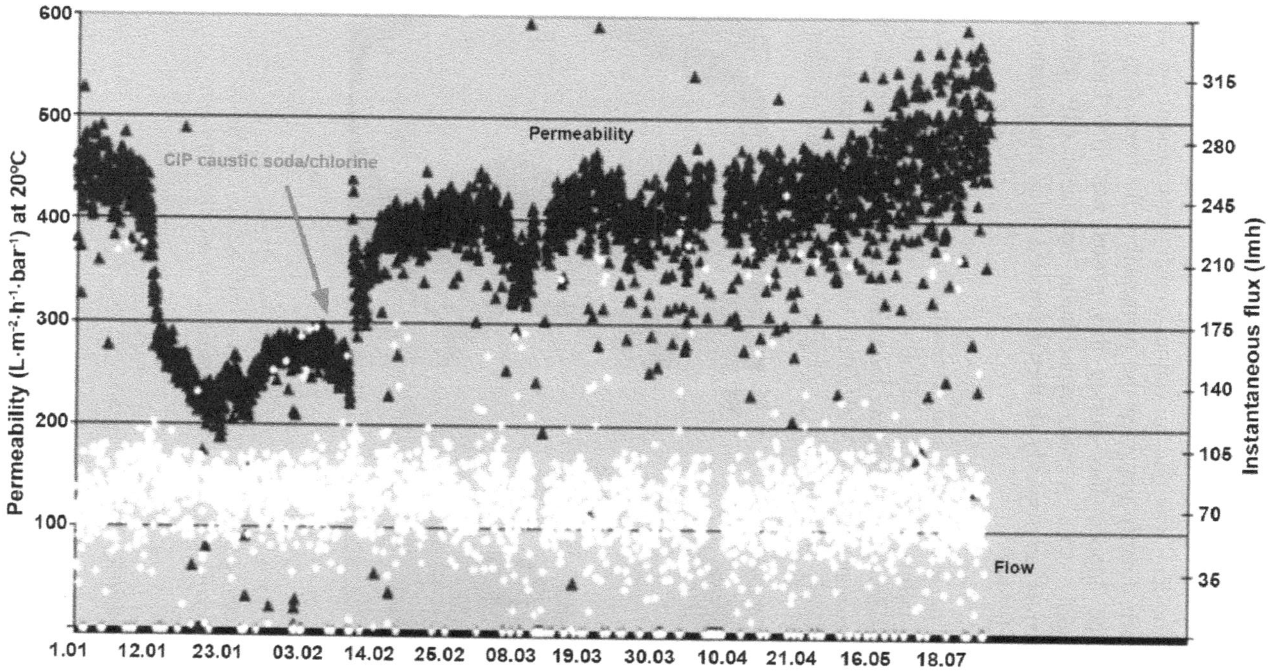

Figure 17.31. *Drop in permeability after clogging and its recovery after chemical cleaning. For a color version of this figure, see www.iste.co.uk/gaid/watertreatment4.zip*

Figure 17.31 shows a clogging incident on a plant treating surface water, in which the UF membranes are placed in refining after clarification and advanced treatment (O_3-CAG). The applied flux is around 90 Lmh.

The clogging incident was due to various constituents present on the GAC filtered water, which directly impacted the permeability for more than 3 weeks, especially since the applied flux during this period was 100 Lmh and the temperature was only 3°C. A soda/hypochlorite and acid CIP allowed the permeability to recover very quickly. It then recovered to its steady state between 400 and 500 $L \cdot m^{-2} \cdot h^{-1} \cdot bar^{-1}$.

Chemical cleaning with oxalic acid (concentration 0.5%, T 35°C and soaking 2 h) recovered the permeability from 220 $L \cdot m^{-2} \cdot h^{-1} \cdot bar^{-1}$ to 420 $L \cdot m^{-2} \cdot h^{-1} \cdot bar^{-1}$. Figure 17.32 shows membrane clogging by a colloid-$Fe(OH)_3$ aggregate that was cleaned with Hydrex reagent No. 4921 (Veolia), which is more effective than oxalic acid and allows recovery of permeability to higher values (550 $L \cdot m^{-2} \cdot h^{-1} \cdot bar^{-1}$).

a) b)

Figure 17.32. *UF module clogged with Fe (a) before and (b) after CIP with Hydrex No. 4921 (Veolia). For a color version of this figure, see www.iste.co.uk/gaid/watertreatment4.zip*

17.9.5. *Membrane integrity*

An MF-UF membrane is a barrier against micro-organisms, if it is integrated. This monitoring parameter is thus essential on MF or UF plants.

The monitoring of this parameter can be direct or indirect.

Indirect monitoring of integrity is done through measurements on the permeate. Direct integrity monitoring, which is much more accurate and essential in drinking water, is carried out in a punctual way by a test most often based on the bubble point principle. The most commonly used direct test is the PDT.

17.9.5.1. *Indirect integrity monitoring*

Indirect integrity monitoring is done through permeate analyses, such as particle counts or turbidity measurements.

The advantage of indirect monitoring is that it is performed continuously on an easily monitored parameter, such as turbidity. Particle counts can be used to detect changes in permeate quality. For example, if the feed water to the membranes is loaded with particles, these will be more easily counted in the permeate in the case of damaged fibers. Turbidity monitoring requires the use of very sensitive turbidimeters (<0.05 NTU).

17.9.5.2. *Direct integrity monitoring*

Direct integrity monitoring is complementary to indirect monitoring. It is usually performed by a test based on the bubble point principle. There are different tests based on the latter:

– PDT;

– vacuum decay test;

– diffusion flow test.

17.9.5.2.1. Bubble point principle

The bubble point is based on the fact that a liquid is held in a membrane pore by surface tension and capillary forces. The minimum air pressure to force the liquid through a pore (bubble point pressure, P_{PB}) is a function of the pore size and is calculated as follows:

$$P_{PB} = \frac{4\,\sigma\, f\, \cos\theta}{D}$$

With:

– P_{PB}: bubble point pressure (Pa);

– σ: surface tension of the liquid = 73,000 (k Dynes);

– f: corrective coefficient of the pore geometry (take max value of $f = 1$);

– θ: angle of liquid-membrane contact (take $\theta = 0$ for hydrophilic membrane);

– D: pore diameter of the membrane (μm).

With f = 1 and θ = 0, we get:

$$P_{PB} = \frac{4\sigma}{D}$$

This represents the surface tension force divided by the area to which it is applied:

$$P_{PB} = \frac{\pi D \sigma}{\pi \left(D^2/4\right)} = \frac{4\sigma}{D}$$

It is not possible to perform such a test at the bubble point of a UF membrane with integrity (e.g. pore diameter of 0.02 μm). The pressure would be very high (>100 bar!), and therefore far above the maximum acceptable pressure of the membrane.

Therefore, the pressure drop test is performed at a pressure of 0.7–1.0 bar. This allows us to determine if there are any holes larger than 3–4 μm. The integrity check against parasites is therefore well performed by the bubble point, and not by means of viruses (MS2 phages).

To check the retention of a given membrane against smaller micro-organisms, challenge tests are available. This type of test is performed with calibrated particles, in order to calculate the log rejection value (LRV) of the membrane with respect to the selected particle size. This can occasionally be done on a module with a challenge test with MS2 phages, so as to test the rejection of viruses from a UF membrane in situ, and to verify that there is no modification in time of the membrane cut-off threshold with respect to the virus. The challenge test cannot be implemented on a regular and continuous basis on an installation. It can therefore only be used in addition to the bubble point test to check the behavior of a membrane at a given time with respect to a type of particle.

Regular testing based on the bubble point principle is necessary to monitor the integrity of the system and to detect broken fibers.

The most commonly used indirect test is the PDT, which is described in section 17.9.5.2.2.

17.9.5.2.2. Pressure drop test

The pressure drop test consists of draining the inner channel of the fibers, isolating the skid or cell and pressurizing the inside of the fibers to approximately 1 bar of air. The test consists of a pressure drop measurement over a given time (Figures 17.33 and 17.34).

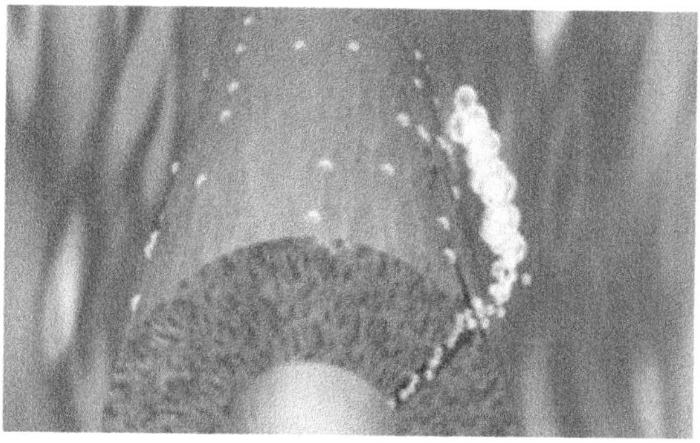

Figure 17.33. *Bubble test on a module with a broken fiber. For a color version of this figure, see www.iste.co.uk/gaid/watertreatment4.zip*

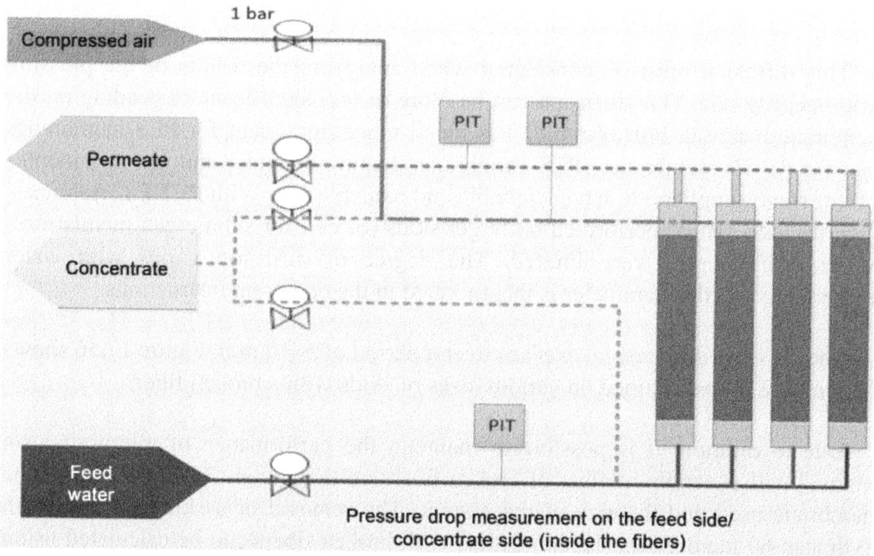

Figure 17.34. *Example of PDT implementation on a skid in an internal–external configuration. For a color version of this figure, see www.iste.co.uk/gaid/watertreatment4.zip*

By applying a pressure of 1 bar inside the fibers, the air passes through holes larger than the diameter corresponding to the applied pressure, and thus lowers the same pressure. This means that fibers are broken in one or more modules.

Air diffusion occurs within each water-filled pore, which naturally generates a small pressure reduction, and very slight bubbling on the permeate side of the membrane while the membrane has integrity. Modeling of integrity test systems is most often implemented on units (Figure 17.35).

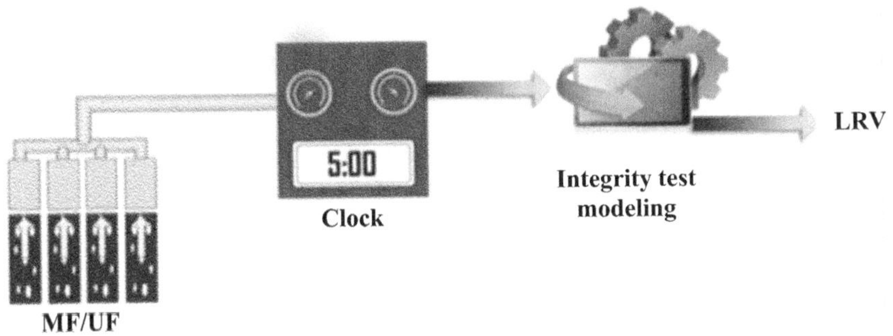

Figure 17.35. *Illustration of the integrity test and its modeling. For a color version of this figure, see www.iste.co.uk/gaid/watertreatment4.zip*

This diffusion must be considered when analyzing the results of the pressure drop integrity test. This diffusion can be more or less significant, depending on the membrane material. For example, it is much more significant for PES membranes than for PVDF membranes. PES membranes require a limited number of modules (60 for some suppliers) to have a reliable and readable test. With PVDF membranes, integrity tests can be performed on larger skids (or cell for submerged membranes) because diffusion is very limited. The degree of diffusion varies with water temperature and this parameter is incorporated in the operating instructions.

The pressure drop test takes place over a period of 5–10 min. Figure 17.36 shows the pressure drops obtained on various sizes of skids with a broken fiber.

Due to dilution, it is possible to maintain the performance of microorganism removal with a certain number of broken fibers on the system. This depends on the membrane used and the size of the system. The removal of a skid equipped with Hydracap 60 membranes (Hydranautics) with broken fibers can be calculated using the following equation:

$$LRV = -log\,(1 - exp^{(-\frac{5.74}{1.5}k)})$$

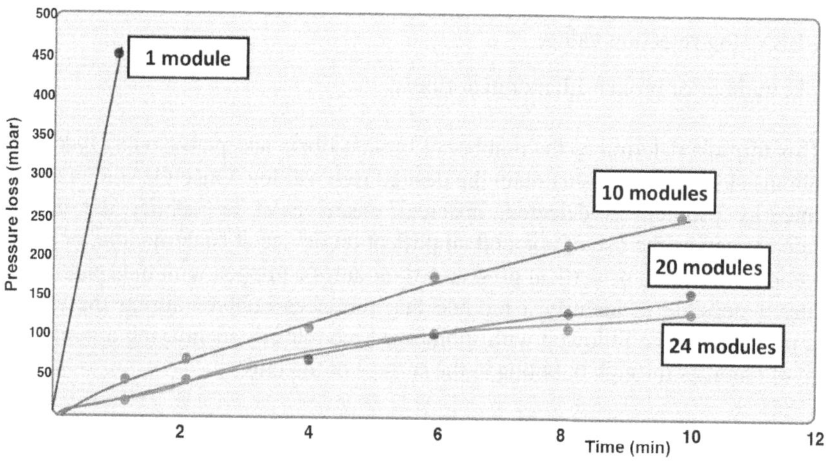

Figure 17.36. *Examples of PDT results with a broken fiber according to the size of the skids. For a color version of this figure, see www.iste.co.uk/gaid/watertreatment4.zip*

a) Evoqua's Memsep system for removing submerged membranes from the tank for a bubble test and repair after identifying the bubbling modules

b) Transparent tube for visual identification

Figure 17.37. *Identification systems of the modules to be repaired with submerged membranes (Evoqua type) in the tank (a) and pressurized membranes (hydranautics) with transparent tube (b). For a color version of this figure, see www.iste.co.uk/gaid/watertreatment4.zip*

With:

– LRV: log rejection value;

– k: number of broken fibers/intact fibers.

This formula determines the number of broken fibers acceptable on a Hydracap 60 membrane skid in order to maintain the desired removal log. Once the skid is tested, if an integrity problem is detected, different means exist to identify the defective module(s) such as the use of a sound amplification device at each module, or a visual inspection in the case of vertical pressurized modules equipped with transparent tubes, making it possible to identify a module that produces bubbles during the test. The same protocol can be followed with submerged modules by identifying a module or a group of modules through bubbling at the surface of the tank.

17.9.5.2.3. Repair of broken fibers (isolation)

The identified module(s) is removed from the skid or cell and installed on a test station. The bubble point test is then performed in reverse, pressurizing the module to 0.7–1 bar on the outside of the fibers using a repair kit. Damaged fibers are identified by their bubbling and are plugged with a needle or an adhesive given by the membrane manufacturer, and approved for use in drinking water (Figure 17.38).

17.10. MF and UF's place in a treatment process

MF and UF's place in a treatment process depends on the quality of the water to be treated and the treatment objectives. They can be placed directly on the raw water, after a settling or flotation stage, after a sand filtration or an activated carbon filtration.

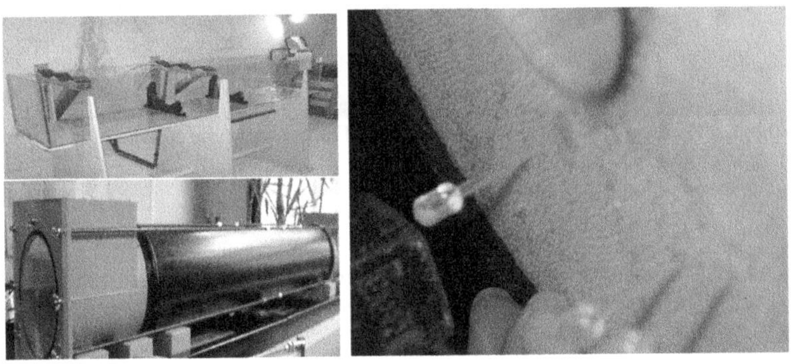

a) Test station for a module

b) Repair of defective fibers

Figure 17.38. *Example of a fiber repair system. For a color version of this figure, see www.iste.co.uk/gaid/watertreatment4.zip*

17.10.1. *Turbidity and SS*

The MF membrane can be placed directly on raw karstic water with turbidity variations up to 800 NTU.

The UF membrane is used in frontal filtration for water with an average turbidity $\leq$ 5 NFU, and occasional peaks up to 15 NFU. For certain types of resources (such as karst water), water with turbidity peaks up to 100 NFU can be treated on UF membranes, but special provisions must be made in the treatment cycles. Often, national regulations specify that a UF membrane can only be used directly on raw water when the suspended solids concentration of the raw water is less than 25 $mg \cdot L^{-1}$ or about 15 NFU.

17.10.2. *TOC (and UV254)*

It is considered that MF and UF membranes do not retain dissolved organic matter. This is the case even if we observe that about 10% of DOC contained in raw water is retained by UF membranes by adsorption. For waters with more than 2 $mg \cdot L^{-1}$ of TOC, this adsorption will induce drops in permeability and it will then be necessary to provide a sufficient frequency of CEB with caustic soda (and even additional chemical cleaning for very difficult waters).

For these reasons, for water loaded with TOC, the UF membrane is installed or positioned after coagulation–flocculation–settling or after coagulation–flocculation–settling–filtration (sand or activated carbon).

17.10.3. *Algae*

Algae are well removed by MF and UF. The corresponding removal is estimated at 5 log. On the other hand, the presence of algae will quickly lead to clogging issues in the membranes. It is assumed that a maximum concentration of 2,000 $cells \cdot mL^{-1}$ of algae in the feed water is acceptable in order to maintain constant water production. For higher concentrations, a suitable pre-treatment such as Spidflow® (Veolia process of flotation) should be provided. In addition, the sizing will take into consideration a higher frequency of chemically enhanced backwash (CEB).

17.10.4. *Iron and manganese*

Membranes only retain these metal ions if they are in particulate form. Special attention must be paid to the presence of dissolved iron and manganese in the raw water. A layer of iron and manganese oxide can be deposited on the membrane during filtration, but also during backwashing. However, to a certain extent, CEB acid can limit this clogging effect. On the other hand, during CEB with chlorine, depending on the content of the water, iron and manganese precipitation can take place outside and inside the membrane (the slower kinetics of manganese oxidation cause a higher risk with the latter).

17.10.4.1. *Treatment of low turbidity water, free of pesticides, micropollutants and organic matter*

In this application, where only turbidity treatment and a physical barrier to microorganisms are required, most types of MF or UF membranes are applicable. Submerged membranes can be used, as well as pressurized membranes with internal–external or external–internal filtration.

Micro-coagulation (Figure 17.39) is sometimes performed upstream of the membranes (addition of a very low dosage of coagulant) to improve the filterability of the water. The coagulant dosage is low (in the range of 0.1–0.5 ppm metal).

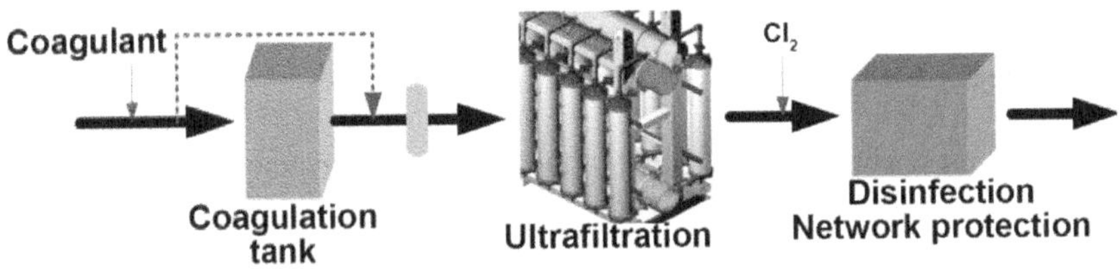

Figure 17.39. *Treatment of low turbidity water with direct UF membranes. For a color version of this figure, see www.iste.co.uk/gaid/watertreatment4.zip*

17.10.4.2. *Treatment of slightly turbid water with low levels of pesticides, micropollutants and organic matter (DOC < 2.5–3 mg·L^{-1})*

This is the Opaline C process (Veolia). The dosages of PAC applied range between 2 and 10 g·m^{-3}, depending on the concentration of pesticides, micropollutants and the DOC.

Since the water is not very turbid, it is possible to use submerged membranes (external–internal filtration) or pressurized membranes with internal–external filtration (Figure 17.40).

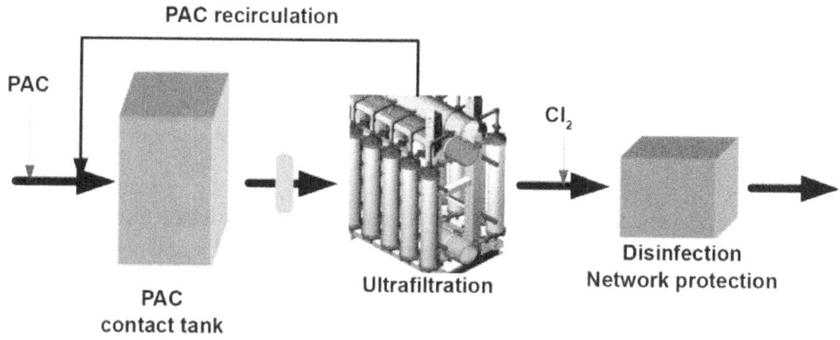

Figure 17.40. *Combination of activated carbon and ultrafiltration membranes. For a color version of this figure, see www.iste.co.uk/gaid/watertreatment4.zip*

The PAC is rejected (sent to sludge treatment) during backwashes, the frequency of which will be adapted to the load brought by the PAC. Tangential filtration is recommended for this application with UF. On a UF membrane, the filtering cycle takes between 30 and 45 min.

GAC filtration is also a solution that is applicable to this type of water, containing a turbidity lower than 2 NTU and a DOC < 2.5–$3\ mg{\cdot}L^{-1}$.

A third combination is possible with ozonation followed by GAC filtration in order to remove pesticides, micropollutants and to biologically treat organic matter. However, beware of the presence of certain types of pesticides, whose biological degradation is known to lead to the formation of other pesticides that are less adsorbable on activated carbon.

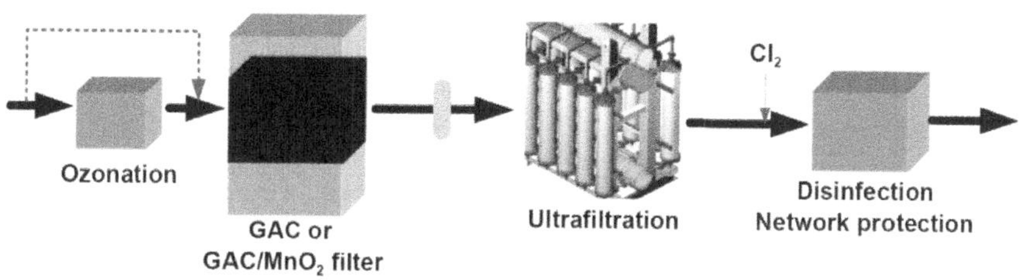

Figure 17.41. *Ultrafiltration placed in refining behind an oxidation–adsorption–biology combination. For a color version of this figure, see www.iste.co.uk/gaid/watertreatment4.zip*

UF is placed in refining. This is the Opaline S process (Veolia). In the presence of manganese or metal ions, the filter can be a GAC-MnO_2 dual media filter in which the manganese dioxide will treat the metal ions (Figure 17.41).

17.10.4.3. *Treatment of slightly turbid water, heavily loaded with pesticides, micropollutants and/or DOC*

The Opaline S process (Veolia) makes it possible to consider different treatment methods for raw water with an average turbidity ≤ 5 NFU and a presence of pesticides, micropollutants and/or a total organic carbon concentration lower than 4–5 $mg{\cdot}L^{-1}$.

In the case of a Multiflo Carb, an Opacarb FL, a Filtraflo Carb, an Opacarb MF or an Opacarb MG, the UF unit can be placed directly behind these technologies (Figure 17.42). The UF's operation is tangential. A micro-coagulation (Al or Fe) and a pre-filter (120–140 μm) are provided upstream of the UF.

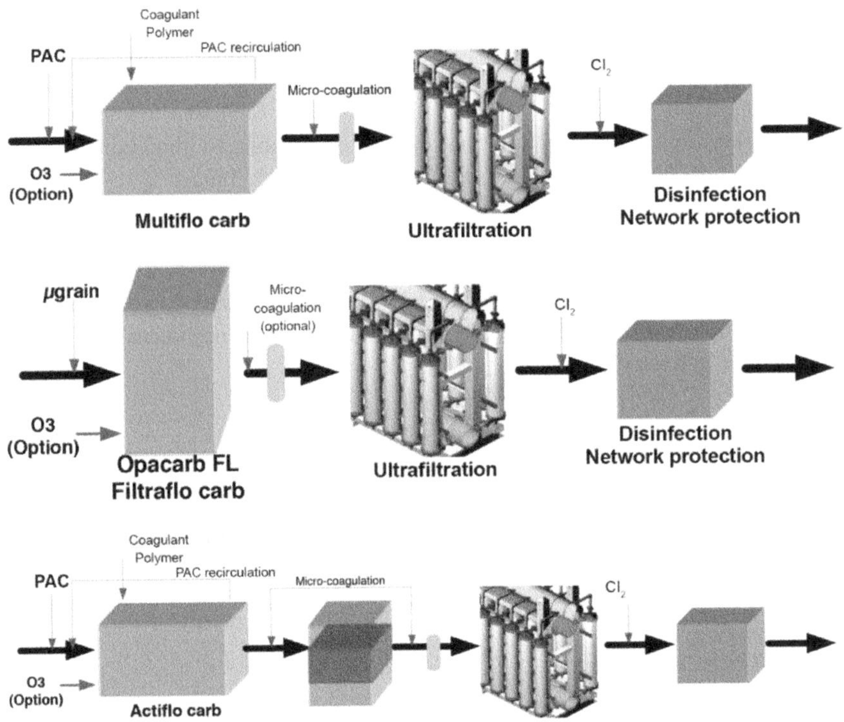

Figure 17.42. *Different possibilities of the Opaline S process. For a color version of this figure, see www.iste.co.uk/gaid/watertreatment4.zip*

In situ ozonation in the activated carbon reactor may be an interesting option to consider in order to increase removal efficiencies and reduce activated carbon dosages.

When an Actiflo Carb is proposed, a sand or dual media filtration must be installed between the settler and the membrane unit. A micro-coagulation (dosage between 0.05 and 0.25 ppm in Fe or Al metal) is installed or positioned upstream of the filters and upstream of the membranes.

After the sand filter, both submerged membranes (MF with external–internal filtration) and pressurized membranes (UF with internal–external filtration) can be used.

17.10.4.4. *Treatment of highly turbid water, heavily loaded with pesticides, micropollutants and (TOC 3–14 mg·L^{-1})*

The treatment file implemented for this type of application consists of three barriers (Figure 17.43): the first barrier is responsible for removing turbidity and some of the organic matter by means of advanced coagulation (see Volume 2, Chapter 7, Section 7.4.2.1). If algae are reported, a settling stage or flotation stage can be proposed. The second barrier is responsible for removing the organic carbon residual, as well as pesticides and micropollutants. The third barrier consists of a polishing of the water quality by removing the residual turbidity and a large number of microorganisms.

When the second stage is a PAC reactor (Actiflo Carb or Multiflo Carb) for the treatment of pesticides, micropollutants and TOC, the PAC is recirculated in the reactor and the purges are re-injected into the first stage, allowing the reactor to use the eventual residual adsorption capacity of the PAC, for the removal of additional pesticides and DOC.

In addition, the new PAC is injected on water that has already been clarified, which reduces the competition between pesticides and suspended solids. The PAC is then more efficiently used than in the case of an injection upstream of the decanter.

When the second stage is a micro-grain activated carbon reactor (Opacarb FL, Filtraflo Carb or Opacarb MG) for the treatment of pesticides, micropollutants and TOC, the used µgrain-activated carbon is extracted from the reactor and is not re-injected upstream, because this activated carbon is regenerable. It is stored on site, then recovered by the supplier in order to regenerate it and restock the site.

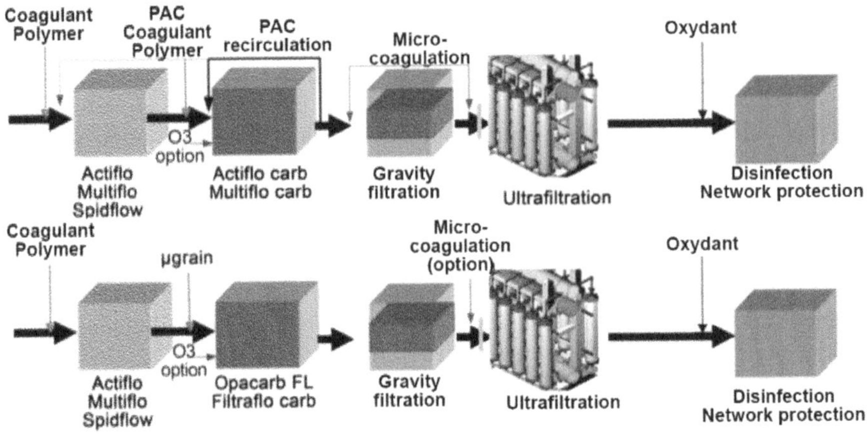

Figure 17.43. *Three-barrier treatment system. For a color version of this figure, see www.iste.co.uk/gaid/watertreatment4.zip*

17.10.4.5. *Treatment of highly turbid water (or algae), moderately loaded with pesticides, micropollutants and/or (COD < 3 mg·L^{-1})*

The processes used for this type of application consist of settling or flotation (if the presence of algae is reported), followed by filtration on granular activated carbon (Figure 17.44). These may be preceded by an ozonation stage. The UF membrane acts as a refiner (turbidity and micro-organisms). A punctual addition of PAC is often provided in the event of peaks in contaminants.

A PAC reactor or a micro-grain carbon reactor can be alternative solutions, but it depends on the concentrations of pesticides and micropollutants.

The economic cost is to be estimated between the GAC lifetime (and thus its renewal) and the dosage of activated carbon (powder or micro-grain) to be injected daily. It is important that new activated carbon is injected continuously (even at a low dosage: 2 g·m^{-3}), in order to maintain an available adsorption capacity in the reactor in case of a sudden peak in organic substances.

With the O_3-GAC solution, a sand filtration is very often placed between the decanter and the ozonation stage. This depends on the quality of the settled water and the installed ozonation system. For example, behind an Actiflo, the turbidity is less than 1.2 NTU on average and less than 2 NTU for the majority of the time. This is acceptable for conventional ozonation systems.

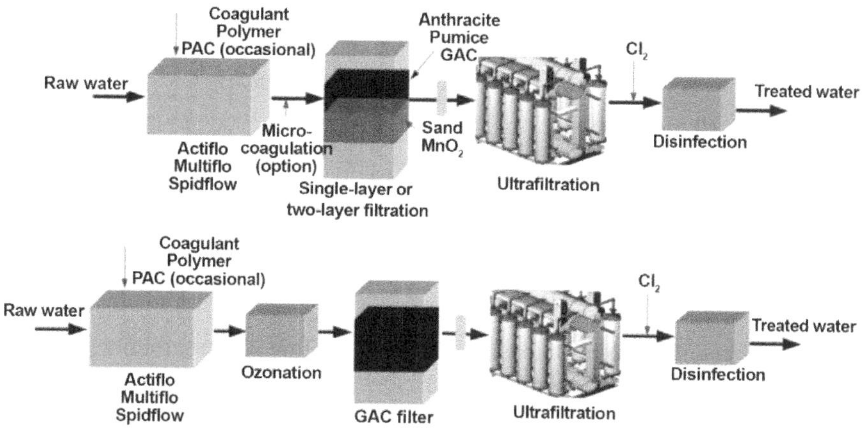

Figure 17.44. *Conventional settling-filtration GAC-ultrafiltration. For a color version of this figure, see www.iste.co.uk/gaid/watertreatment4.zip*

Alternatively, ozonation systems in which ozone is injected into a static mixer and then sent to an ozonation chamber do not have to deal with turbidity.

17.10.4.6. *Treatment of water with variable turbidity (2-500 NTU)*

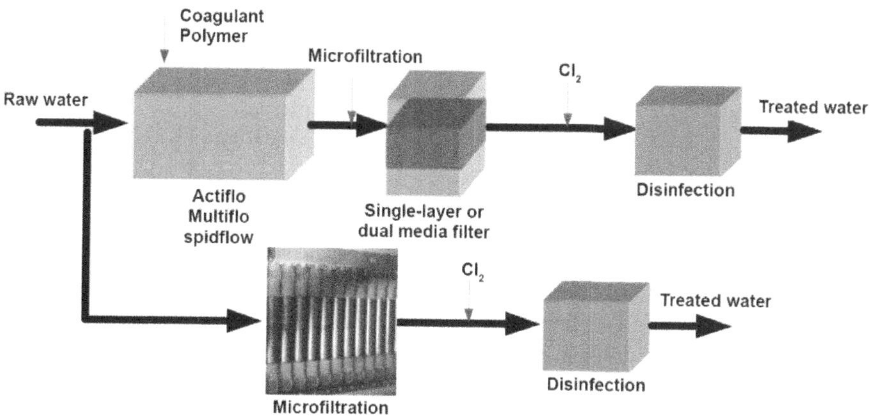

Figure 17.45. *Microfiltration instead of clarification. Top line with DOC, bottom line without DOC. For a color version of this figure, see www.iste.co.uk/gaid/watertreatment4.zip*

In this application, where only turbidity treatment and a physical barrier to micro-organisms are desired, most types of MF membranes are applicable (Figure 17.45), either with submerged membranes or pressurized membranes with external–internal filtration.

A disinfection that is specifically adapted to the viruses is necessary, as well as to the protection of the network.

17.10.4.7. *Treatment of slightly turbid water (<10 NTU on average and <30 NTU in peak), with low levels of pesticide and/or (COD < 3 mg·L^{-1})*

This application concerns karstic waters or reservoirs, where turbidity is low for a large part of the year and where there are few pesticides, micro-pollutants and/or DOC (Figure 17.46).

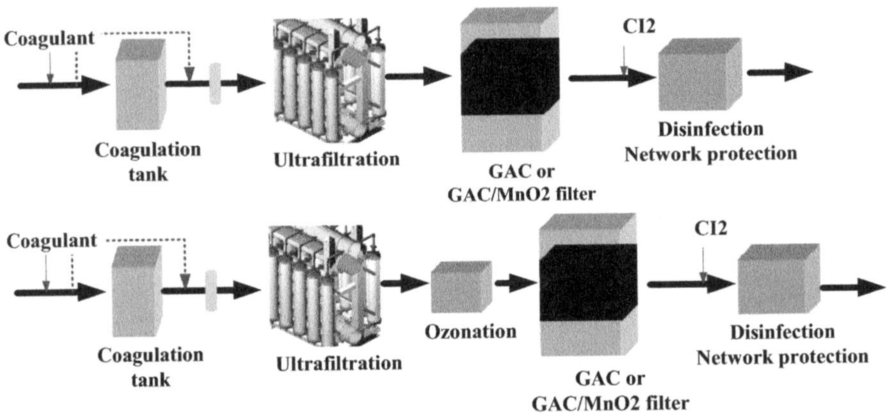

Figure 17.46. *Diagram of a process with a UF at the head, followed by a GAC or O_3-GAC filter. For a color version of this figure, see www.iste.co.uk/gaid/watertreatment4.zip*

The installation of a UF membrane at the head of the filter removes turbidity and micro-organisms. The downstream GAC filter removes organic contaminants. The advantage of such a solution is that the washing frequencies of the GAC filter are low and the issue of fines is less problematic. In addition, the GAC filter can be bypassed during certain periods when the contaminants are low or absent, and still produce good quality water.

In summary, the application of UF to the various compounds and elements present in water can be summarized as shown in Figure 17.47.

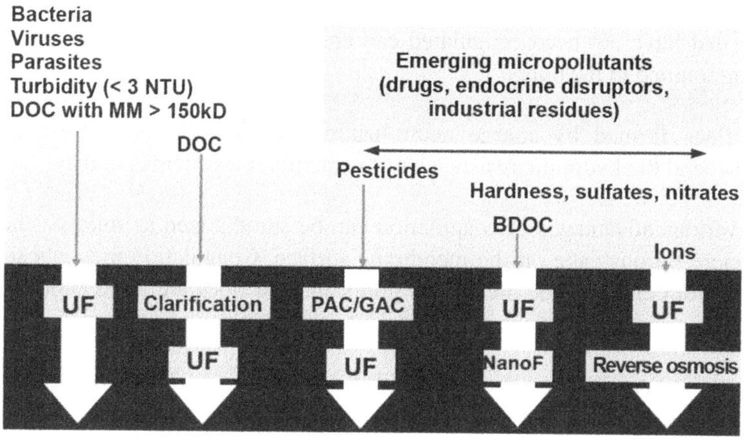

Figure 17.47. *Processes associated with ultrafiltration for the removal of compounds and ions from water. For a color version of this figure, see www.iste.co.uk/gaid/watertreatment4.zip*

17.11. Combination of coagulation and UF membranes

In drinking water production, coagulation as a pre-treatment process for UF membrane filtration improves the applied flow levels, as well as helps to remove some of the natural organic matter (NOM). This is because the decrease in flow of an operating UF membrane can be due to the presence of NOM, which promotes organic clogging, and to turbidity, which leads to mineral clogging.

NOM is a highly heterogeneous mixture of organic substances that vary in acidity, molecular weight, molecular structure and charge density. Since coagulation can effectively remove hydrophobic substances present in NOM (see section 7), it can therefore slow the decrease in permeability and flow. On the other hand, hydrophilic substances that are poorly removed tend to clog membranes and reduce permeability. DOC removal efficiencies are around 30–35% and UV 254 removal efficiencies are greater than 60% for conventional dosages of 30–40 $g \cdot m^{-3}$ of aluminum salt. Figure 17.48 shows the evolution of the TMP when the UF is supplied with raw water, and when coagulation is arranged upstream. With a raw water DOC of 2.6–3 $mg \cdot L^{-1}$, the TMP rose to 400 mbar after 8 days of operation without the addition of coagulant, whereas it remained fairly stable (at 250 mbar) with coagulation. The residual DOC after coagulation was 1.8–2.1 $mg \cdot L^{-1}$. Coagulation reduces reversible clogging by agglomerating small particles into large particles that are then discharged during backwashing. On the other hand, small particles that have not been coagulated can create irreversible particle clogging by remaining trapped in the pores.

The flocs formed by charge neutralization are less compressible and more permeable, and the hydraulic resistance of the membrane system is higher.

The various advantages of coagulation can be summarized as follows: the flocs form a more porous cake on the membrane surface. Coagulation prevents some of the NOM from entering the membrane pores and the flocs are large enough not to settle on the membrane surface due to various re-suspension phenomena.

Coagulation can be done in a tank (with a contact time of 2 min) or in line. In the case of low temperature, coagulation in a tank is recommended. In the case of direct treatment of surface water (with the risk of the presence of algae), the coagulant dosage upstream of the membranes is between 1 and 4 mg Al or Fe/mg of TOC.

Figure 17.48 illustrates the effect of micro-coagulation installed on settled water, to which excess polymer has been added. Permeability stability is observed with a coagulant dosage of 0.2 ppm Al. The permeability drops by 5.2 $L \cdot m^{-2} \cdot h^{-1} \cdot bar^{-1}$ for a very low dosage (0.03 ppm Al), which does not manage to neutralize the residual polymer, impacting the permeability.

Micro-coagulation is recommended upstream of the UF membrane when the feed water has undergone a pre-treatment, during which a polymer residual may be present. For example, downstream of an Actiflo or Actiflo Carb, dosages of 0.05–0.25 ppm in Al or Fe are recommended.

It also appears that the choice of coagulant is a parameter to consider. Indeed, the addition of ferric chloride ($FeCl_3$, 6 H_2O at 41%) implies a lower decrease in permeability than that observed with aluminum sulfate as coagulant. The reason for this is probably the formation of flocs with different characteristics, with iron hydroxide flocs being larger than aluminum hydroxide flocs and of lower density. As a result, the cake layer formed on the membrane surface is more complex and permeable. The higher the dosage of $FeCl_3$, the lower the decrease in permeability, which is explained by a greater removal of dissolved organic matter and colloidal matter. The floc cake created on the membrane retains part of the coagulated NOM, as well as small flocs. The compression of the cake is reversible because as the TMP decreases, the specific resistance of the cake also decreases.

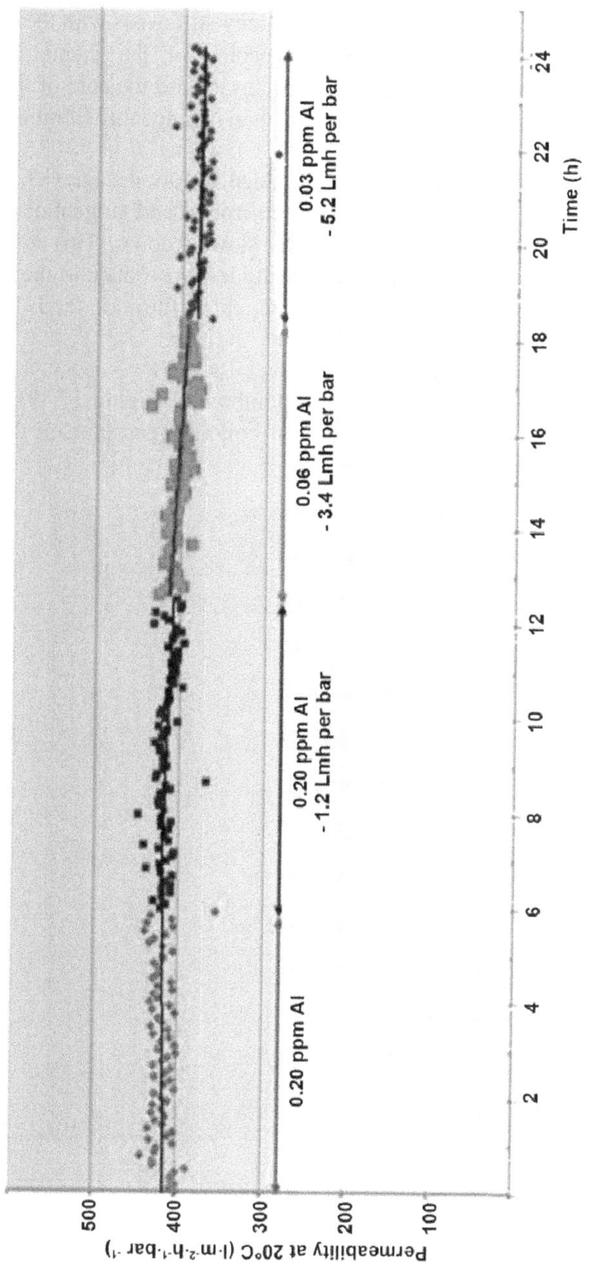

Figure 17.48. *Effects of micro-coagulation upstream of an ultrafiltration membrane. For a color version of this figure, see www.iste.co.uk/gaid/watertreatment4.zip*

In frontal filtration, an increasing dosage of coagulant implies a progressive decrease in clogging capacity. In tangential filtration, the impact of coagulation is less pronounced, although the decrease in clogging capacity still exists with the addition of coagulant. This is because the continuous sweeping of the membrane surface accelerates the back-transport of colloids and particles toward the core of the flow. As a result, clogging is more rapid in frontal filtration than in tangential filtration.

When coagulant such as ferric chloride is added at low dosage (<1 mg Fe/mg TOC), the difference in clogging kinetics between frontal and tangential filtration is reduced because clogging in frontal filtration is slowed down. This is due to the formation of a more porous cake, therefore offering less resistance to the passage of water. In addition, the recirculation of water in tangential filtration tends to break up the flocs or prevent their growth.

The same observations are made with MF membranes (Figure 17.49), for which adsorption of humic substances and retention of colloidal particles on the external wall participate in organic and mineral clogging.

a) b)

Figure 17.49. *Microfiltration module (a) without coagulation and (b) after Al coagulant injection. For a color version of this figure, see www.iste.co.uk/gaid/watertreatment4.zip*

The addition of an aluminum-based organic coagulant (WAC HB) at dosages ranging from 40 to 60 g·m^{-3} allowed the membrane to be cleaned and the permeability to be restored.

17.12. Combination of PAC and UF

UF coupled with PAC adsorption, in an integrated configuration according to the implementation of the Opaline C process, allows the removal of organic matter, pesticides, micropollutants, molecules responsible for tastes and odors, as well as micro-organisms. This Opaline C process developed and implemented by Veolia is applied in the case of slightly turbid waters with variable concentrations of pesticides and micropollutants, the TOC of raw water being on average between 2 mg·L^{-1} and 3 mg·L^{-1}, with peaks of 4 mg·L^{-1}.

The combination of PAC and UF membranes involves a vertical configuration of the modules. Depending on the quality of the raw water and the amount of activated carbon to be maintained in the PAC reactor, which is then applied to the membranes, the operation can be frontal or tangential.

Frontal filtration consists of a circulation of water perpendicular to the membrane. All the retained matter accumulates on the membrane. It is frequently evacuated by hydraulic backwashes.

In tangential filtration, retained substances are captured on the membrane surface and can be removed from the module continuously by sweeping the water that flows parallel to the tangential membrane. The Opaline C process offers many advantages, such as high efficiency, a small footprint and flexible operation.

PAC adsorbs some of the various organic substances and reduces the potential for organic clogging of the membranes. When the concentration of organic substances is high, which implies high dosages of PAC, the membrane filtration must be tangential in order to avoid an accumulation of matter on the membrane, which in the long term makes the clogging irreversible. The PAC deposited on the membrane surface can

– decrease the efficiency of organic substances removal by reducing its concentration in suspension in the interstitial water;

– promote biological clogging through the development of biomass on the PAC;

– indirectly participate in the increase of the organic substances–membrane surface interactions;

– increase the risk of fiber breakage by creating tension forces on the membrane and reducing permeability.

It appears that in the absence of organic matter (such as in groundwater), the PAC is deposited on the surface of the membrane and forms a layer that is easily removed by backwashing. Pesticides and micro-pollutants have no influence here, because they are at concentrations that have little impact on the interactions with the membrane.

Frontal mode UF of water containing natural organic matter (2–4 $mg \cdot L^{-1}$), and a calcium concentration around 40–50 $mg \cdot L^{-1}$ (100–125 $mg \cdot L^{-1}$ $CaCO_3$), promotes the formation of a complex layer on the membrane that accumulates over time, despite backwashing. This heterogeneous layer has a low porosity, because the accumulation of organic substances can reduce the interstitial spaces between the activated carbon particles. The permeability decreases over time. The same water treated in a tangential filtration mode regularly sweeps the membrane surface and avoids deposits that are detrimental to the proper functioning of the UF. A recirculation loop is created between the concentrate outlet and the raw water inlet. This loop creates a constant flow along the fibers of the membranes and prevents their early clogging. The skids are sized to handle the instantaneous raw water flow, plus a concentrate recirculation flow (up to 25% of the raw water flow to the membranes). The concentrate recirculation loop used to ensure tangential operation of the membranes controls the PAC concentration in the contact tanks, located upstream of the UF.

A PAC matter balance on the recirculation loop quantifies the PAC concentration in the reactor.

17.13. Performance and guarantees

The performance and guarantees are linked to the membrane used, and the guarantees given by the supplier. In addition, Veolia provides additional guarantees based on its experience and feedback from the field.

17.13.1. *Turbidity*

With MF or UF in the final stage, the turbidity of the treated water is below 0.2 NFU (100% of the time). Conventional systems are already capable of delivering water with a turbidity of less than 0.5 NFU at the sand filter outlet in the first stage, and even less than 0.3 NFU at the granular activated carbon outlet in the second stage. However, in the case of strong degradation of the feed water, these performances can be altered.

MF and UF offer an additional guarantee of quality. Indeed, the quality of the water at the outlet is not very sensitive to the quality of the water at the inlet.

Figure 17.50 illustrates the performance of a pressurized MF membrane on karstic water (Veolia site), which is characterized by sudden and high turbidity variations. Thus, the MF membrane treats turbidity both as low as 10 NTU or 500 NTU, with an output turbidity lower than 0.1 NTU.

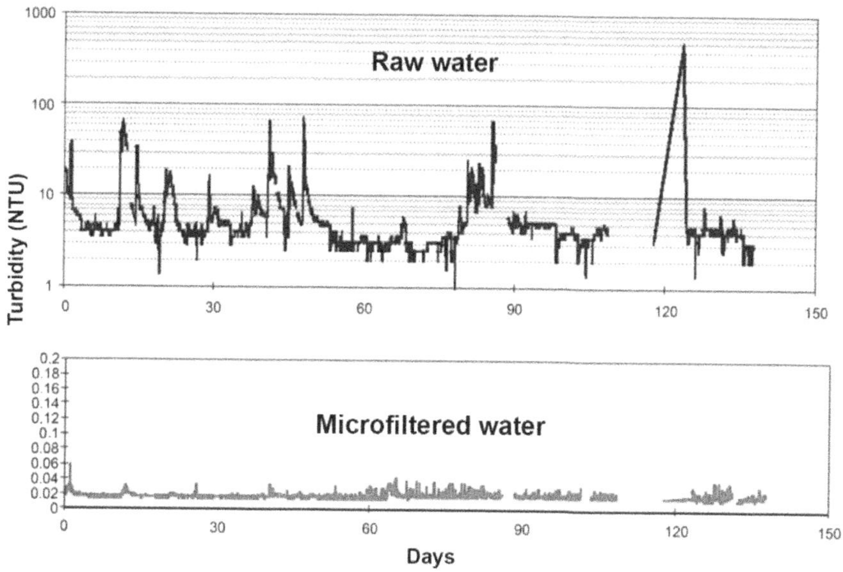

Figure 17.50. *Turbidity variation and its removal by means of microfiltration. For a color version of this figure, see www.iste.co.uk/gaid/watertreatment4.zip*

Therefore, when these membranes are used for polishing after conventional filters, their role with respect to turbidity is to ensure that the quality of the treated water is stable at less than 0.2 NFU (100% of the time) and 0.1 NFU 95% of the time.

When used on raw or settled water, MF and UF membranes can produce treated water at less than 0.2 NFU, in a very stable way.

Although the turbidity of the water obtained by UF is less than 0.2 NTU (100% of the time) and fully meets the objectives, direct use on surface water is only possible in the case of a resource with a low organic content. This limitation is due to the fact that the clogging phenomenon is exacerbated in the presence of natural organic matter, and that the regulations concerning the formation of disinfection by-products cannot be complied with, since the measured UV and TOC reductions are too low and therefore insufficient to avoid the presence of these by-products.

17.13.1.1. *Algae impact and removal efficiency*

A UF membrane can accept 10^6 units·L^{-1} of micro-algae without the permeability being degraded. For higher concentrations, rapid clogging of the membranes occurs according to a pore blocking mechanism with a very rapid drop in permeability, requiring very frequent chemical cleaning, or even a stop in production (Figure 17.51).

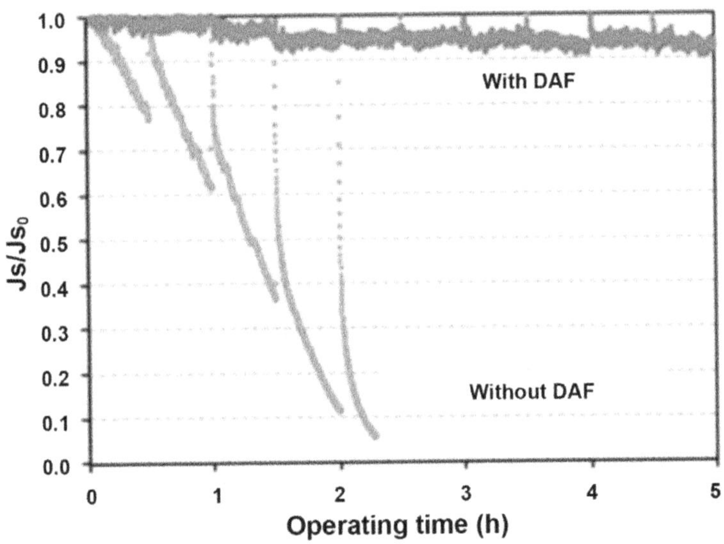

Figure 17.51. *Flow drop of a UF membrane with water containing 5 × 107 units·L^{-1}*

The implementation of dissolved air flotation (DAF) stabilizes the flux, while a flux shutdown can be observed in less than 2 h by feeding the membranes directly with raw water. In fact, the algae cells themselves are not responsible for the potential for membrane clogging, but rather for the exo-polysaccharides (EPSs) they secrete or for a combination between EPS, microalgae and organic substances. Living microalgae cells settle on the membrane surface and release extracellular material during filtration. The released polysaccharides can bind to other organic species and increase the filtration resistance. In addition, cake formation and partial blockage of the pores (Figure 17.52) reduce their diameter, and are other mechanisms involved in the decrease in flow and clogging irreversibility.

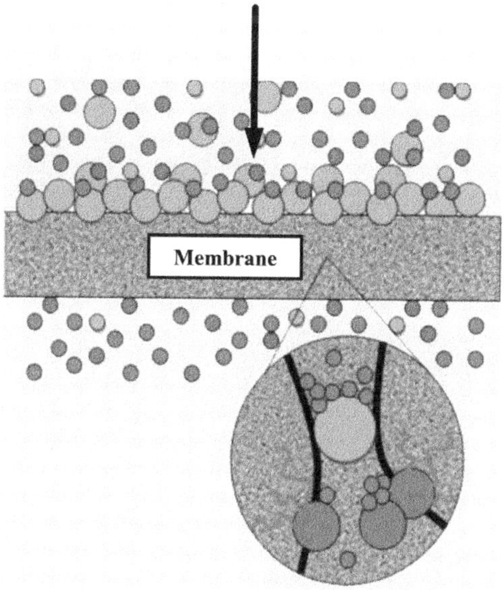

Figure 17.52. *Clogging by steric exclusion. For a color version of this figure, see www.iste.co.uk/gaid/watertreatment4.zip*

Organic matter also participates in clogging through its adsorption on the membrane surface. Figure 17.53 shows an electron microscope view of a clean MF membrane (0.2 μm), clogged with microalgae. The micro-algae cells that form the cake on the membrane are about 10 μm in diameter, and the binder between these cells is also present in the cake.

General speaking, EPS concentrations are often related to algae or chlorophyll concentrations. Chlorophyll measurement can be a good predictive indicator. Increasing EPS in the UF feed water most often results in a linear increase in TMP, through clogging potential and an exponential increase in irreversible clogging.

When the chlorophyll concentration exceeds 5 $\mu g \cdot L^{-1}$, special attention must be paid to the operation of the membranes. Above 10 $\mu g \cdot L^{-1}$, emergency operating conditions are applied, such as a reduction in flow, the implementation of coagulation (Figure 17.54) and higher cleaning frequencies. However, these conditions are restrictive for the operator and the reduction in flux means a reduction in production. This can be a real problem with regard to guarantees given to the customer.

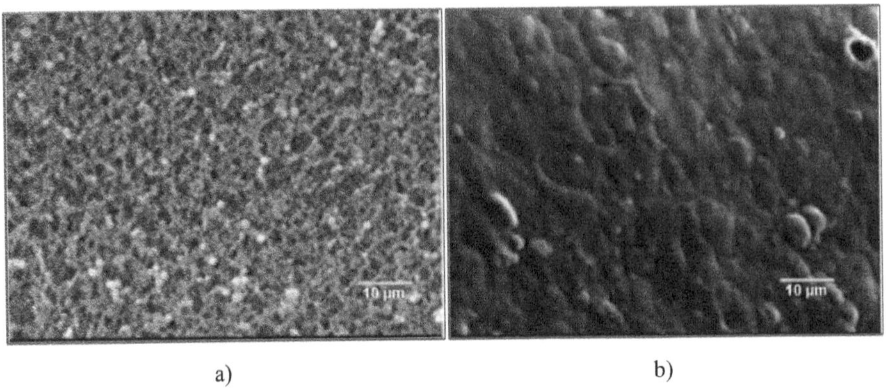

Figure 17.53. *Electron microscope view of (a) a clean MF membrane (0.2 µm) and (b) one clogged with micro-algae*

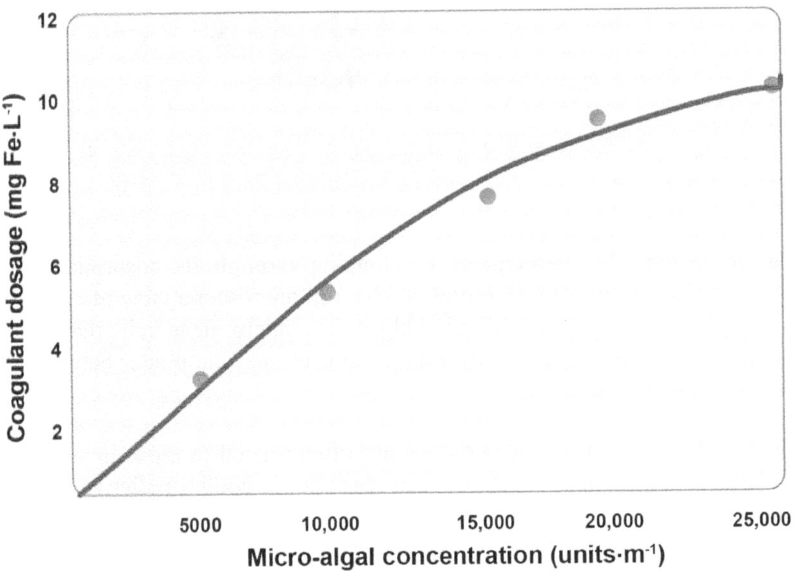

Figure 17.54. *Coagulant (Fe) dosage according to the micro-algae concentration*

Coagulant dosage ahead of the UF can reduce the number of EPS that feed the UF, while simultaneously reducing its rapid clogging (Figure 17.55).

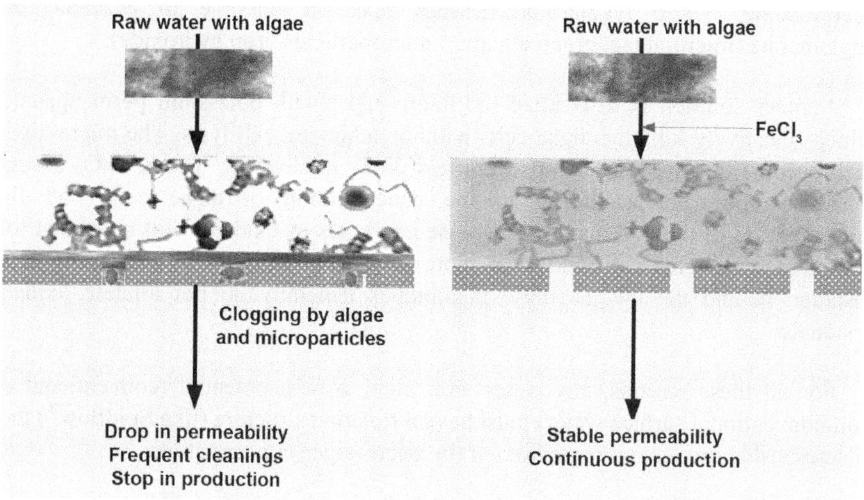

Figure 17.55. *Clogging of UF membranes with micro-algae. For a color version of this figure, see www.iste.co.uk/gaid/watertreatment4.zip*

The MFI_{UF} indicator ($s \cdot L^{-2}$), which corresponds specifically to clogging by cake formation on the membrane surface, is given by the following relation:

$$MFI\ (UF) = \frac{\mu 20°\ I}{2\ \Delta Po\ Ao^2}$$

With:

– μ: dynamic viscosity of water ($kg \cdot m^{-1} \cdot s^{-1}$ corrected at 20°C);

– I: index related to the properties of the particles to form a layer on the membrane, by creating hydraulic resistance (m^{-2});

– ΔPo: reference TMP (bar);

– Ao: membrane surface area (m^2) used for water filtration with a cut-off of 0.45 μm ($Ao = 13.8 \times 10^{-4}\ m^2$).

MFI_{UF} is correlated with EPS and chlorophyll concentration.

In-line coagulation with ferric chloride, which traps micro-algae and organic matter, stabilizes permeability. The coagulation consists of flocculating the materials deposited on the surface, and in forming a cake that can be easily evacuated by backwashing. These regular backwashes make it possible to evacuate the agglomerate (micro-algae, organic matter, microparticles, iron hydroxide).

Another solution is to oxidize the micro-algae with potassium permanganate, which selectively kills the algae cells without achieving cell lysis. The micro-algae then become inactive particles that are less prone to clogging. The $KMnO_4$ dosage must be optimized according to the concentration of algae cells and the concentration of organic matter, otherwise an excess of oxidants will be present on the membrane. This can lead to a more serious problem if reverse osmosis is installed behind the UF, as these membranes generally do not tolerate oxidant residuals.

For all these reasons, any water with algal growth potential (conventional or non-conventional surface water) must have a flotation structure (like Spidflow®) that ensures stable production regardless of the micro-algae concentration.

17.13.1.2. *Downstream chlorination*

The fact that almost all micro-organisms and parasites are retained on the membranes leads to a much lower consumption of chlorine, whose addition is only intended to protect the network.

A final chlorine dosage is, however, required in order to ensure a reliable quality of produced water (principle of multiple barriers), as well as the protection of the water against bacterial revival during its transport in distribution network.

The potential for the formation of chlorination by-products is therefore very low. This is an important point in the quality of the water distributed, both at the chemical level and other levels of more subjective parameters such as taste and odor, which remain among the best parameters for consumer appreciation.

17.13.1.3. *Reducing water losses*

In conventional systems, the recycling of decanter purge water and filter washing water at the head of the plant is subject to specific regulations in each country. The overall aim is to guarantee that potential parasites (*Cryptosporidium* and *Giardia*) do not escape as the water is recirculated, and that no risk reaches the production of drinking water. In this respect, several recommendations mention a reduction of the turbidity of the water to be recirculated must be less than 2 NTU, or disinfection with UV or ozone, a filtration-disinfection, or an MF or UF membrane.

As MF and UF are considered an ultimate barrier to these substances, the risk is greatly reduced. The supernatant from the sludge treatment stage (such as the thickener) can thus be recirculated upstream of the treatment process.

The cleaning water can also be treated by a dedicated system with membranes in the second stage, followed by a recirculation of the permeate at the top, with appropriately sized flux for the quality of the water to be treated. This make is possible to strongly reduce water losses, which are then only limited to the dirty water coming from the second stage, and from the chemical backwash and cleaning in place (which is generally well below 1%).

17.13.1.4. *Removal of micro-organisms*

MF and UF membranes provide removal of micro-organisms and parasites of about 6 log. UF offers a 3–4 log barrier to viruses.

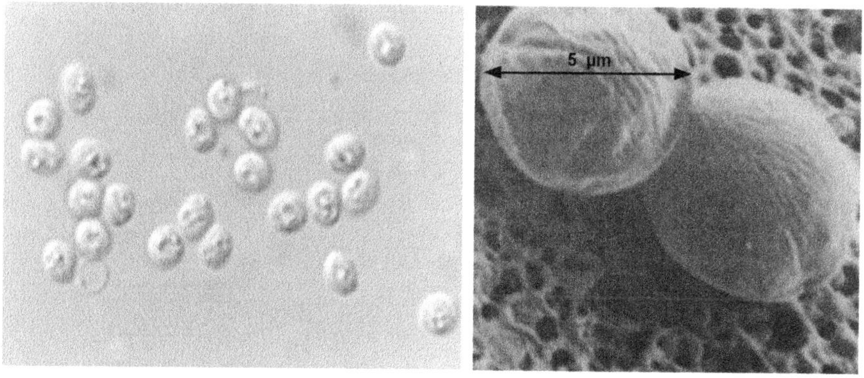

a) b)

Figure 17.56. *Cryptosporidium parvum protozoa (a) in water and (b) on an MF membrane. For a color version of this figure, see www.iste.co.uk/gaid/watertreatment4.zip*

Virus retention with low-pressure membranes (MF and UF) depends on membrane characteristics, module design and mode of operation. The efficiency of virus removal can be expressed by a log reduction value (LRV), which is defined as the ability of a filtration system to remove particles (in this case, particles > 0.2 µm). The LRV is given by:

$$LRV = \log(\frac{Cin}{Cout})$$

With:

– C in: concentration in the feed water;

– C out: concentration in the filtered water.

Viruses are very small (10–400 nm) and should be well retained (around 4 log) by an integrated UF membrane that is characterized by a MWCO < 150 kDa. Although this parameter is a good indicator, retention cannot be only predicted on the basis of nominal pore size or by MWCO.

More efficient and reliable retention is achieved with coagulation pre-treatment. Bacteriophage MS2 is most often used as a reference virus because it is similar in size, shape and nucleic acid to hepatitis A and enteric poliovirus. Various laboratory tests (VERI-Veolia sources) have shown, for example, that an MF membrane (0.2 μm) has less than 0.5 log MS2 removal, whereas coagulation with PAX or aluminum sulfate gave a removal rate of more than 3 log. The UF membrane (100 kDa) shows 4 log without a coagulant, and over 7 log with a coagulant.

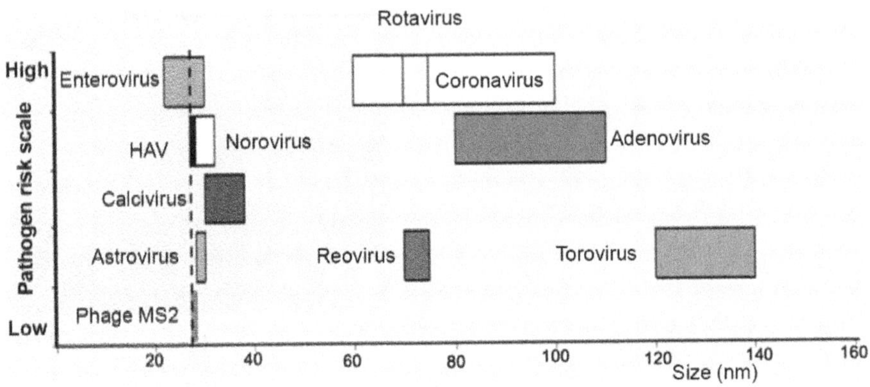

Figure 17.57. *Size of viruses present in water and associated pathogen risk. For a color version of this figure, see www.iste.co.uk/gaid/watertreatment4.zip*

For comparison, an LRV of 2–4 can be achieved with a coagulation–flocculation–settling treatment stage. The possible aggregation of the viruses between one another is not possible because the operating pH is between 6.5 and 7.2, and aggregation only occurs when the pH is close to the isoelectric point (3.9).

17.13.1.5. *Performance and guarantees on micro-organisms*

For the other parameters in dissolved substances (DOC, Fe, Al, Mn, etc.), there are no changes between the entry and exit of the membranes.

	Microfiltration (100% of the time)	**Ultrafiltration (100% of the time)**
Turbidity	≤0.2 NTU	≤0.2 NTU
Bacteria	≥5 log removal	≥5 log removal
Cryptosporidium	≥5 log removal	≥5 log removal
Giardia	≥5 log removal	≥5 log removal
Viruses	1 log	≥4 log removal

Table 17.10. *Examples of process guarantees in MF-UF*

17.13.1.6. *Expected performance with submerged membranes*

Applications	**Application of the submerged membrane (CMF-S® example)**
Surface water	Removal of turbidity and suspended solids Removal of bacteria Removal of parasites Removal of algae combined with the addition of a coagulant Removal of a part of DOC and color with the addition of a coagulant Injection of powdered activated carbon (PAC) for punctual or continual removal of pesticides and micropollutants Removal of oxidized iron and manganese (PVDF membranes)
Groundwater	Removal of turbidity and suspended solids Removal of bacteria Removal of parasites Injection of powdered activated carbon (PAC) for punctual or continual removal of pesticides and micropollutants Removal of oxidized iron and manganese (PVDF membranes)
Sea water	Removal of micro-particles including bacteria, colloidal particles and other suspended solids to ensure optimum operation of reverse osmosis membranes positioned downstream Removal of hydrocarbons and organic matter by possible addition of of PAC Makes it possible to obtain an SDI < 3

Table 17.11. *Applications of submerged membranes*

17.13.2. *Supplier warranty on the life of the membranes*

For drinking water applications, the lifetimes of organic hollow fiber membranes are in the range of 7–8 years. The mechanical warranty of 2 years is given by the supplier (100%) and the remaining number of years is often *prorata temporis*.

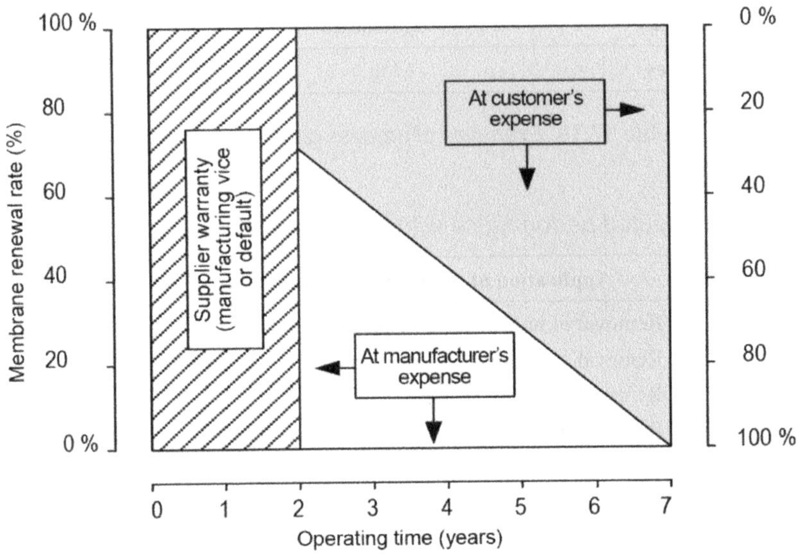

Figure 17.58. *Prorata temporis*

The supplier warranty covers:

– a guarantee in terms of water quality;

– a guarantee of hydraulic performance (production and water losses in the defined conditions of flow and washes);

– a guarantee on fiber breakage.

Certain restrictive clauses of the supplier impose access to the recording of the operating data, the respect of hydraulic constraints (rate of pressurization, etc.) and a particular maintenance of the modules.

Veolia has a tool for monitoring hydraulic parameters (Veolink), which is installed at many drinking water plants and provides access to the required information at any given time.

17.14. Advantages of MF and UF

Advantages of microfiltration	Advantages of ultrafiltration
Removal of micro-organisms including *Cryptosporidium* and *Giardia* Stabilization of turbidity to less than 0.2 NFU Removal of dissolved matter (including organic matter) is zero	Removal of micro-organisms, including *Cryptosporidium* and *Giardia* Removal of viruses Stabilization of turbidity to less than 0.2 NFU Reduction in disinfection reagents Reduction in THM formation Generally speaking (for ultrafiltration with a cut-off of 100–150 kDa), the reduction of dissolved matter is nil and the reduction of organic matter is very low (around 10%). Removal of organic matter can be observed when it is made up of organic matter with a very high molecular weight. However, this is a specific case for certain surface waters (highly colored waters).

Table 17.12. *Advantages of microfiltration and ultrafiltration for drinking water applications*

17.15. Veolia's experience

Veolia has been building and operating plants with membranes for more than 40 years, for drinking water, wastewater and industrial water. In drinking water, Veolia has participated in the construction and/or operation of more than 200 plants, and the extensive information reported from the various sites and their association with its research and development center (VERI) enable it to compare the performance of the various membranes on the market, to optimize the sizing of a drinking water plant, to give performance guarantees that exceed those of suppliers and to ensure optimized operating costs.

Parameters	Nitto Denko Hydranautics		BASF Inge		Cabot-Norit	
	HYDRAcap 60	HYDRAcap 80	Dizzer XL0.9MB60_W	Dizzer XL1.5MB40_W	SXL-225	SXL-55
Types of membrane	UF	UF	Multi-bore UF	Multi-bore UF	UF	UF
Fiber materials	Modified PES	Modified PES	Modified PES	Modified PES	Modified PES	Modified PES
Pore size, µm	0.02–0.025	0.02–0.025	0.02	0.02	0.01-0.05	0.01–0.05
Cut-off points, kDa	100–150	100–150	100	100	200	200
Fiber inner diameters, mm	0.8	0.8	0.9 × 7 capillaries	1.5 × 7 capillaries	0.8	0.8
Fiber outer diameters, mm	1.3	1.4	4	6	1.3	1.3
Horizontal or vertical configurations	Vertical	Vertical	Vertical	Vertical	Vertical or horizontal	Vertical or horizontal
Module dimensions: **– diameter, mm** **– total length, m**	250 1.708	250 2.216	250 1.68	250 1.68	200 1.537 (excluding Aquaflex fittings)	200 1.537 (excluding Aquaflex fittings)
Surface area of a module, m^2	46	62.2	38, 45, 50 and 68	25 and 40	40	55
Applications	Refining (behind sand filter or GAC), reverse osmosis pre-treatment, direct application for low turbidity water	Refining (behind sand filter or GAC), reverse osmosis pre-treatment, direct application for low turbidity water	More loaded water, recycling of dirty water from an MF/UF	Xiga: refining (behind sand filter or GAC), Aquaflex: direct application for low turbidity water or refining Seaguard: reverse osmosis pre-treatment on low turbidity water Seaflex: reverse osmosis pre-treatment on more loaded or pre-treated water	–	–

Parameters	Nitto Denko Hydranautics		BASF Inge		Cabot-Norit	
	HYDRAcap 60	HYDRAcap 80	Dizzer XL0.9MB60_W	Dizzer XL1.5MB40_W	SXL-225	SXL-55
Operating modes	**Hydracap 60**		**Dizzer**		**SXL 225: SXL 55**	
Chlorine tolerances (during cleaning)	100 ppm (exceptional max 200 ppm)		200 ppm		200 ppm	
Temperatures	≤40°C		≤40°C		1–40°C	
pH (during cleaning)	1.5–13		1–13		2–12 (max 1-13)	
Permeabilities in clear water $L \cdot m^{-1} \cdot bar^{-1}$ at 20°C	Very high--PES modified to make it hydrophilic > 600 Lmh/bar		Very high--PES modified to make it hydrophilic > 800 Lmh bar		High--PES modified to make it hydrophilic	
Usual types of filtration	Frontal, usually tangential for turbidity peaks		Frontal		Frontal	
Backwash flux (BW) $L \cdot m^{-1} \cdot h^{-1}$	250		230–250		–	
Air scouring (during washing)	No--possible air shocks (pressurization to 1 bar)		No		–	

Table 17.13. *Membranes on the market operating under internal–external pressure*

Parameters	Pall-Asahi	Polymem	Omexell-DOW			Evoqua	
	Microza UN620/A	**UF 120 S2**	**SFX-2680**	**SFX-2860**	**SFX-2880**	**S10T**	**S10V**
Types of membrane	MF	UF	UF	UF	UF	Submerged MF	Submerged MF
Fiber materials	PVDF	PS	H-PVDF	H-PVDF	H-PVDF	PP	PVDF
Pore size, um	0.1	0.01	0.03	0.03	0.03	0.2	0.04
Fiber inner diameters, mm	0.7	0.4	0.7	0.7	0.7	0.39	0.5
Fiber outer diameters, mm	1.2	0.72	1.3	1.3	1.3	0.65	0.8
Horizontal or vertical configurations	Vertical	Vertical	Vertical	Vertical	Vertical	Vertical	Vertical
Module dimensions: **- diameter, mm** **- total length, m**	 165 2.338	 315 0.925	 165 2.36	 225 1.86	 225 2.36	 131 1.186	 131 1.186

Parameters	Pall-Asahi	Polymem	Omexell-DOW			Evoqua	
	Microza UN620/A	UF 120 S2	SFX-2680	SFX-2860	SFX-2880	S10T	S10V
Surface area of a module, m^2	50	114	44	51	77	31.09	27.9
Chlorine tolerances	NEP to 250–500 ppm	200	2,000	2,000	2,000	No chlorine tolerated	200
Temperatures	<40°C	<35°C	1–40°C	1–40°C	1–40°C	< 45°C	<45°C
pH	1–10	2–12	1–11	2–11	2–11	2–14	2–10
Operating mode							
Usual types of filtration	Frontal	Frontal	Frontal	Frontal	Frontal	–	–
Backwash flux (BW): **- air/water phase** **- in water phase only**	36 Lmh (1.8 m^3/h per module) pdt 60 sec 80 Lmh (4 m^3/h per module) pdt 45 s	87–131 Lmh	100–150 Lmh	100–150 Lmh	100–150 Lmh	2.25 m^3/h/mod (72 Lmh)	2.25 m^3/h/mod (72 Lmh)
Air scouring (in BW)	7 Nm^3/h per module	Yes	Yes 7 Nm^3/h/module to 1 bar	Yes 12 Nm^3/h/module to 1 bar	Yes 12 Nm^3/h/module to 1 bar	12 Nm^3/h/module	12 Nm^3/h/module

Table 17.14. *External–internal pressurized and submerged membranes on the market*

Figure 17.59 summarizes the ever-present balance between performance, capital costs and operating costs.

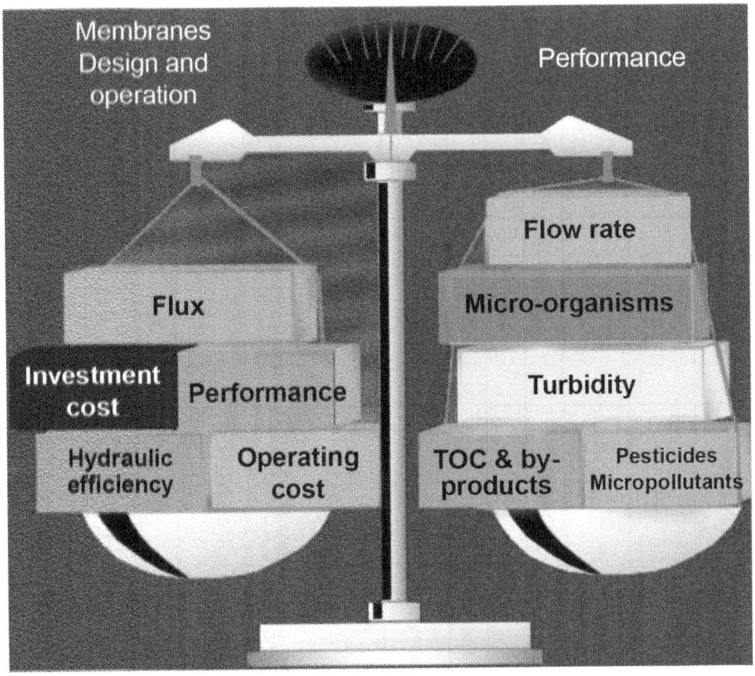

Figure 17.59. *Balance between performance, capital costs and operating costs. For a color version of this figure, see www.iste.co.uk/gaid/watertreatment4.zip*

Veolia has always analyzed and compared suppliers' membranes, whether they are on the market or under development. Given its long and rich experience, as well as the data bank from operating sites, many suppliers (Asia, the United States, Europe) ask Veolia to help them improve the performance of their membranes. This is why Veolia has references with many suppliers in MF and UF (Evoqua, Pall, Cabot-Norit, Hydranautics, Nitto-denko, Koch, Asahi, DOW, Samsung, Toray, etc.). Figure 17.60 illustrates some examples of membranes used and/or tested.

17.16. Appendix: sheets

The data sheets presented in Figure 17.61 summarize examples of MF and UF membrane implementations at various sites and in different configurations. This explains the different ways in which membranes can be integrated into a treatment system, depending on the quality of the water to be treated and the treatment objectives. For each project, various considerations are taken into account:

– membrane type and selection of the most suitable membrane;

– module configuration;

– pre-treatment;

– applied flows;

– reversible and irreversible clogging;

– cleaning conditions and frequency;

– life expectancy;

– integrity tests;

– outcome of the concentrates;

– investment and operating costs, etc.

Figure 17.60. *Some examples of MF-UF membranes used or tested. For a color version of this figure, see www.iste.co.uk/gaid/watertreatment4.zip*

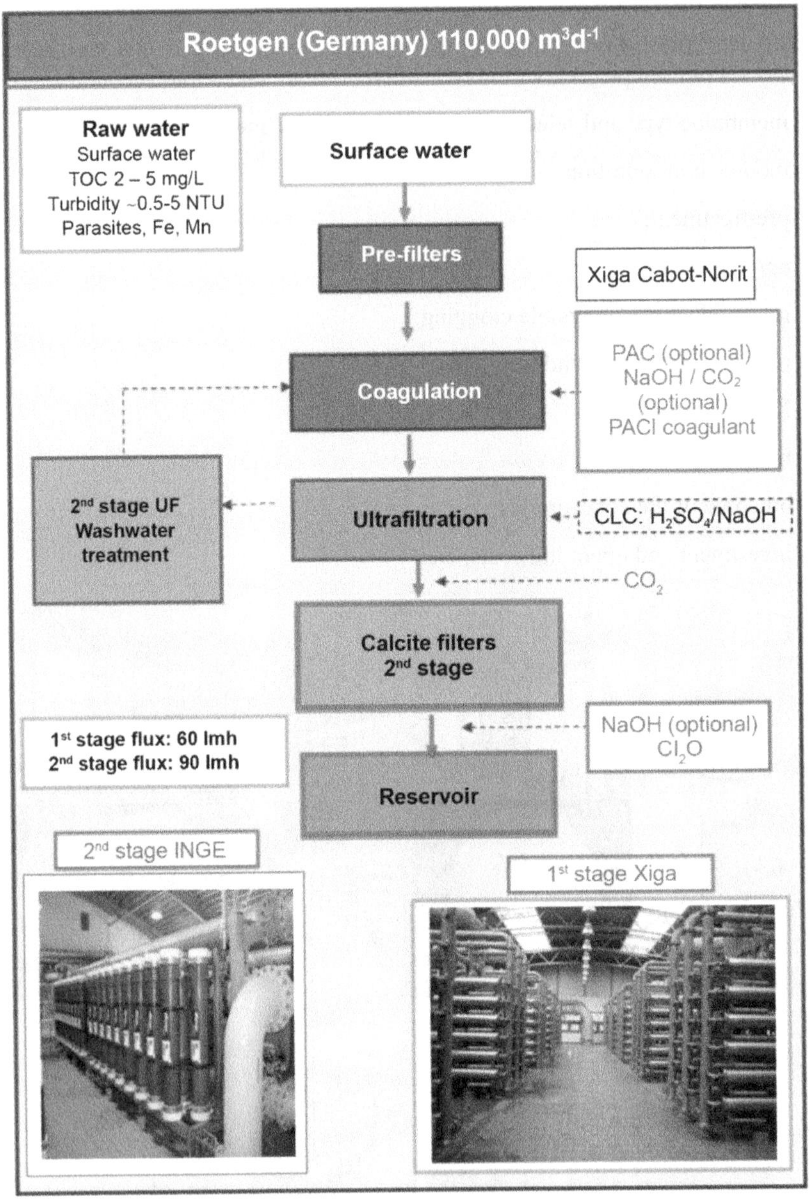

a)

Figure 17.61. *a) Some examples of MF and UF membrane implementations on different sites and in different configurations. For a color version of this figure, see www.iste.co.uk/gaid/watertreatment4.zip*

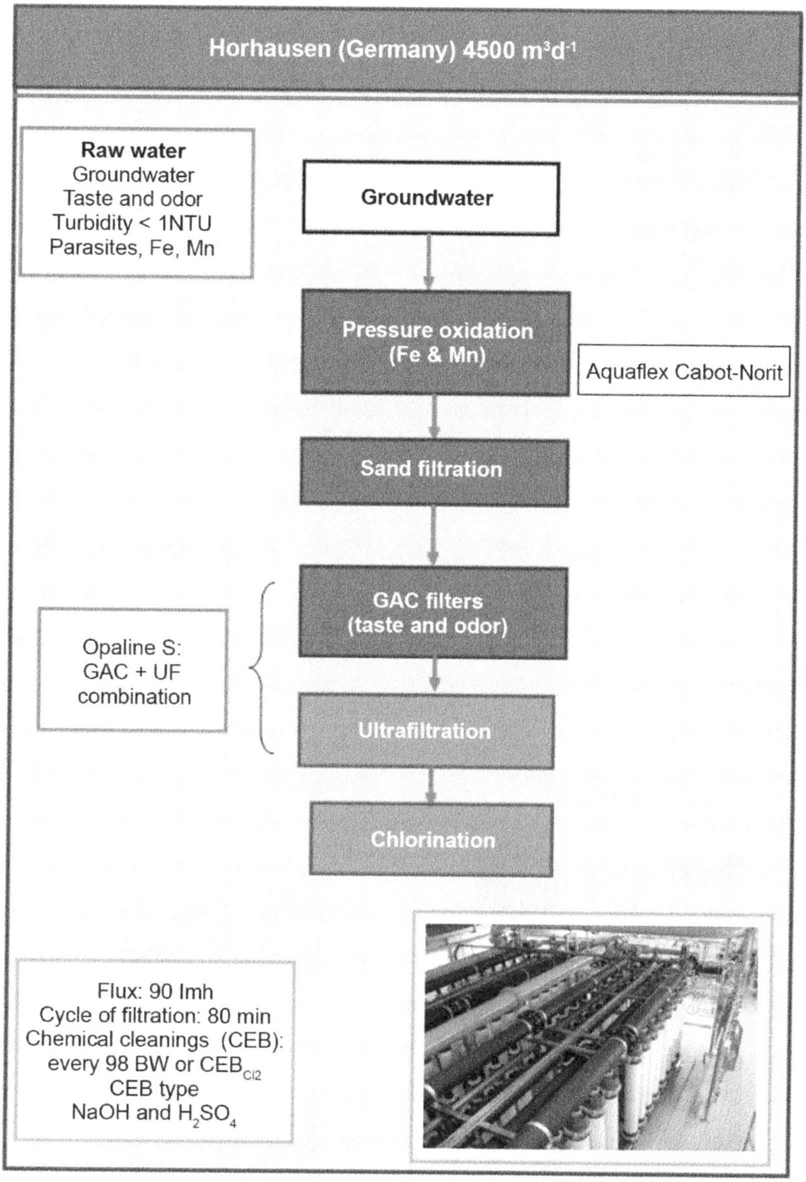

b)

Figure 17.61. *b) Some examples of MF and UF membrane implementations on different sites and in different configurations (continued). For a color version of this figure, see www.iste.co.uk/gaid/watertreatment4.zip*

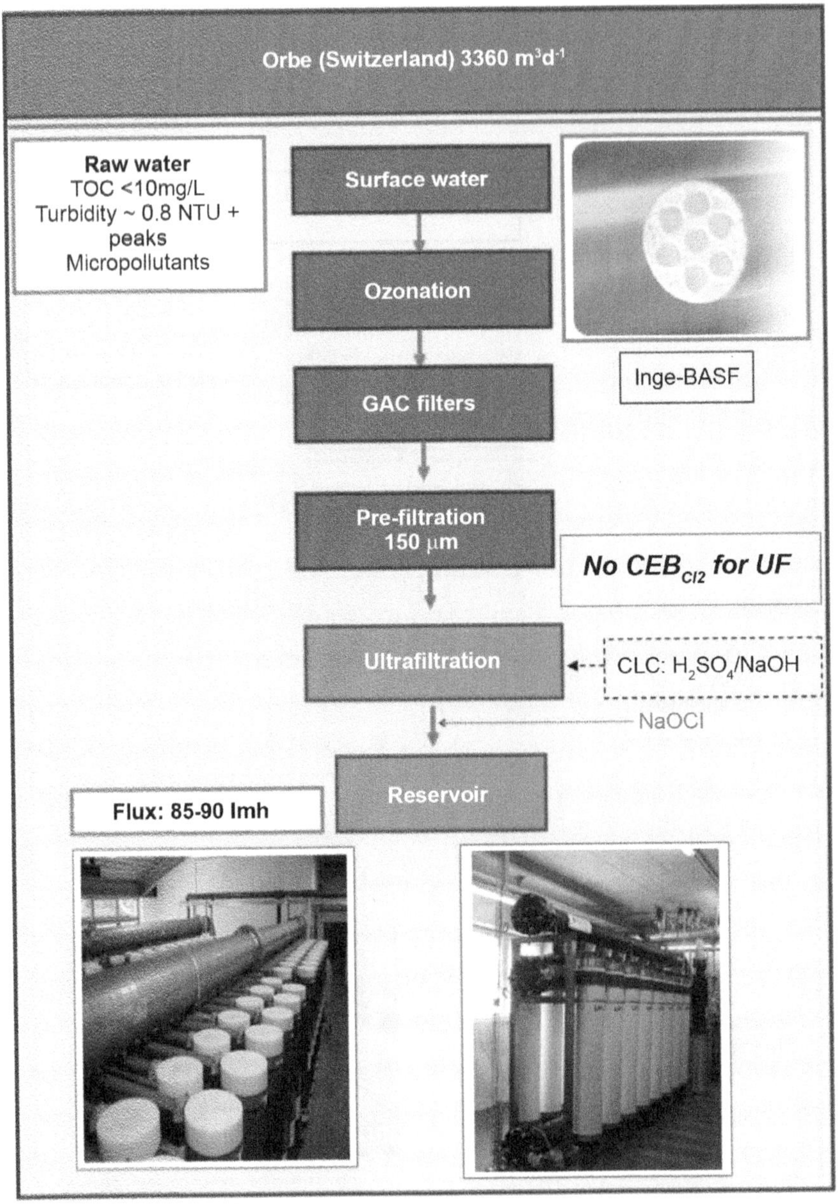

c)

Figure 17.61. *c) Some examples of MF and UF membrane implementations on different sites and in different configurations (continued). For a color version of this figure, see www.iste.co.uk/gaid/watertreatment4.zip*

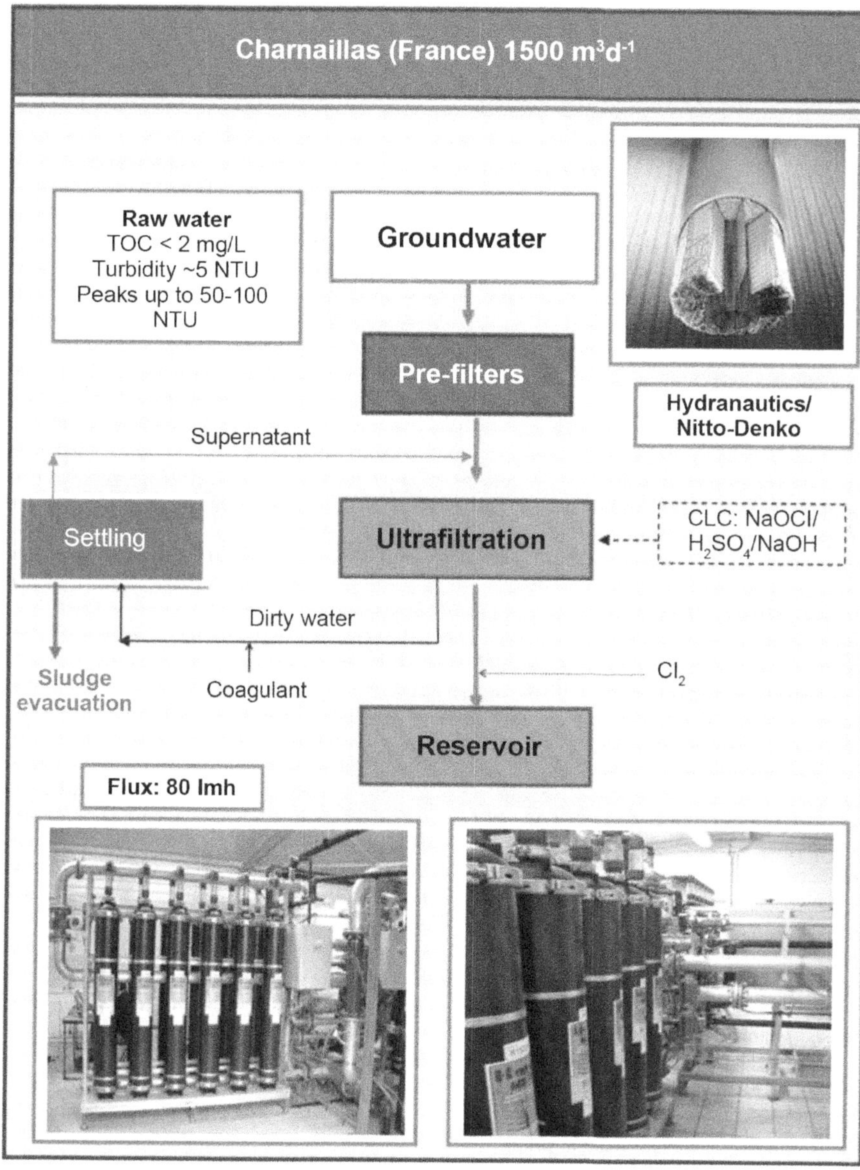

d)

Figure 17.61. *d) Some examples of MF and UF membrane implementations on different sites and in different configurations (continued). For a color version of this figure, see www.iste.co.uk/gaid/watertreatment4.zip*

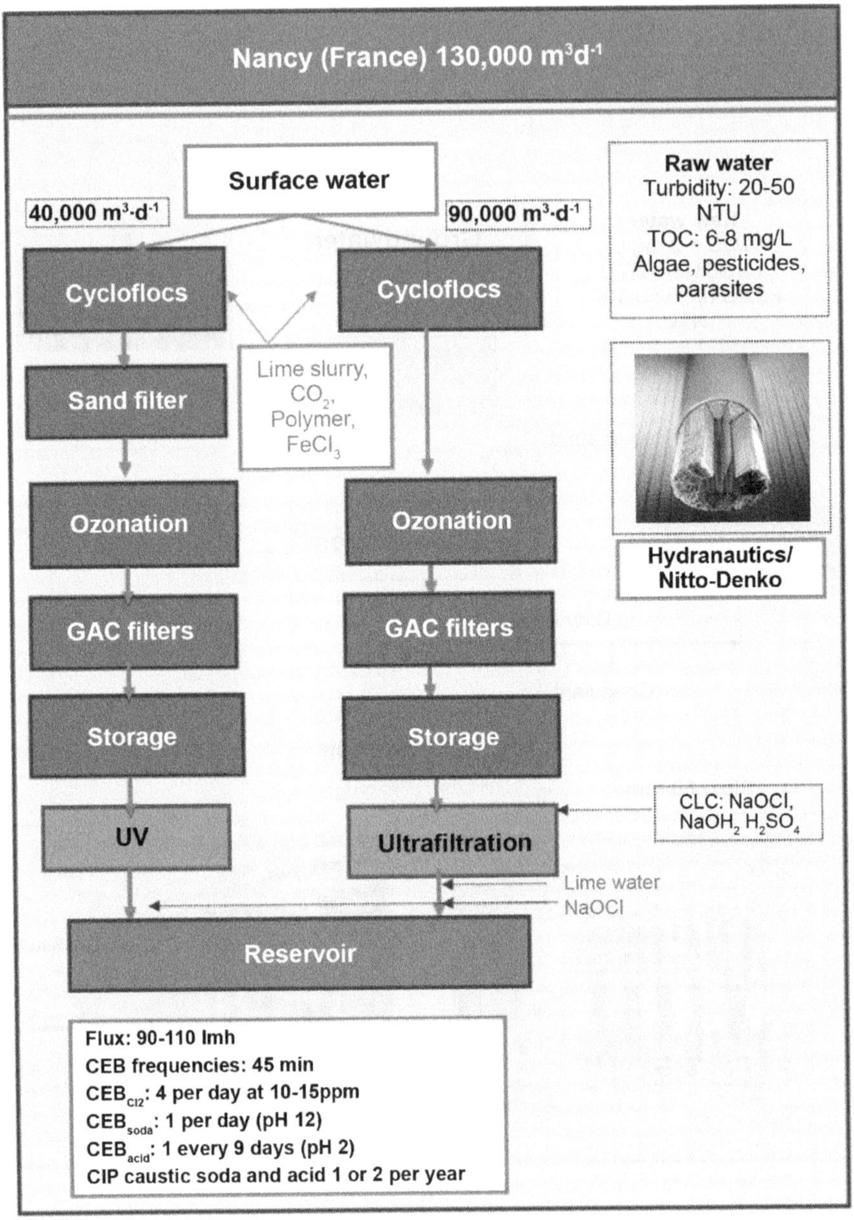

e)

Figure 17.61. *e) Some examples of MF and UF membrane implementations on different sites and in different configurations (continued). For a color version of this figure, see www.iste.co.uk/gaid/watertreatment4.zip*

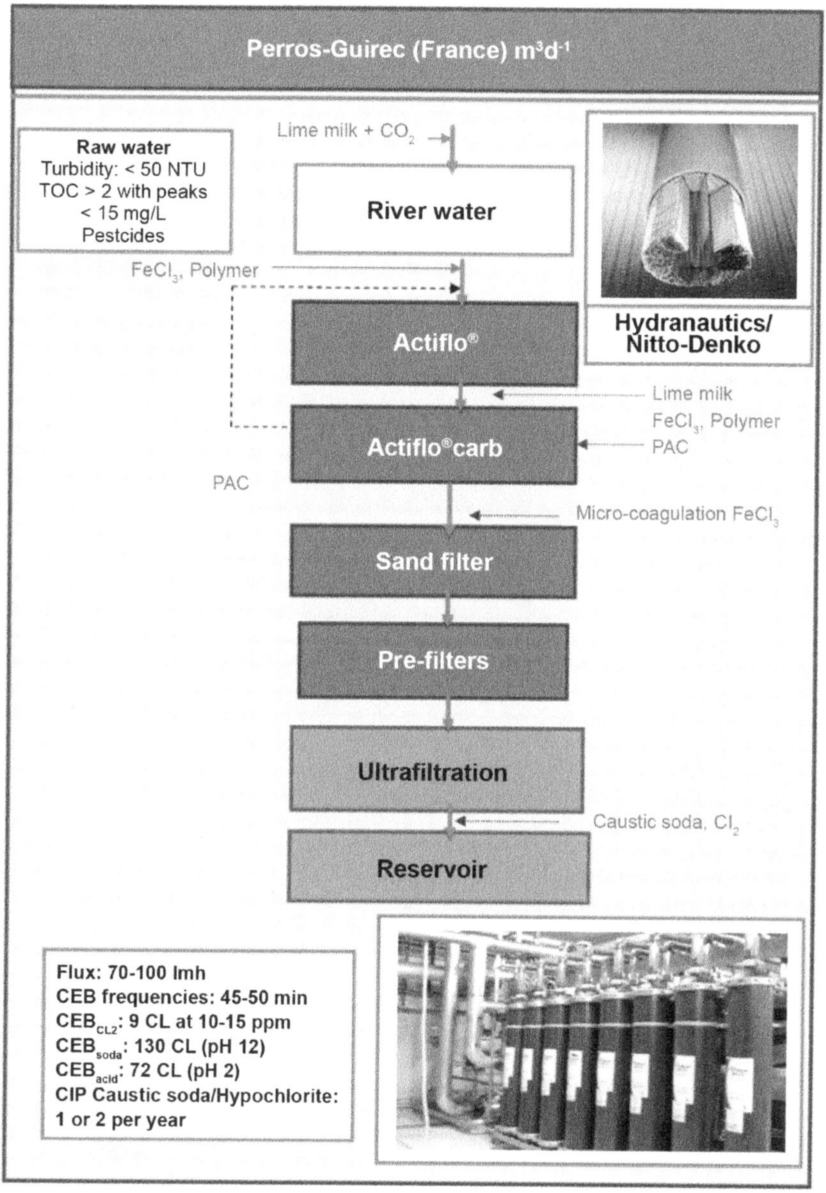

f)

Figure 17.61. *f) Some examples of MF and UF membrane implementations on different sites and in different configurations (continued). For a color version of this figure, see www.iste.co.uk/gaid/watertreatment4.zip*

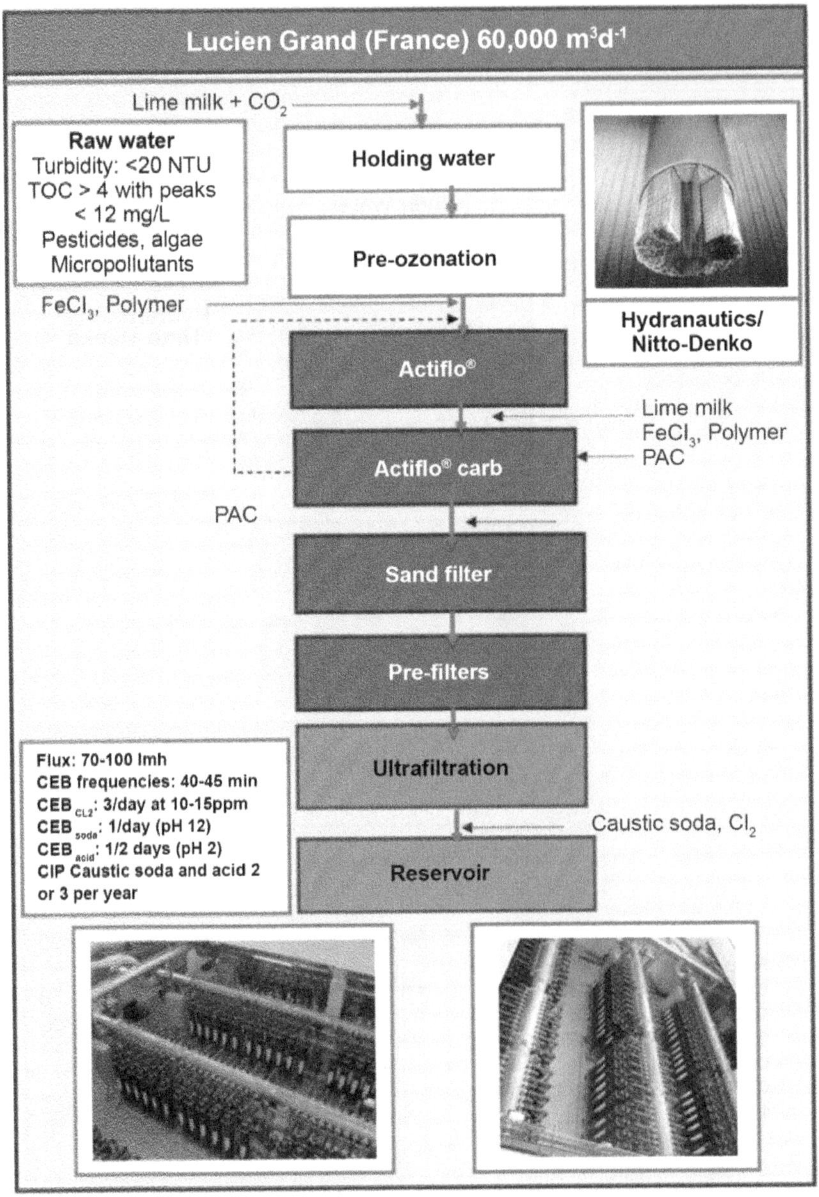

g)

Figure 17.61. *g) Some examples of MF and UF membrane implementations on different sites and in different configurations (continued). For a color version of this figure, see www.iste.co.uk/gaid/watertreatment4.zip*

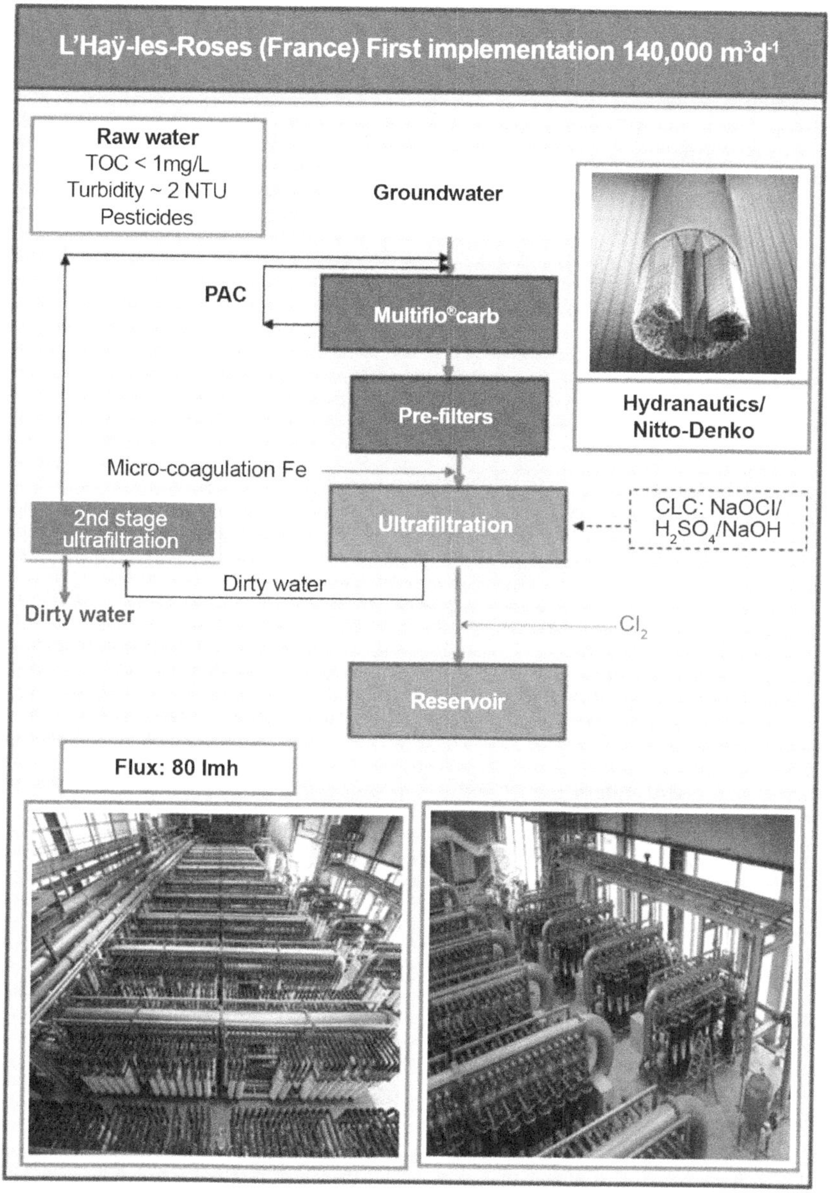

h)

Figure 17.61. *h) Some examples of MF and UF membrane implementations on different sites and in different configurations (continued). For a color version of this figure, see www.iste.co.uk/gaid/watertreatment4.zip*

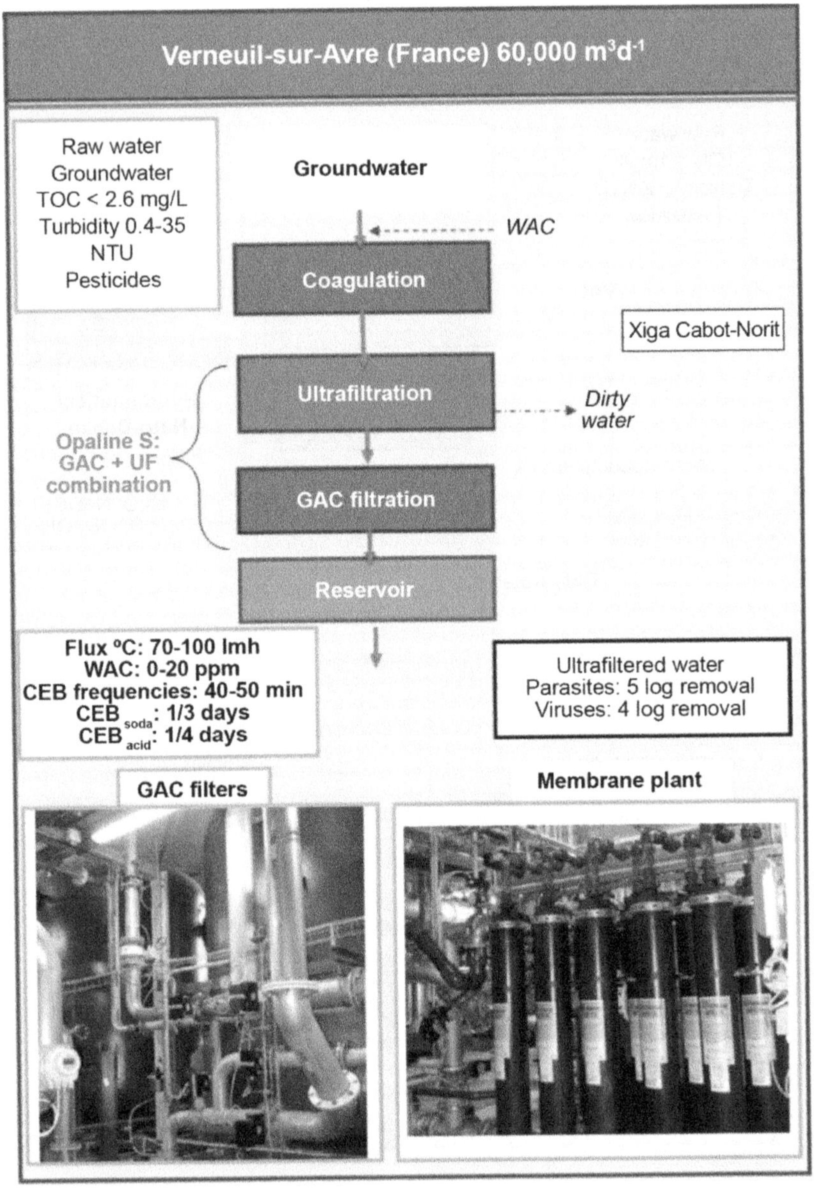

i)

Figure 17.61. *i) Some examples of MF and UF membrane implementations on different sites and in different configurations (continued). For a color version of this figure, see www.iste.co.uk/gaid/watertreatment4.zip*

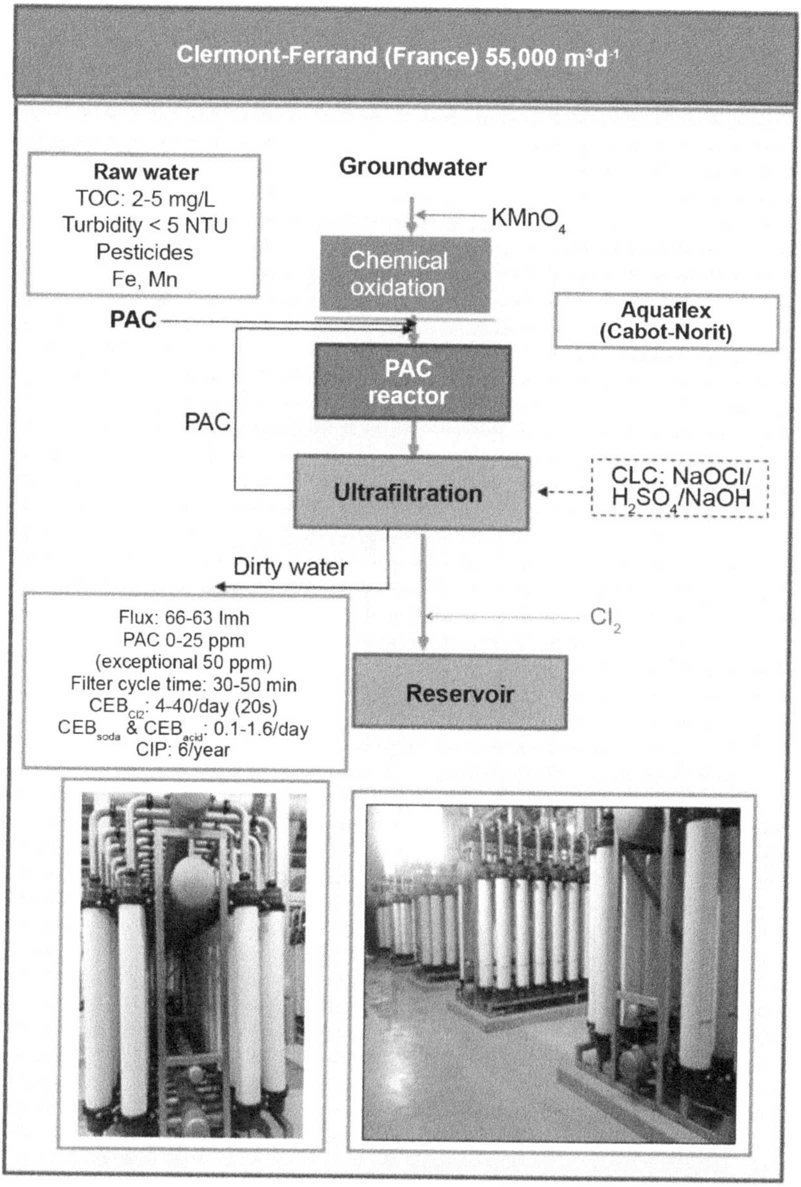

j)

Figure 17.61. *j) Some examples of MF and UF membrane implementations on different sites and in different configurations (continued). For a color version of this figure, see www.iste.co.uk/gaid/watertreatment4.zip*

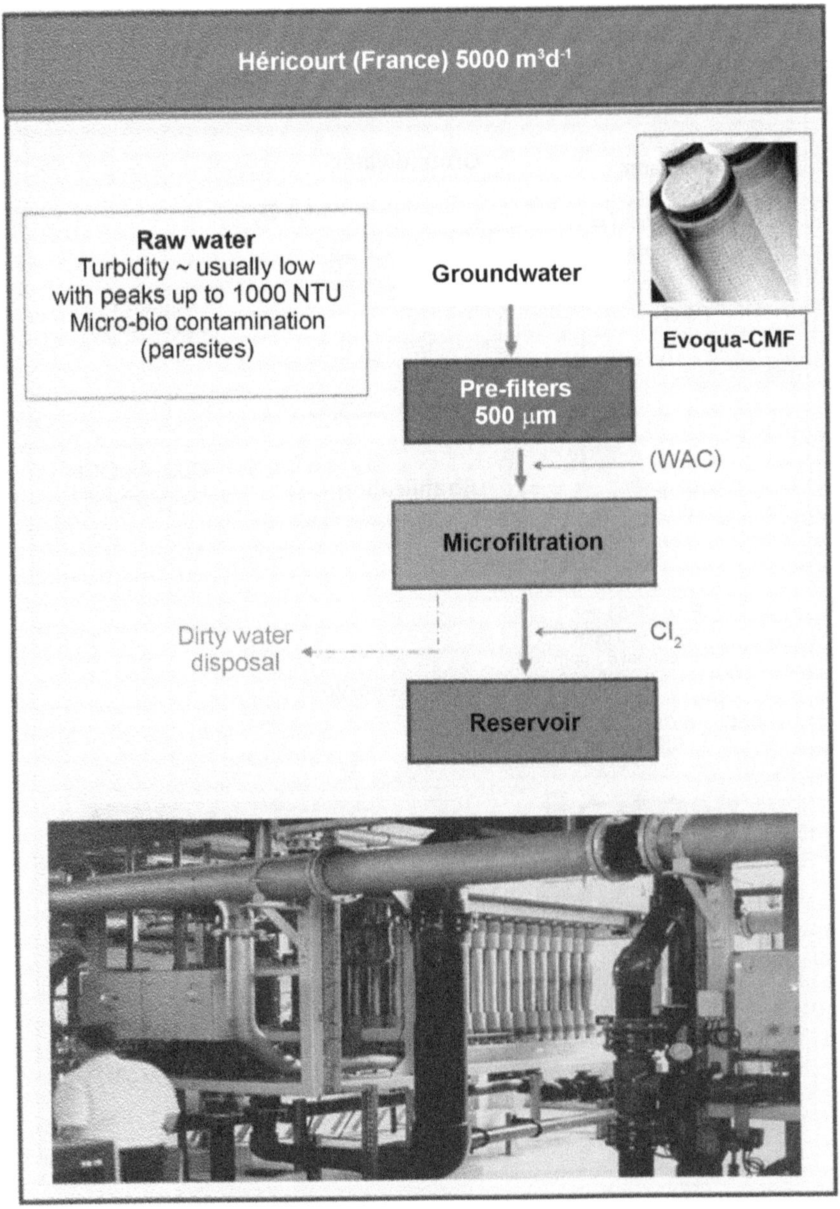

k)

Figure 17.61. *k) Some examples of MF and UF membrane implementations on different sites and in different configurations (continued). For a color version of this figure, see www.iste.co.uk/gaid/watertreatment4.zip*

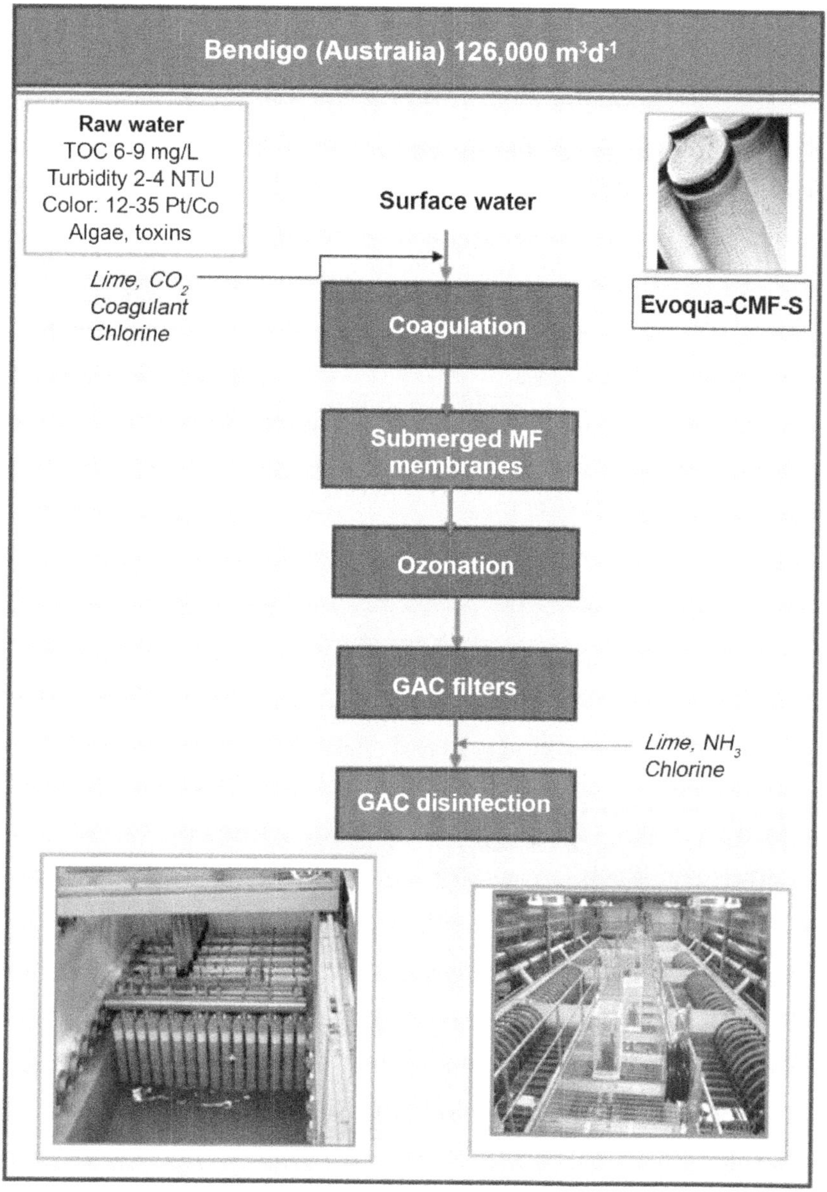

l)

Figure 17.61. *l) Some examples of MF and UF membrane implementations on different sites and in different configurations (continued). For a color version of this figure, see www.iste.co.uk/gaid/watertreatment4.zip*

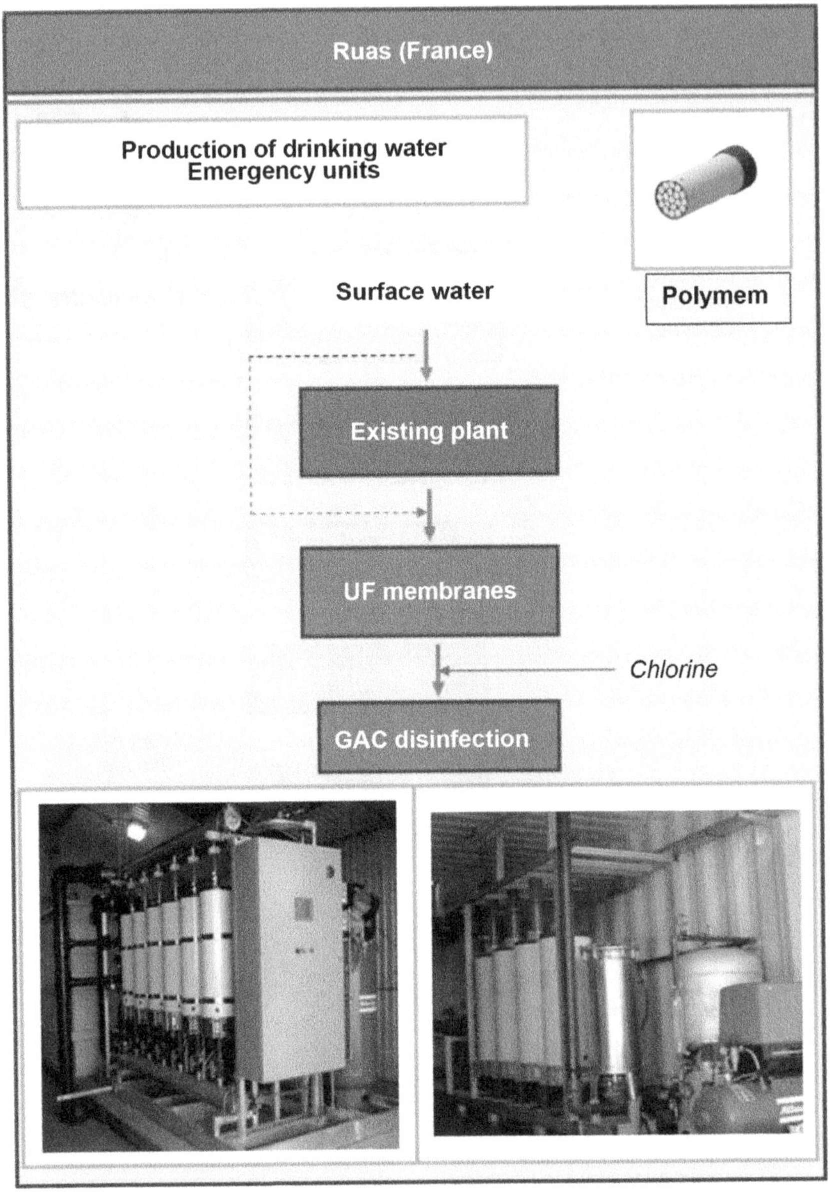

m)

Figure 17.61. *m) Some examples of MF and UF membrane implementations on different sites and in different configurations (continued). For a color version of this figure, see www.iste.co.uk/gaid/watertreatment4.zip*

17.17. References

Altmann, J. and Ripperger, S. (1989). Experimental and theoretical study of the influence of particle interaction forces in crossflow micro- and ultrafiltration. *Sep. Sci. Technol.*, 24, 383–398.

Aoustin, E., Schafer, A.I., Fane, A.G., Waite, T.D. (2001). Ultrafiltration of natural organic matter. *Sep. Purif. Technol.*, 22–23, 63–78.

Bacchin, P. (1994). Formation et résistance au transfert d'un dépôt de colloïdes sur une membrane d'ultrafiltration. PhD Thesis, Université Paul Sabatier, Toulouse.

Bacchin, P. (2005). Principes de base des technologies à membranes. PhD Thesis, 2e École d'Été Franco Maghrébine Sciences et Technologies à Membranes, Monastir.

Bacchin, P., Meireles, M., Aimar, P. (2002). Modelling of filtration: From the polarised layer to deposit formation and compaction. *Desalination*, 145, 139–146.

Bai, R. and Tien, C. (2005). Further work on cake filtration analysis. *Chem. Eng. Sci.*, 60, 301–313.

Bessiere, Y. (2005). Filtration frontale sur membrane : mise en évidence du volume filtré critique pour l'anticipation et le contrôle du colmatage. PhD Thesis, Université Paul Sabatier, Toulouse.

Best, G., Singh, M., Mourato, D., Chang, Y.J. (2001). Application of immersed ultrafiltration membranes for organic removal and disinfection by-products reduction. *Wat. Sci. Tech.: Wat. Supp.*, 5/6, 221–231.

Bian, R., Watanabe, Y., Tambo, N., Ozawa, G. (2000). Removal of humic substances by UF and NF membranes systems. In *Proc. International IAWQIWSA Joint Specialist*, Trondheim.

Bowen, W.R. and Jenner, F. (1995). Dynamic ultrafiltration model for charged colloidal dispersions: A Wigner-Seitz cell approach. *Chem. Eng. Sci.*, 50, 1707–1736.

Cabane, B. and Hénon, C. (2003). *Liquides. Solutions, dispersions, émulsions, gels.* Belin, Paris.

Campos, C., Marinas, B.J., Snoeyink, V.L., Baudin, I., Laine, J.M. (2000). PAC membrane filtration process I: Model development. *J. Environ. Eng.*, 126, 97–103.

Carroll, T., King, S., Gray, S.R., Bolto, B.A., Booker, N.A. (2000). The fouling of microfiltration membranes by NOM after coagulation treatment. *Water Res.*, 34(11), 2861–2868.

Chellam, S. and Jacangelo, J.G. (1998). Existence of critical recovery and impact of operational conditions on potable water microfiltration. *J. Environ. Eng.*, 124, 1211–1219.

Cho, J., Amy, G., Pellegrino, J. (2000). Membrane filtration of natural organic matter: Factors and mechanisms affecting rejection and flux decline with charged ultrafiltration (UF) membrane. *J. Membr. Sci.*, 164, 89–110.

Choi, Y.J. (2003). Critical flux, resistance and removal of contaminants in ultrafiltration of natural organic materials. PhD Thesis, University of Pennsylvania, Philadelphia.

Choi, Y.J. and Dempsey, B.A. (2004). In-line coagulation with low-pressure membrane filtration. *Water Res.*, 38, 4271–4281.

Crozes, G., Anselme C., Mallevialle, J. (1993). Effect of adsorption of organic matter on fouling of ultrafiltration membranes. *J. Membr. Sci.*, 84, 61–77.

Espinasse, B. (2003). Approche théorique et expérimentale de la filtration tangentielle de colloïdes : flux critique et colmatage. PhD Thesis, Université Paul Sabatier, Toulouse.

Espinasse, B., Bacchin, P., Aimar, P. (2002). On an experimental method to measure critical flux in ultrafiltration. *Desalination*, 146, 91–96.

Farahbakhsh, K., Svrcek, C., Guest, R.K., Smith, D.W. (2004). A review of the impact of chemical pretreatment on low-pressure water treatment membranes. *J. Environ. Eng. Sci.*, 3, 237–253.

Fan, L., Harris, J.L., Roddick, F.A., Booker, N.A. (2001). Influence of the characteristics of natural organic matter on the fouling of microfiltration membranes. *Water Res.*, 35, 4455–4463.

Goosen, M.F.A., Sablani, S.S., Al-Hinai, H., Al-Obeidani, S., Al-Belushi, R., Jackson, D. (2004). Fouling of reverse opsmosis and ultrafiltration membranes: Critical review. *Sep. Sci. Technol.*, 39(10), 2261–2298.

Goslan, E.H. (2003). Natural organic matter character and reactivity: Assessing seasonal variation in a moorland water. PhD Thesis, Cranfield University.

Guigui, C. (2000). Procédé hybride de coagulation/ultrafiltration pour la potabilisation des eaux de surface. PhD Thesis, Université Paul Sabatier, Toulouse.

Hirata, T. and Hashimoto, A. (1998). Experimental assessment of the efficacy of microfiltration and ultrafiltration for Cryptosporidium removal. *Water Sci. Technol.*, 38, 103–107.

Hong, S. and Elimelech, M. (1997). Chemical and physical aspects of natural organic matter (NOM) fouling of nanofiltration membranes. *J. Membr. Sci.*, 132, 159–181.

Howe, K.J. (2001). Effect of coagulation pretreatment on membrane filtration performance. PhD Thesis, University of Illinois, Springfield.

Jacangelo, J.G., Rhodes Trussell, R., Watson, M. (1997). Role of membrane technology in drinking water treatment in the United States. *Desalination*, 113, 119–127.

Jucker, C. and Clark, M.M. (1994). Adsorption of aquatic humic substances on hydrophobic ultrafiltration membranes. *J. Membr. Sci.*, 97, 37–52.

Judd, S.J. and Hillis, P. (2001). Optimisation of combined coagulation and microfiltration for water treatment. *Water Res.*, 35, 2895–2904.

Jung, C.W., Son, H.J., Kang, L.S. (2006). Effects of membrane material and pretreatment coagulation on membrane fouling: Fouling mechanism and NOM removal. *Desalination*, 197, 154–164.

Kabsch-Korbutowicz, M. (2006a). Removal of natural organic matter from water by in-line coagulation/ultrafiltration process. *Desalination*, 200, 421–423.

Kabsch-Korbutowicz, M. (2006b). Impact of pre-coagulation on ultrafiltration process performance. *Desalination*, 194, 232–238.

Kecili, K. (2006). Étude et caractérisation de membranes synthétiques organiques immergées de microfiltration et d'ultrafiltration employées pour la production d'eau destinée à la consommation humaine : intensification des opérations de déconditionnement et de nettoyage. PhD Thesis, Université Pierre et Marie Curie, Paris.

Kennedy, M., Kim, S.M., Mutenyo, I., Broens, L., Schippers, J. (1998). Intermittent crossflushing of hollow fiber ultrafiltration systems. *Desalination*, 118, 175–187.

Kim, J.S., Akeparathumchai, S., Wickramasinghe, S.R. (2001). Flocculation to enhance microfiltration. *Journal of Membrane Science*, 182, 161–172.

Kulovaara, M., Metsamuuronen, S., Nystrom, M. (1999). Effects of aquatic humic substances on a hydrophobic ultrafiltration membrane. *Chemosphere*, 38, 3485–3496.

Kwon, B., Park, N., Cho, J. (2005). Effect of algae on fouling and efficiency of UF membranes. *Desalination*, 179, 203–214.

Laabs, C., Amy, G., Jekel, M. (2004). Organic colloids and their influence on low-pressure membrane filtration. *Water Science and Technology*, 50, 311–316.

Laine, J.-M., Vial, D., Moulart, P. (2000). Status after 10 years of operation – Overview of UF technology today. *Desalination*, 131, 17–25.

Lee, N., Amy, G., Habarou, H., Schrotter, J.C. (2002). Identification and control of fouling of low pressure (MF and UF membranes) by drinking-water. Report, Natural Organic Matter, Proceedings-IWA, Membranes in Drinking and Industrial Production (M DI W).

Lee, N., Amy, G., Croue, J.P., Buisson, H. (2004). Identification and understanding of fouling in low pressure membrane (MF/UF) filtration by natural organic matter (NOM). *Water Research*, 38, 4511–4523.

Liang, H. (2007). Study on algae monitoring/prediction and its control mechanisms. PhD Thesis, Harbin Institute of Technology, Nangang.

Liang, H., Gongb, W.J., Chena, Z.L., Tiana, J.Y., Qia, L., Lia, G.B. (2009). Effect of chemical preoxidation coupled with in-line coagulation as a pretreatment to ultrafiltration for algae fouling control. *Desalination and Water Treatment*, 9, 241–245.

Lin, C.F., Huang, Y.J., Hao, O.J. (1999). Ultrafiltration processes for removing humic substances: Effect of molecular weight fractions and PAC treatment. *Water Research*, 33, 1252–1264.

Lin, C.F., Lin, T.Y., Hao, O.J. (2000). Effects of humic substance characteristics on UF performance. *Water Res.*, 34, 1097–1106.

Maartens, A., Swart, P., Jacobs, E.P. (2000). Membrane pretreatment: A method for reducing fouling by natural organic matter. *J. Colloid Interface Sci.*, 221, 137142.

Martin-Mousset, B., Croue, J.P., Lefebvre, E., Legube, B. (1997). Distribution and characterization of dissolved organic matter of surface waters. *Water Res.*, 31, 541–553.

Matsui, Y., Yuasa, A., Ariga, K. (2001a). Removal of a synthetic organic chemical by PAC-UF systems I: Theory and modeling. *Water Res.*, 35, 455–463.

Matsui, Y., Colas, F., Yuasa, A. (2001b). Removal of a synthetic organic chemical by PAC-UF systems II: Model application. *Water Res.*, 35, 464–470.

Miljatovic, I., Matosic, M., Hajduk Cerneha, B., Bratilic, D. (2004). Removal of natural organic matter by ultrafiltration and nanofiltration for drinking water production. *Desalination*, 169, 223–230.

Mo, L. and Huang, X. (2003). Fouling characteristics and cleaning strategies in coagulation-microfiltration combination process for water purification. *Desalination*, 159, 1–9.

Mulder, M. (1996). Polarisation phenomena and membrane fouling. In *Basic Principles of Membrane Technology*, Mulder, M. (ed.). Kluwer Academic Publishers, Dordrecht.

Park, P.K., Lee, C.H., Choi, S.J., Choo, K.H., Kim, S.H., Yoon, C.H. (2002). Effect of the removal of DOMs on the performance of a coagulation – UF membrane system for drinking water production. *Desalination*, 145, 237–245.

Rickman, M., Pellegrino, J., Davis, R. (2012). Fouling phenomena during membrane filtration of microalgae. *J. Membr. Sci.*, 423–424, 33–42.

Serra, C. (1996). Ultrafiltration frontale par fibres creuses : dimensionnement des modules et simulation du procédé en clarification des eaux. PhD Thesis, Université Paul Sabatier, Toulouse.

Wetterau, G.E., Clark, M.M., Anselme, C. (1996). A dynamic model for predicting fouling effects during the ultrafiltration of a groundwater. *J. Membr. Sci.*, 109, 185–204.

Yuan, W. and Zydney, A.L. (2000). Humic acid fouling during ultrafiltration. *Environ. Sci. Technol.*, 34, 5043–5050.

Zhang, X., Hu, Q., Sommerfeld, M., Puruhito, E., Chen, Y. (2010). Harvesting algal biomass for biofuels using ultrafiltration membranes. *Bioresour. Technol.*, 101, 5297–5304.

18

Nanofiltration and Reverse Osmosis

While MF and UF membrane technologies are based on processes that use microporous membranes operating under pressure to remove particles (including turbidity and micro-organisms) via a size exclusion mechanism, nanofiltration (NF) and reverse osmosis (RO) membrane technologies are based on the use of semi-permeable membranes that also operate under pressure in order to mainly target the removal of dissolved contaminants (organic substances and ions) via a controlled diffusion separation process.

When a high pressure greater than the osmotic gradient is applied to the feed side of the high-pressure membrane, water is forced through the membrane surface. Contaminants are then rejected, even though some may diffuse through these membranes. Since the mass transfer rate of these contaminants is slower than that of water, the result is a permeate that is largely free of these compounds.

The amount of energy (hydraulic pressure) required to drive the feed water through the membrane depends on the membrane material and thickness, as well as the osmotic pressure of the feed water. Osmotic pressure is the pressure created by the natural process of water flowing from a dilute solution (low compound concentration) through a semi-permeable membrane to a more concentrated solution (higher compound concentration). Energy in the form of hydraulic pressure is required to overcome both the physical resistance of the membrane and the osmotic pressure. Since the pressure applied is against the osmotic pressure, the treatment process is called “reverse osmosis”. Under given operating conditions, the feed pressure mostly depends on the total dissolved solids concentration (TDS) and the temperature of the feed water. Thus, the osmotic pressure (π) is the pressure that, if applied to the most concentrated solution, would prevent water transport across the membrane.

NF membranes were industrialized in the early 1980s as an alternative to RO membranes. Indeed, these membranes operate at high pressures while delivering a very good water quality, but at a high energy cost. The alternative was to manufacture a membrane operating under pressure with properties between RO and ultrafiltration (UF), with lower rejection characteristics (removal %) for monovalent ions such as sodium and chloride. The latter was done while remaining high for organic substances and micropollutants, in addition to offering a higher water permeability.

These membranes, initially called "low-pressure reverse osmosis membranes", have become NF membranes. The term "nanofiltration" means that compounds of nanometric dimensions are separated through NF membranes. As these membranes require a lower feed pressure, the NF membrane is well suited for applications such as the removal of natural organic substances, pesticides, micropollutants, taste and odor causing substances, and where monovalent ions are not the main concern with respect to water quality objectives.

Moreover, these performances are obtained with lower energy consumption than with RO. For example, the removal of dissolved organic carbon (DOC) and micropollutants only requires a pressure of less than 12 bar with NF membranes.

NF and RO membranes have thus shared the market according to the applications (surface water, groundwater, brackish water, seawater), with different quality objectives.

18.1. Membranes

18.1.1. *Materials*

The two types of membranes currently used are asymmetric homogeneous membranes and composite membranes. The two main materials used to manufacture RO and NF membranes are cellulose acetate (CA; and its derivatives) and various polyamides (PAs), especially aromatic PAs.

NF membranes are organic and composite; in other words, these membranes are composed of several distinct layers:

– A porous polyester textile support (approximately 120 µm thickness). This support gives the membrane its mechanical resistance.

– A layer of microporous polysulfone (PS), with a thickness of about 40 µm. This layer offers an extremely smooth surface to allow the adhesion of the third layer.

– A very thin skin (ultrafine barrier) of about 0.2–2 µm thickness, which gives the membrane its specific rejection properties of the different compounds. The composition of this layer varies from one membrane to another (polypiperazine-amide, PS, etc.).

The water to be filtered is in contact with the last layer and flows from the thinnest layer to the loosest layer. Due to this particular composite structure, it is not possible to reverse the direction of the water flow (from the permeate to the concentrate), as this would cause the different layers of the membrane to become detached. Therefore, it is necessary to always maintain a positive pressure difference between the concentrate and the permeate. The surface or skin gives the membrane its rejection characteristics, and the porous substrate supports this skin to withstand pressure differences of over 70 bar in some seawater RO applications.

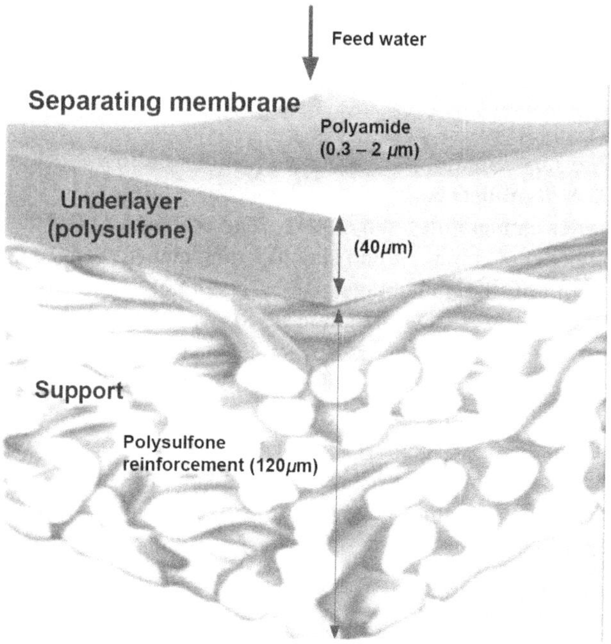

Figure 18.1. *Cross-section of a membrane. For a color version of this figure, see www.iste.co.uk/gaid/watertreatment4.zip*

A variety of polymers are used for the barrier layer, including fully aromatic PAs, the most common of which are aryl-alkyl PAs and polyureas. NF uses modified PAs and piperazine-amides for the barrier layer.

Commonly used support materials for thin film composite membranes are PS, polyethersulfone (PES), polyvinylidene fluoride (PVDF), polyacrylonitrile (PAN) and polyether ether ketone.

Polymeric membranes represent the largest proportion of insulated membranes currently in use. Several different polymers are used to match the required molecular weight cut-off (MWCO) or to achieve the desired resistance to clogging.

For example, the NF-270 membrane is a polypiperazine membrane, while the polymeric active layer of the NF-90 is PA. The active layer is supported by a porous PS layer, reinforced with a non-woven polyester layer.

18.1.1.1. *Cellulose acetate*

CA is one of the first polymer membranes used for water-based separation, for example, RO and UF techniques. Being hydrophilic, CA offers good resistance to fouling. CA membranes have a high water permeability and are also inexpensive and easy to manufacture.

CA hydrolyzes to cellulose over time and results in a loss in salt rejection. The hydrolysis of CA is influenced by both pH and temperature. The latter accelerates hydrolysis at water temperatures above 35°C. The optimum pH of the feed water is about 5 for cellulosic membranes, and should be maintained between 4.5 and 6.5 to maximize the life of the membrane. As a result, chemical cleaning must be performed in a relatively narrow pH range (4–8), which may limit its effectiveness in some cases. Cellulosic membranes are biodegradable in the presence of micro-organisms capable of producing cellulosic enzymes, even if preventive actions are performed by chlorination (1 $mg \cdot L^{-1}$) of the feed water. On the other hand, it is no longer necessary to control the chlorine and neutralize it as it is done for other types of membranes. The asymmetric structure of CA membranes makes them sensitive to co-packing under high operating pressures, especially at high temperatures, leading to a reduction in product flow.

18.1.1.2. *Aromatic PAs*

PA membranes overcome some of the problems associated with CA membranes. They have better resistance to hydrolysis and biological attack than cellulosic membranes.

Aromatic PA membranes tolerate pH ranges from 2 to 12 without hydrolysis, and are thus wider than cellulosic membranes. These membranes are also not susceptible to biological attack, but the material is sensitive to oxidants, which can degrade the membrane and cause loss of rejection capabilities. They require strict control of chlorine residue.

Due to the higher rejections and lower operating pressures, thin film composite membranes are the most commonly used in water treatment applications. Asymmetric membranes, regardless of the base material, are subject to compaction, which occurs primarily in the porous spongy support and results in a loss of productivity over time. The degree of compaction is influenced by the applied pressure, which is especially observed in seawater desalination applications.

18.1.2. *Membrane element configurations*

A membrane by itself is fragile and must be carefully supported in order to be integrated into the pressure tubes. The membrane is conditioned to minimize hydraulic pressure losses, while allowing sufficient flow velocities to keep the membrane surface clean of concentrated salts and particles. The manufacturing principle consists of wrapping the flat membrane around a perforated tube, which is used to collect the permeate.

A membrane used in a spiral configuration is manufactured in sheet form on a backing material such as a web fabric (for cellulosic membranes) or a non-woven polyester sheet (for composite membranes).

The membranes are packaged in the form of spiral modules: the manufacturing principle consists of gluing together two membrane sheets with a permeate spacer between them. They are wrapped around a perforated tube, which serves as a permeate collector. Lines of glue seal the inner (permeate) side of the sheet against the outer (feed-concentrate) side. The open side of the sheet is sealed against the central part of a perforated tube, which is the permeate collector for all sheets. The sheets are rolled up with a spacer sheet that is interspersed between the permeate and concentrate sheets, so as to improve water circulation. The presence of these spacers makes it possible to maintain the spacing between the membrane sheets, as well as to create turbulences on the feed side.

Figure 18.2 shows the assembly method of the membrane divided into two flat films of equal surfaces, which are assembled by gluing, and enclose a porous support which will serve as a collector for the permeate.

During operation, the feed water enters from the front of the module through the feed spacer channels and spirals toward the center. As it passes through the membrane sheets, it is stripped of ions and organic substances. It exits at the opposite end as a concentrate (Figure 18.3). Thus, one part ends up in the central permeate collector, while the other part flows through the module and exits under pressure through the concentrate pipes.

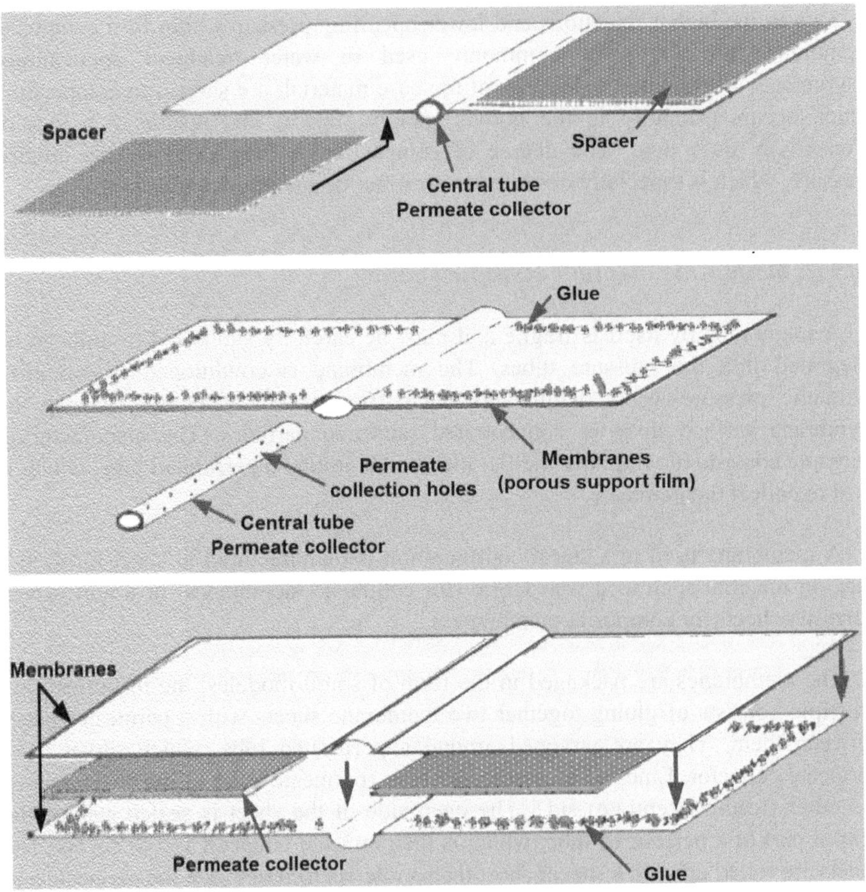

Figure 18.2. *Membrane assembly method. For a color version of this figure, see www.iste.co.uk/gaid/watertreatment4.zip*

The resulting spiral is then inserted into a module, which is itself inserted into a pressure tube. Several modules (from 6 to 8) are integrated in series in a pressure tube (Figure 18.4).

Although RO and NF membranes are manufactured in several physical configurations, the spiral configuration is the only one that is widely used in municipal water treatment applications.

Spirally wound designs offer many advantages over other module designs. A module must balance several objectives:

– maximize the surface area per module;

– ensure proper flow distribution within the module;

– minimize concentration polarization and accumulation of suspended matter on the membrane surface;

– minimizing energy consumption.

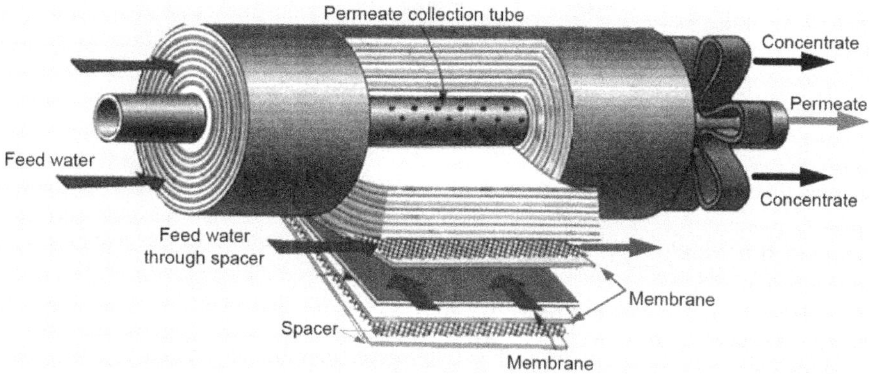

Figure 18.3. *Description of a spiral module. For a color version of this figure, see www.iste.co.uk/gaid/watertreatment4.zip*

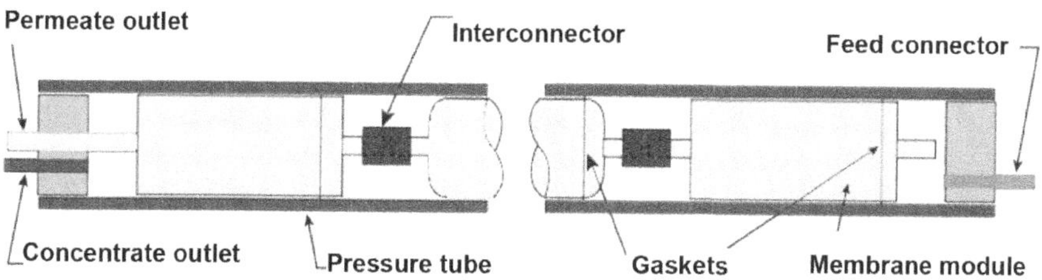

Figure 18.4. *Assembly in a pressure tube. For a color version of this figure, see www.iste.co.uk/gaid/watertreatment4.zip*

Tubular designs provide excellent flow distribution and a high velocity sweep, which minimizes concentration bias and pretreatment conditions, but causes high pumping costs. For this reason, tubular modules are typically used for specialized operations at low flow rates.

Industrial-type modules have a diameter of 203.2 mm and a length of 1,016 mm, and are characterized by a hydraulic efficiency (conversion rate) of approximately 15%.

Figure 18.5. *Interconnectors in a high-pressure membrane. For a color version of this figure, see www.iste.co.uk/gaid/watertreatment4.zip*

Each pressure tube consists of six to seven membranes linked together by a permeate connector, connecting the permeate collectors together (Figure 18.5). The tubes are then closed by components known as "bottom plates". These are connected to the high-pressure collector (feed and concentrate) by Victaulic connectors. The pressure tubes are made of stainless steel and have a side port connection, with the tube being fed from the side (Figure 18.6).

a) b) c)

Figure 18.6. *Various pressure tube components: bottom plate a); Victaulic b) and feed c). For a color version of this figure, see www.iste.co.uk/gaid/watertreatment4.zip*

The spiral membranes are standardized and allow interchangeability from one supplier to another.

18.2. Principles of operation and separation

18.2.1. *Conceptual principle*

In membrane processes, a membrane separates two phases acting as a selective barrier, regulating the transport of substances between the two compartments. The structure of the membrane barrier can be classified according to its porous nature. The membrane surface can be a dense selective skin, allowing only diffusive transport, or a porous skin allowing permeate flow. For porous membranes, the transport rate and selectivity are mainly influenced by viscous flow and steric exclusion. Nevertheless, interactions of solutes with the porous membrane surface can alter membrane performances.

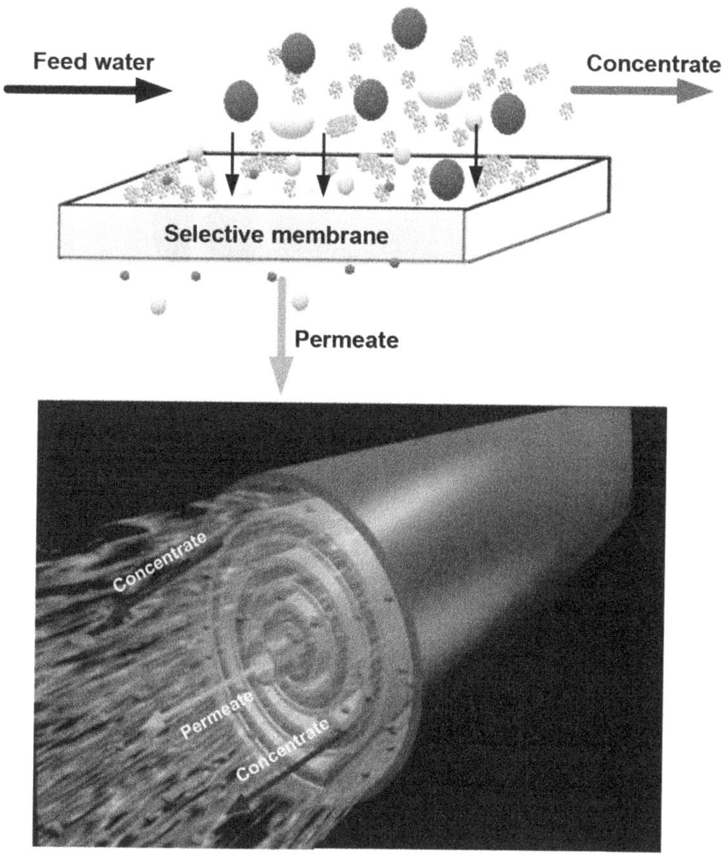

Figure 18.7. *Simplified conceptual diagram of a membrane separation. For a color version of this figure, see www.iste.co.uk/gaid/watertreatment4.zip*

The main advantages of membrane technology over other unit operations are related to this unique separation principle, namely the transport selectivity of the membrane. Separations with membranes do not require additives, and they can be performed isothermally at low temperatures and then compared to other energy-efficient thermal separation processes (Figure 18.7).

Membrane processes, such as NF and RO, use the pressure difference between the feed and permeate sides as the driving force to transport water across the membrane. Dissolved compounds are partially or fully retained based on properties such as size exclusion, charge repulsion or dielectric exclusion. Separation can also be due to chemical interactions such as hydrophobic interactions or hydrogen bonds. A schematic representation of a membrane process is given in Figure 18.8. The feed flow is divided into two flows: the permeate flow and the concentrate flow.

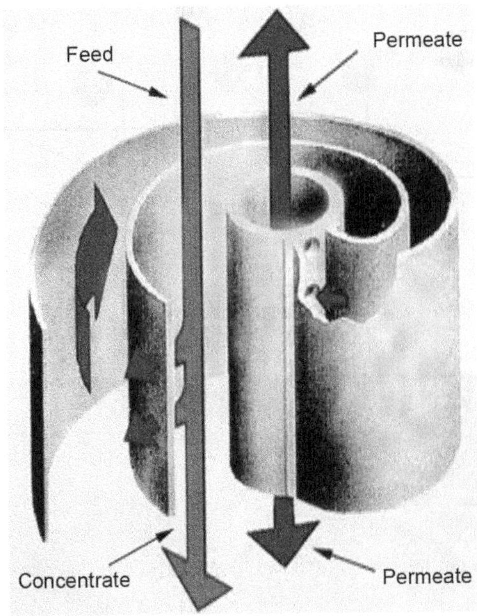

Figure 18.8. *Separation into two flows, permeate and concentrate (reject). For a color version of this figure, see www.iste.co.uk/gaid/watertreatment4.zip*

Thus, the effect of the molecular size involves most organic compounds with a molecular weight (MW) greater than the MWCO that are rejected by the membrane. Those with a MW below the MWCO of a membrane easily penetrate through the membrane. In addition, Van der Waals interactions are mainly responsible for the retention of uncharged molecules, and their transport is by convection due to a pressure difference, as well as by diffusion due to a concentration gradient across the membrane.

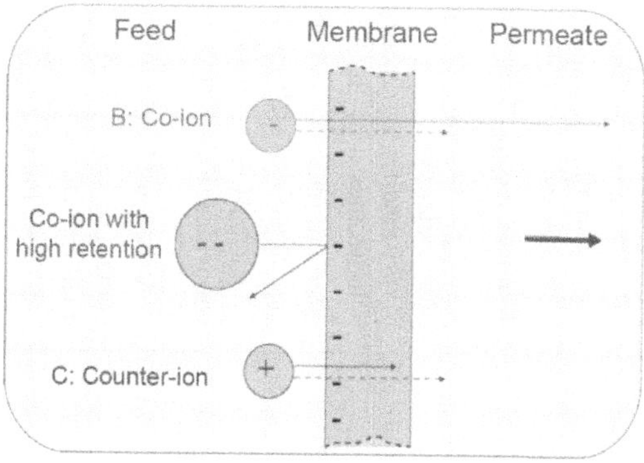

Figure 18.9. *Donnan effect. For a color version of this figure, see www.iste.co.uk/gaid/watertreatment4.zip*

For charged compounds, size exclusion and electrostatic interactions are responsible for the separation.

Given that most NF membranes are charged (mainly negatively), an electrostatic interaction takes place between the compound and the membrane. The separation of ions resulting from electrostatic interactions between the ions and the membrane surface charge is based on the Donnan exclusion mechanism.

According to Donnan's theory, for a negatively charged membrane, an increase in co-ion charge and a decrease in counter-ion charge result in an increase in salt discharge. The sequence of salt discharge should be $MgCl_2 < NaCl < Na_2SO_4$.

Thus, the co-ions (which have the same charge as the membrane) are repelled by the membrane surface. In order to satisfy the electroneutrality condition, an equivalent number of counter-ions are retained, resulting in salt retention. In addition, the Donnan effect (Figure 18.9) results in a difference in retention according to the ionic charge. Multivalent ions have a higher NF rejection than monovalent ions, because the charge interactions are greater. For ions of the same valence, the rejection is affected by the difference in ionic diffusivity. That is to say, an ion is better retained if it has a smaller diffusivity and a larger Stockes radius.

Membranes	Operating modes	Types of separation	Porous structures	Cut-off characteristics (µm)	Permeate compositions	Elements and compounds removed
Nanofiltration	Hydrostatic pressure difference	Sieving + diffusion-exclusion	Micropores (2 nm)	0.001–0.01	Very small molecules Monovalent ions	Small and high molecules, color, medium hardness, bacteria, parasites, viruses
Low-pressure reverse osmosis	Hydrostatic pressure difference	Sieving + diffusion-exclusion	Dense micropores	0.0001–0.01	Low content of very small molecules Medium ionic composition Partially monovalent ions	Very small and high molecules, color, ions, hardness, bacteria parasites, viruses
Reverse osmosis	Hydrostatic pressure difference	Sieving + diffusion-exclusion	Dense (<2 nm)	0.0001–0.001	Very low content of very small molecules Very low ionic composition	Very small and high molecules, color, medium hardness, ions, bacteria, parasites, viruses

Table 18.1. *General characteristics of nanofiltration and reverse osmosis membranes*

Ion rejection by NF membranes at a given volume flow is related to the partition coefficient between the volume and the pore (as a function of steric exclusion, electrical and dielectric interactions), and ion transport within the pore, which is controlled by electro-migration and ionic diffusivity. Divalent anions are strongly rejected by a negatively charged NF membrane due to the Donnan exclusion.

On the other hand, the selectivity between divalent and monovalent cations, assuming that they can freely enter the nanopores, is not as pronounced because the mobility of divalent cations is much lower than that of monovalent cations ($1.333 \times 10^9\ m^2 \cdot s^{-1}$ for Na^+ and $0.705 \times 10^9\ m^2 \cdot s^{-1}$ for Mg^{2+}).

The diffusion coefficient is related to the Stokes radius. The Stokes–Einstein radius is given by:

$$rs = \frac{kT}{6\pi\mu Di}$$

With:

– k: Boltzmann constant ($J \cdot mol^{-1} \cdot K^{-1}$);

– μ: viscosity ($kg \cdot m^{-1} \cdot s^{-1}$);

– T: temperature (K);

– Di: diffusion coefficient ($m^2 \cdot s^{-1}$).

Ions	**Stokes radius (nm)**	**Hydration energy ($kJ \cdot mol^{-1}$)**
Na^+	0.184	407
Cl^-	0.121	376
F^-	0.117	515
NO_3^-	0.128	329
SO_4^{2-}	0.231	1,138
Ca^{2+}	0.310	1,548
Mg^{2+}	0.341	2,018

Table 18.2. *Stokes radius and hydration energy for various ions*

Table 18.2 gives the Stokes radius and hydration energy of the different ionic types. Ions with high hydration energy are better retained, because hydration is considered to be an energy of transfer of ions from water to the membrane pores.

Compared to MF and UF, hydrodynamic resistance increases with processes such as NF and RO and, therefore, higher driving forces are required. On the other hand, the production rate decreases across the membrane. It is determined by the applied pressure and the membrane resistance. In addition, the size of the retained compounds is also smaller.

18.2.2. *Molecular weight cut-off*

NF membranes can be characterized according to their cut-off: opened NF (200–400 Da and a pore size of 0.8–1.0 nm) and tight NF (100–200 Da and a pore size of 0.6–0.7 nm). They have an electrostatic surface charge, which gives them a high selectivity toward ions or charged molecules. The NF membrane can be used to remove small organic molecules (MW > 200) and due to its electrostatic surface properties, it retains multivalent ions well, while monovalent ions are retained much less. The NF membrane offers several advantages, such as a lower operating pressure (5–20 bar) than RO, a high flow (20–30 $L{\cdot}m^{-2}{\cdot}h^{-1}$) and a lower operating and maintenance cost than RO.

Figure 18.10. *Steric exclusion for a high-pressure membrane. For a color version of this figure, see www.iste.co.uk/gaid/watertreatment4.zip*

Properties	Tightly packed nanofiltration	Nanofiltration	Opened nanofiltration	Reverse osmosis	Reverse osmosis
	NF-90 polyamide	**NF-270 polypiperazine**	**NF-200 polypiperazine**	**BW30-400 polyamide**	**XLE 440 polyamide**
KW pure water ($l \cdot m^{-2} \cdot h^{-1} \cdot bar^{-1}$)	6.4–6.6	14.7–15.0	10.8–11.0	2.8–3.0	2.8–3.0
MWCO (Da)	120–180	220–250	250–300	<100	<60
Pore radius (nm)	0.32–0.35	0.43–0.45	0.45–0.60	<0.3	<0.30
		Salt rejection (%)			
NaCl	88 ± 2	35–45	35–40	99.5	99.0
$MgCl_2$	94 ± 1	63	63	–	–
Na_2SO_4	98 ± 1	96	96	99.0	99.0
$MgSO_4$	98 ± 1	96	96	99.5	99.5
Active layer thickness (nm)	120–230	50–100	50–100	160–180	130–230

Table 18.3. *Characteristics of some NF and RO membranes (DOW Filmtec supplier)*

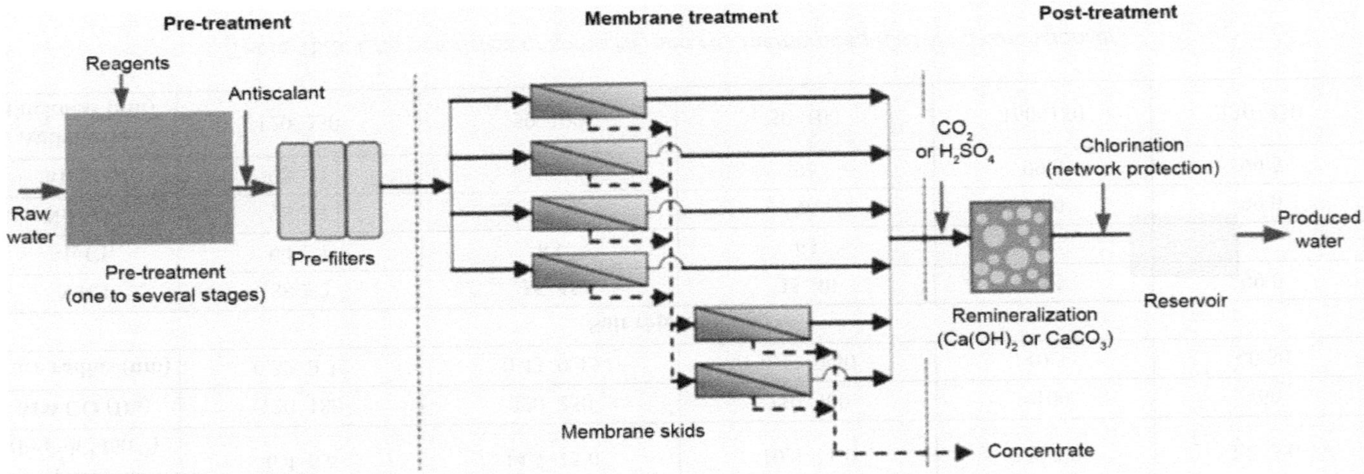

Figure 18.11. *Nanofiltration or reverse osmosis treatment process. For a color version of this figure, see www.iste.co.uk/gaid/watertreatment4.zip*

Lastly, the relatively small pore radius (tight NF) implies rapid clogging by pore blockage, instead of a possible progressive fouling (by adsorption or cake layer formation). Pore blockage could indeed restrict solute and water transport.

The rejection characteristic of an NF membrane is often quantified by the MWCO. This is defined as the MW of a compound that is 90% rejected. The MWCO concept is based on the fact that molecules become larger as their mass increases.

As a result, sieving effects due to steric exclusion (Figure 18.10) increase, and the molecule is then rejected by the membrane more often than a smaller molecule. The MWCO can also be related to diffusion, as a large molecule diffuses more slowly than a smaller molecule.

The choice of a membrane process depends on the objective to be achieved such as hardness reduction, removal of organic substances and micropollutants, removal of heavy metals or desalination.

The MWCO and pore radius of NF-90 (0.32–0.35 nm) are lower than those of NF-270 (0.43–0.45 nm) and NF-200 (0.45–0.60 nm). The pore size of a membrane correlates well with the results of salt discharges such as Na_2SO_4 and $MgSO_4$, which are well retained by these membranes.

18.3. Treatment process including high-pressure membranes and parameters to be considered

A treatment process that includes NF or RO membranes consists of three distinct stages: pre-treatment, membrane treatment and post-treatment. Figure 18.11 illustrates a typical surface water treatment process with NF or RO. In all cases, three water flows must be considered: the raw water flow, the produced water flow (permeate) and the concentrate flow.

18.3.1. *Particulates and SS*

NF acts on dissolved substances and is not the right process to retain particulate pollution. Therefore, particulate pollution should be treated as much as possible before the membranes, so as to minimize clogging. Turbidity can give a first indication, but is limited to assess the quality of the pre-treatment with respect to the NF membranes.

18.3.2. *Particle count*

Most counters count particles as small as 1.5 µm, with several channels indicating a "particle size distribution". Some counters "go down" to 0.5 µm. However, in use, this sensitivity has proven to be too high most of the time for this application.

18.3.3. *Conductivity*

The conductivity of water is a measure of the solution's ability to conduct an electric current. As the electric current is carried by the ions present, the conductivity increases when the concentration of ions increases. It is expressed in millisiemens per m ($mS{\cdot}m^{-1}$) and micromhos per cm ($\mu mho{\cdot}cm^{-1}$), with 1 $mS{\cdot}m^{-1}$ = 10 $\mu mho{\cdot}cm^{-1}$.

A quick estimate gives: conductivity ($\mu mho{\cdot}cm^{-1}$) = TDS ($mg{\cdot}L^{-1}$)/0.60.

18.3.4. *The SDI or MFI: clogging indices*

Two indices are commonly used to estimate the clogging capacity of membrane feed waters: the silt density index (SDI) and the modified fouling index (MFI). Both indices are based on the principle of Darcy's law applied to cake filtration. They are measured by following the flow loss through a flat membrane of low porosity under a pressure of 2.07 bar. High indices are generally associated with a rapid decrease in membrane performance.

The difference between MFI and SDI is the fact that the former is directly related to the concentration of clogging particles and is informative about cake-like clogging, while the latter is more global and is not solely related to cake-like clogging.

18.3.5. *The SDI*

This is the clogging index most often used by the water industry and membrane suppliers. It is easier to implement. The SDI is representative of a particle density, which makes it possible to qualify the clogging nature of water. It is obtained from the time required to filter a given volume of water through a flat membrane with a pore diameter of 0.45 µm (microfiltration) and a filtering surface (1.73×10^{-4} m^2) under a pressure of 2.07 bar. The duration of the experiment varies from a few minutes to an hour, depending on the nature of the water. The flow on the flat membrane is frontal. The principle of the measurement consists of timing how long it takes (t_1) to filter 500 mL of water through a new membrane. Then, after a filtration time between 1 and 60 min, timing how long it takes again (t_2) to filter another 500 mL.

The SDI is expressed in %/min. It is calculated by:

$$SDI = \frac{100}{tf}\left(1 - \frac{t_1}{t_2}\right)$$

With:

– t_1 : time needed to collect the first 500 mL of the sample (min);

– t_2 : time needed to collect the last 500 mL of the sample (min);

– tf: total filtration time (min).

This index only provides an indication of particulate clogging and a static measure of filtration resistance. The more the water clogs, the shorter the overall filtration time because the membrane is quickly clogged.

The ranges of validity of the measurements are therefore as follows: SDI_{15}: "0" to 6.6; SDI_{10}: "0" to 10; SDI_5: "0" to 20; SDI_3: "0" to 33.3. A reference time of 15 min for tf is often used for pre-treated water. For example, tests for two different types of water are the given results in Table 18.4.

Time (min)	Volume of filtered water (mL)	Sample type A collection time (s)	Sample type B collection time (s)	SDI
t_1 Sample A	500	9	–	–
t_2 Sample A	500	14	–	2.4
tf Sample A	15 min			
t_1 Sample B	–	–	65	–
t_2 Sample B	500	–	165	20.2
tf Sample B	3 min			

Table 18.4. *Examples of SDI results for two types of water*

Type A water has a $SDI_{15} < 3$ and is a good quality water that may be either groundwater, spring water or pre-treated water.

Type B water has an SDI_3 equal to 20.2, which is a highly contaminated raw water that requires an efficient pre-treatment.

The SDI measurement does not consider the different phases of membrane clogging. Indeed, it is considered that, first, there is a pore blockage, then the formation of a cake. It all depends on the water that is being tested. The SDI measurements can be done partially or totally before the formation of a cake, which can in some cases complicate the interpretation of the results.

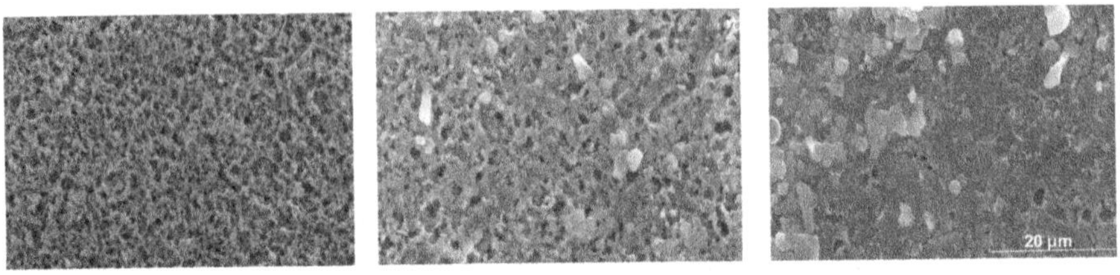

a) b) c)

Figure 18.12. *Electron microscope observations of (a) clean, (b) SDI 2.5 and (c) SDI 4.8 membranes (0.45 µm)*

For an NF or RO system, the objective is to feed the membranes with water characterized by an $SDI_{15} > 5$ (100% of the time).

Types of water	SDI
Groundwater	<2
Surface water	>6
Feed water for a spiral module	<5

Table 18.5. *Typical SDI values*

18.3.6. *The MFI*

Compared to SDI, MFI physically represents the resistance of a unit of deposit mass per unit area of filter membrane. The clogging of a membrane in frontal filtration, where the pressure gradient is constant, takes place in three stages: pore blockage, formation of an incompressible cake and formation of a compressible cake. The denser the deposit formed, the more difficult it is for the fluid to be filtered. The MFI is therefore high.

When an incompressible cake is formed, a linear relationship exists between t/V and V, as shown in Figure 18.13. The MFI indicator specifically corresponds to the clogging by cake formation on the surface of the membrane, and is represented by the slope of the linear part of the curve t/V = f(V).

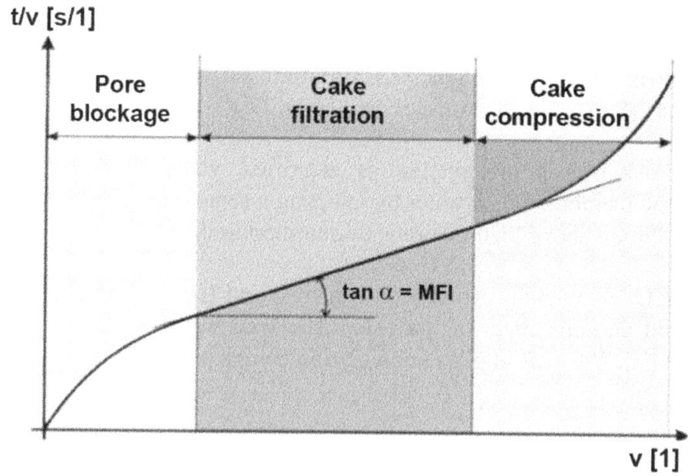

Figure 18.13. *Experimental determination of MFI. For a color version of this figure, see www.iste.co.uk/gaid/watertreatment4.zip*

$$MFI0.45 = \frac{\mu 20}{\mu T}\left(\frac{\Delta P}{210}\right) tan\ \alpha$$

V, t, ΔP and μ represent filtrate volume, filtration time, applied transmembrane pressure and dynamic viscosity of water, respectively. The MFI value is taken at a temperature of 20°C and a pressure of 210 kPa. In practice, the determination of the MFI is similar to that of the SDI. The same experimental setup is used. The tests consist of a series of time and cumulative permeate volume measurements that allow t/V to be plotted against V.

18.3.7. *Salts and metals*

NF and RO are processes that concentrate salts that tend to precipitate once their solubility product (Ksp) is exceeded. The higher the conversion rate, the higher the salt concentration.

In this case, the risk of exceeding the solubility products of the salts is high. The concentration of salts and metals retained by the membranes increases from upstream to downstream in the concentrate. This can lead to precipitation and clogging.

Some salts present in groundwater or brackish water can limit the conversion rate of the plant, especially for high sulfate concentrations. These are calcium fluoride (CaF_2), barium sulfate ($BaSO_4$), strontium sulfate ($SrSO_4$), calcium sulfate ($CaSO_4$), silica (SiO_2) and calcium carbonate ($CaCO_3$).

When a risk of salt precipitation is identified, water conditioning must be provided in the membrane feed water by injecting a sequestering agent, which keeps the salts in solution above their theoretical saturation limit.

With $CaCO_3$, the action of the sequestering agent (antiscaling chemical) is not sufficient, and an adjustment of the pH of the feed water is also necessary. The sizing of a NF or RO unit implies knowing the complete ionic balance of the water to be treated.

Groundwater can also contain sulfides, ferrous iron and manganese. These different substances do not cause problems in dissolved form and are fairly well retained by the membranes (except for sulfides, which pass through). However, if they are oxidized, they can precipitate and clog the membranes and pre-filters.

An oxidation and filtration step should be performed, so as to remove them before passage to the membranes. It is also necessary to remove the residual oxidant before passage to the membranes in order to avoid their degradation.

18.3.8. *Biological clogging*

Algae and bacteria can, in some cases, adhere to the membrane and develop in the NF modules, producing a viscous gel consisting of exo-polysaccharides that create a "biofilm" that is difficult to remove. The parameters that influence such developments are mainly:

– the concentration of bacteria or algae in the feed water;

– the temperature of the water;

– the operating mode of the installation: prolonged shutdowns promote proliferation. Appropriate operating procedures can limit the risks (maximum downtime, rinsing, conditioning and operation);

– the design of the installation: for example, transparent pipes must be banned because algae can grow inside;

– pre-chlorination of the water is possible, but not sufficient to prevent biofilm development. Moreover, chlorination promotes the breaking down of organic substances into biodegradable substances that participate in the formation of biofilm.

In all cases, a neutralization of the residual oxidant is essential for the feeding of the membranes.

Algae need to be very well removed during pre-treatment, because its removal on pre-filters and membranes is restrictive.

18.3.9. *Undesirable substances*

A number of substances are considered to be incompatible with the membranes by the manufacturers. These are mainly oils, greases, non-ionic and cationic detergents, and hydrocarbons. Most often the required limits are for hydrocarbons (<0.1 mg·L^{-1}) and for non-ionic and cationic detergents (<0.1 mg·L^{-1}). Lastly, the use of cationic polymers should be avoided (at the pre-treatment level) because, in case of leakage, a reaction is possible with the sequestering agent and can lead to the clogging of the membranes.

18.3.10. *Limit values of compounds at the inlet of high-pressure membranes*

Most of the compounds responsible for clogging, as presented above, can be removed by chemical cleaning with the exception of algae, calcium sulfate, barium sulfate and strontium sulfate. For these salts, the addition of a particular sequestrant is necessary, but may not be sufficient.

Parameters	Reference limit values	Imperative limit values
Turbidity (NTU)	<0.2	<0.5
Particles > 1.5 µ	<200	<2,000
SDI	Around 3–4	<5
Total coliforms (nb/100 mL)	<50	–
Total germs (nb·mL^{-1})	<50	<10
Algae	Absence	Absence
PO_4 (mg·L^{-1})	<0.1 mg·L^{-1}	<0.5
Al (µg·L^{-1})	<20 (detection limit)	<50
Fe (µg·L^{-1})	<20 (detection limit)	<50
Mn (µg·L^{-1})	<20 (detection limit)	<50

Table 18.6. *Limit values of some acceptable elements and substances for the feeding of high-pressure membranes*

Table 18.6 roughly defines different cases for the feeding of high-pressure membranes. Its values must be interpreted as average values. Indeed, a short overrun may require cleaning without any significant impact on the performance of the membranes or on the annual operating cost.

18.4. Sizing parameters

The maximum permeate flow that can be produced depends on the potential for membrane clogging and scaling. A two-stage membrane system that treats hard groundwater is typically between 70% and 80%. For both surface water and groundwater, the issue of concentrate must be considered in order to determine the outcome. The concentrations of salts and other contaminants in the concentrate are calculated based on the conversion rate. The assessment of capital and operating costs must take into account the physicochemical quality of the raw water, the costs of reagents (including pre-treatment, membrane station and post-treatment), energy costs, maintenance and membrane renewal costs.

18.4.1. *Temperature*

Temperature is the parameter that has the greatest influence on performance, both on water quality and inlet pressure. As surface waters have significant temperature variations during the year, the sizing must consider the minimum temperature, which implies calculating the maximum pressure, the maximum rejection of the membranes and the conditions of equilibrium recovery (pH of saturation, the pH of the water if it were saturated and at equilibrium with calcium carbonate and caustic soda consumption).

18.4.2. *Implementation configuration*

Given that RO and NF are pressure-driven processes, the required membrane feed pressure is a critical hydraulic design parameter. Membrane feed pressure directly influences membrane flux and permeate water quality. Five parameters influence the efficiency of the RO/NF system:

– osmotic pressure;

– pressure loss through the membrane;

– permeate backpressure;

– pressure loss along the feed channel of multiple membrane elements configured in series within a pressure vessel;

– friction losses in the feed piping between membrane feed stages.

The choice of the flow must be assessed on a compromise between an aggressive sizing that leads to excessive clogging and the investment costs by considering a flow that is too low.

For fresh and brackish waters (total dissolved solids < 12,000 mg/L), average osmotic pressure within a membrane stage is closely approximated by the following equation:

$$\Delta\Pi = \frac{TDSin + TDSc}{2} - TDSp$$

where:

– $\Delta\Pi$ = osmotic pressure;

– TDSin = total dissolved solids concentration in the feed water, mg/L;

– TDSc = total dissolved solids concentration in the concentrate water, mg/L;

– TDSp = total dissolved solids concentration in the reverse osmosis permeate water, mg/L.

The permeate flow (Jw, $L \cdot m^{-2} \cdot h^{-1}$) is the ratio of the permeate flow rate (Q in $L \cdot h^{-1}$) to the total membrane surface (St in m^2). It is also proportional to the pressure gradient across the membrane (transmembrane pressure) and to the permeability of the membrane (Lp in $L \cdot m^{-2} \cdot h^{-1} \cdot bar^{-1}$):

$$Jw = \frac{Qp}{St} = Lp\ (\Delta Pt - \Delta\Pi\ ) = Lp\ (\Delta Peffective)$$

With:

$$\Delta Pt = [\frac{Pe+Pc}{2} - Pp]$$

or:

$$\Delta Peffective == \left[\frac{Pin\ e+Pc}{2} - Pp\right] - [\frac{\Pi in\ e+\Pi c}{2} pf - \Pi p]$$

with:

– $\Delta\Pi$: osmotic pressure difference on either side of the membrane, which ranges from 0.1 to 1 or 2 bar depending on the nature of the quality of the water to be treated (water can be more or less concentrated in salts). $\Delta\Pi$ (bar) includes the osmotic pressures of the feed water, permeate and concentrate;

– Pin, Pc, and Pp (bar): inlet pressure, concentrate pressure, and permeate pressure, respectively. A temperature coefficient is included to establish the correction to a reference temperature;

– Pf: polarization factor.

The attack pressure (Pin, bar) gradually increases when there is a decrease in flow, due to a drop in temperature or clogging.

A compromise must therefore be found, depending on the quality of the feed water, between the conversion rate and the clogging. The latter also depends on the permeate flow. The permeability *Lp* depends on the temperature and the degree of clogging of the membrane. The higher it is, the better the hydraulic performance of the membrane.

The flow rate produced (Qp, $m^3 \cdot d^{-1}$) determines the number of modules to be implemented.

The recovery rate (Y %) (also called *feed water recovery*) is the hydraulic efficiency of the system and represents the ratio of the produced flow (Qp) to the input flow (Qin):

$$Y\,\% = \frac{Qp}{Qin}$$

The recovery of a membrane system is the fraction (%) of feed water volume that passes through the membranes as permeate. System recovery for surface water, brackish water and groundwater, RO and NF applications generally ranges between 70% and 85%. Due to the greater osmotic pressures associated with seawater's higher salinity, seawater RO applications are usually limited to recoveries of approximately 42–50%.

For 8-inch membrane elements used in large-scale systems, maximum recovery for individual membrane elements is generally limited to approximately 15% for surface water, brackish water and groundwater applications, and between 13% and 15% for seawater applications. The design engineer should consult the membrane manufacture's recommendations regarding maximum recovery for specific membrane elements. In brackish water applications, recovery is typically limited by the solubility of a "limiting salt". Since a greater fraction of feed water is converted to permeate at higher recoveries, the concentration of dissolved salts in the feed stream also increases.

The higher salt concentration in the feed stream results in a higher rate of salt diffusion across the membrane with a corresponding increase in permeate salinity at a given flux rate, a greater feed stream osmotic pressure with a corresponding reduction in permeate flux at a given pressure and an increased potential for precipitation of dissolved salts.

For a constant permeate flux, the net effects of operating an RO or NF membrane system at higher recoveries are an increase in required feed water pressure, an increase in permeate salinity and an increased potential for membrane scaling.

It is reasonable to assume that the higher the conversion rate, the lower the concentrate flow rate (and thus the tangential sweeping effect) and the more likely the membrane is to clog. From an economic point of view, this should be as high as possible. From a technical point of view, the choice of the conversion rate is essentially a function of the salinity of the water to be treated.

The total membrane area (St) for a train is determined from a produced water flow rate and the selected flow rate:

$$St\ (m^2) = \frac{Qp}{Jw}$$

The minimum number of modules per train is obtained by the ratio between the total membrane surface of a train and the unit surface of a module (Sm):

$$Number\ of\ modules/train = \frac{St}{Sm}$$

The salt passage rate, which is the ratio of the compound concentration in the permeate (Cp) to its concentration in the feed water (Cin), is:

$$X\,\% = 100\,\frac{Cp}{Cin}$$

The rejection rate (R%), which represents the ability of the membrane to reject unwanted salts, is:

$$R\,\% = 1 - X$$

For certain compounds such as natural organic substances, pesticides and micropollutants, a high rejection rate is sought. For ions, the desired rejection depends on the treatment objectives and their initial concentration in the water. For a given membrane, rejection can be influenced by clogging, temperature, hydraulic conditions (flow, conversion rate) and solution composition (coupling effect, etc.).

The salt flow (Js, $mS \cdot m^{-2} \cdot h^{-1}$) is a function of the concentration upstream and downstream of the membrane, and is independent of the applied pressure:

$$Js = Jw\ (Cin - Cp)$$

The flow is inversely proportional to the viscosity. The latter varies according to the temperature of the water. The following relation indicates the variation of μ according to T °C:

$$\mu\ (mPa \cdot s^{-1}) = 1.777 - 0.052\ T\ °C + 6.25 \times 10^{-4}\ T\ °C$$

The ratio between a flow at 15°C and a flow at 20°C is therefore given as: at 15°C, μ15 is equal to 1.14 $mPa \cdot s^{-1}$ and at 20°C, μ20 is 0.987 $mPa \cdot s^{-1}$:

$$\frac{J20}{J15} = \frac{\mu 15}{\mu 20} = \frac{1.14}{0.987} = 1.155$$

The flow at 20°C is 15.5% higher than the flow at 15°C.

18.4.3. *Calculation of the osmotic pressure*

It is given by the Van Hoff relation:

$$\Pi = zi\ Ci\ R\ T/Mi$$

with:

– zi: ionic valence of the ion;

– Ci: concentration of the ion;

– T: temperature in K;

– R: constant of perfect gases (8.314 $J \cdot K^{-1} \cdot mol^{-1}$);

– Mi: molar mass of the ion ($mol \cdot g^{-1}$).

For a solution with 230 mg $Na^+ \cdot L^{-1}$ and 355 mg $Cl^- \cdot L^{-1}$, at a temperature of 20°C, we obtain the following:

$$\Pi = \left(\frac{230}{23} + \frac{355}{35.5}\right) 8.314 \times 293$$

which gives Π = 48,720 Pa = 0.48 bar. In the presence of a matrix of ionic salts, the osmotic pressure is the sum of the contributions of each ion:

$$\Pi = \sum (i\Phi\ Ci\ R\ T/Mi)$$

Pressure loss that is generated as water passes through a membrane as permeate is dependent on the water mass transfer characteristics of the membrane. This pressure loss is also referred to as net applied pressure.

Net applied pressure (Np) is given by the following equation:

$$Np = \frac{Qp}{\Phi} = \Delta P - \Delta \Pi$$

where:

– Qp: permeate flux;

– Φ: water mass transfer coefficient of the membrane;

– ΔP = transmembrane pressure differential;

– ΔΠ = osmotic pressure.

The pressure loss generated as water passes through a membrane (net applied pressure) is directly proportional to permeate flux and inversely proportional to the membrane's water mass transfer coefficient.

18.4.4. *Mass flow diagram*

Figure 18.14 shows a mass balance of flows on a membrane module. In terms of flow rate, the feed flow rate Qin is the sum of the other two flows (Qp permeate and Qc concentrate):

Qin = Qp + Qc

In terms of mass flows, we get:

Qin Cin = Qp Cp + Qc Cc

with Cin, Cp and Cc being, respectively, concentrations of salts, ions or organic compounds in the feed water, permeate and concentrate. For a recovery rate given by the relation:

Y = Cp/ Cin

The concentration of various elements in the permeate can be deduced.

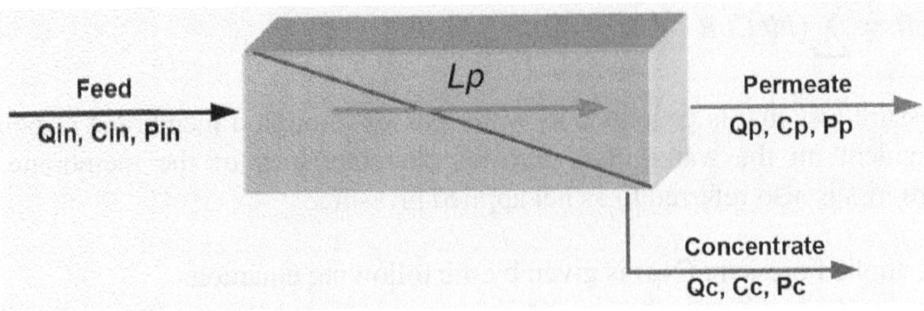

Figure 18.14. *Mass flow diagram. For a color version of this figure, see www.iste.co.uk/gaid/watertreatment4.zip*

18.4.5. *Salt passage*

Salt passage is the opposite of salt rejection and represents the amount of salt (expressed as a percentage) that passes through the elements. The lower the salt passage, the better the performance of the system. A high salt passage may suggest that the membranes need cleaning or replacement:

$$\text{Salt passage (\%)} = 1 - \text{salt rejection (\%)}$$

18.4.6. *Concentration factor*

The concentration factor is related to the conversion rate (Y %) and is an important parameter for system design. The more water recovered as permeate (high conversion rate), the more salts and contaminants are concentrated in the concentrate. This can lead to a saturation potential on the membrane surface when the concentration factor is too high:

$$\text{Concentration factor} = \frac{1}{(1 - Y)}$$

For example, with a recovery rate of 75%, the concentration factor is 4. This means that the concentrations of ions and compounds in the concentrate are four times higher than those in the feed water. Thus, if the feed water has 500 ppm TDS (total dissolved solids), the TDS concentration in the concentrate is 500 × 4 = 2,000 ppm TDS.

18.4.7. *Hydraulic pressure loss*

Permeate backpressure is the result of pressure losses (friction) within the membrane permeate channels and collection tubes, membrane array permeate header losses and pressure resulting from friction losses and elevation changes within a treatment facility's permeate piping network. Feed channel pressure losses are generated as a result of friction that occurs when feed water flows through the feed channel of each membrane element configured in series within a pressure vessel.

Hydraulic losses occur in the feed water (inlet) and the concentrate outlet due to friction on the walls. The friction loss in spiral-wound membranes can be calculated by:

$$\frac{dPhyd}{dx} = \frac{\lambda \rho v^2}{2\ dH}$$

with:

– λ: friction factor;

– dH: hydraulic diameter (m);

– v: circulation velocity ($m \cdot s^{-1}$).

λ is given by: 6.23 $Re^{-0.3}$ for $100 < Re < 1{,}000$.

For example, in a pressure tube comprising six 1-m long membrane modules with a flow velocity of 0.3 $m \cdot s^{-1}$, and a hydraulic membrane diameter of 0.9 mm, the hydraulic pressure loss at 20°C is:

$$Re = \frac{0.3 \times 0.9\ 10^{-3}}{10^{-6}} = 270$$

with η (viscosity) being equal to $10^{-6}\ m^2 \cdot s^{-1}$. We deduce:

$$\lambda = 6.23\ (270^{-0.3}) = 1.161$$

$$\Delta Phyd = 1.161\ \frac{1}{2 \times 0.9 \times 10^{-3}} 1{,}000 \times 0.3^2 = 50{,}000\ \text{kPa} = \ 0.50\ \text{bar}$$

18.4.8. *Pressure tubes and number of modules per tube*

The number and size of the membrane trains should be suitable for a non-binding operation. The choice can be between using multiple membrane trains with an alternate N–1 operation, or with reduced capacity membrane trains with N operation, or with a limited number of higher capacity trains in the plant. The number and size of each train should facilitate startup and shutdown, so as to meet demand. Where appropriate, a train should meet the minimum daily demand. Redundancy may be imposed by the specification in order to ensure that the maximum amount of water produced is met with one train out of service and/or one train being cleaned. Often, a large storage tank is suggested, so as to compensate for the lower water production during cleaning or maintenance periods when the latter does not last longer than 8 h.

Determining the number of pressure tubes and the number of elements per tube is generally a process guided by an economic objective, as well as by the limits imposed by the inlet and concentrate flow rates.

The number of elements per pressure tube is determined according to a minimum flow rate of concentrate specified by the manufacturer, which is produced by the last element in the tube. This flow rate depends on the recovery rate and the number of pressure tubes in parallel.

An economic assessment of the different possibilities takes into account the capital costs of pressure tubes and associated equipment. The energy costs associated with the higher inlet pressures, as the number of elements in a pressure tube increases, are also considered.

Determining the number of pressure tubes involves estimating the number of tubes in parallel and the number of stages needed to achieve the desired recovery rate.

Basic rules exist, but each project must be considered as a specific case:

– for systems operating up to 75% recovery rate, it is common to have two stages with pressure tubes, with six elements per tube;

– for systems operating between 76 and 87% recovery rate, it is common to have two stages with pressure tubes, with seven elements per tube (although a third stage is also often offered);

– for systems with recoveries > 88%, it is common to have three stages with pressure tubes, with six elements per tube.

When the number of elements per pressure tube is chosen, the minimum number of pressure tubes required can be calculated by:

$$\mathrm{NTP} = \frac{\mathrm{NMT}}{\mathrm{NMt}}$$

with:

– NTP: minimum number of pressure tubes required per train;

– NMT: number of modules per train;

– NMt: number of modules per pressure tube.

18.5. Chemical conditioning of pre-treated water

The purpose of conditioning the pre-treated water is to prevent the formation of precipitates and their deposition on the membranes. The concentration of the different salts in the concentrate depends on their content in the inlet water, their rejection by the membrane and the conversion rate.

It is possible to calculate the concentration of the salt(s) in the concentrate from a mass balance between the inlet and the permeate:

$$Cc = Cin\frac{(1 - YX)}{(1 - Y)}$$

with:

– Y: recovery rate of the installation;

– X: salt passage rate;

– Cin: salt concentration in the inlet water;

– Cc/Cin: concentration factor.

If the salt is 100% retained by the membrane (X = 0, case of sulfates), the formula becomes:

$$Cc = Cin\frac{1}{(1 - Y)}$$

Thus, for a Y recovery rate of 85%, a concentration factor of 6.67 is obtained, whereas for a recovery rate of 75%, this factor is only 4.

18.5.1. *Calculation of saturation indices and antiscalant dosage*

The solubility of a salt represents the degree to which it may be dissolved in water at a specific temperature. The solubility of a salt at a specific temperature is given by its solubility product at that temperature, defined as:

$$Ksp = [A^{b+}]^a \, [B^{a-}]^b$$

where:

– Ksp: solubility product of the salt;

– A: cation constituent of salt;

– B: anion constituent of salt;

– a: stoichiometric coefficient of cation in the salt's equilibrium equation, charge of anion;

– b: stoichiometric coefficient of anion in the salt's equilibrium equation, charge of cation.

Ksp varies with temperature. In general, salts become more soluble at higher temperatures.

Given that dissolved salts are rejected by RO, and some of them by NF membranes as water passes through the membrane as permeate, the concentration of these salts in the feed stream rises along the membrane feed path. If a salt's concentration exceeds its solubility product at the system's operating temperature, the salt will begin to precipitate out of the solution. Precipitation of dissolved salts in a membrane system can lead to membrane scaling and reduced membrane performance. Control of mineral scaling is an essential component of the design of an RO or NF system treating brackish surface or groundwaters.

Saturation indices (SI) are defined as the log of the ratio of the ionic activity product (Ip) and the solubility product salt Ksp:

$$\mathrm{IS} = \log \frac{\mathrm{Ip}}{\mathrm{Ksp}}$$

A real solution may not be in the state of equilibrium. This non-equilibrium state is described by the ion activity product (Ip). It has the same form as the equilibrium constant K_{sp}, but involves the actual activities, $\{A\}_{actual}$ and $\{B\}_{actual}$:

$$\text{Real solution} \qquad \mathrm{Ip} = \{A\}^{a}_{actual}\ \{B\}^{b}_{actual}:$$

The value of the product of ion activity of a salt is determined by the ion strength, pH and the ion affinity.

The saturation index is a useful quantity to determine whether the water is saturated, undersaturated or supersaturated with respect to the given mineral:

$$\mathrm{SI} = 0 \quad \mathrm{Ip} = \mathrm{Ksp} \quad \text{saturated (in equilibrium)}$$

$$\mathrm{SI} < 0 \quad \mathrm{Ip} < \mathrm{Ksp} \quad \text{undersaturated}$$

$$\mathrm{SI} > 0 \quad \mathrm{Ip} > \mathrm{Ksp} \quad \text{supersaturated}$$

Computer models incorporate the principles of equilibrium chemistry and can predict with reasonable accuracy the occurrence of scaling based on feed water chemistry and target recovery values. Salts such as calcium sulfate, barium sulfate, strontium sulfate, silica, calcium fluoride and calcium phosphate are well described.

The computer model will provide a solubility warning for common limiting salts, and may recommend a target feed water pH and the use of a scale inhibitor or a mineral acid. Scaling can thus be prevented by removal of seeding material and by not exceeding the solubility product.

With calcium carbonate ($CaCO_3$), the saturation index (Is, also called LSI, the Langelier saturation index) in the concentrate is defined differently from other indices by the formula:

$$\text{Is} = \text{pH concentrate} - \text{equilibrium pH of concentrate} = \text{pH} - \text{pHs}$$

If Is < 0, the concentrate is aggressive, and if Is > 1, the concentrate is scaling. The calculation of Is requires knowing the pH, TH, TDS and the TAC of the concentrate. The equilibrium pH is then calculated using commercial software. Veolia has its own software.

Depending on the sequestrant used, the limits for Is are given in Table 18.7.

Sequestrants	**Is limits calculated with Argo software**
HMPS (20 $mg \cdot L^{-1}$ in the concentrate)	1
Hypersperse or permatreat	1.5
No sequestrant	0

Table 18.7. *Example of a sequestrant used for calcium carbonate*

To lower the LSI below these limits, it may be necessary to acidify the inlet water. So, for example, for conventional surface water (hardness = 300 $mg \cdot L^{-1}$ $CaCO_3$, alkalinity = 200 $mg \cdot L^{-1}$ $CaCO_3$) and a recovery rate of 85%, the approximate acidification pHs needed to maintain an LSI below 1.5 are shown in Table 18.8.

The use of a low salt rejection membrane such as NF-200 or Toray SU 610 requires less acidification at the inlet, as the salts are less concentrated in the concentrate. The final permeate obtained under the best operating conditions (NF-RO membrane, feed pH and recovery rate) is characterized by its total dissolved solids, conductivity, arsenic, sulfate, chloride, calcium and magnesium concentrations.

pH	Membranes
6.5	Membrane type NF-70 Hardness rejection: around 90% Alkalinity rejection: around 85%
7	Membranes NF-200, Toray SU 610 or similar Hardness: around 50% Alkalinity rejection: around 40%

Table 18.8. *pH adjustment according to water hardness and membrane type*

18.5.2. *Choice and implementation of the antiscalant*

Table 18.9 shows the maximum SI theoretically possible in the presence of the anti-precipitant. This calculation is based on the expected recovery rate for the plant (85%), a pH of 7 at the membrane inlet and a maximum sulfate concentration of 120 mg·L^{-1} in the inlet water.

	Saturation indexes		
Salt	**Summer 25°C**	**Winter 3°C**	**Anti-precipitant limits**
$BaSO_4$	14.73	13.47	40
$CaSO_4$	0.52	0.57	2
$SrSO_4$	0.26	0.23	30
SiO_2 (mg·L^{-1})	14	14	21
CaF_2	1.5	1.5	100
$CaCO_3$ (Langelier index)*	1.5	1.38	1.5

* Calculation in the worst case; in other words, 350 mg·L^{-1} calcium hardness at the inlet, and minimum membrane calcium passage rate.

Table 18.9. *Maximum saturation indices for various salts*

These limits, as well as the antiscalant dosage, are recommended by the manufacturer. However, it is difficult to model the chemical and hydraulic interactions at the surface of the membrane, in particular the polarization phenomena (local overconcentration of salts), which are very dependent on the membrane itself. An experimental approach was therefore implemented to verify the limits given by the manufacturers. In addition, the action of the antiscalant is limited on some salts that present a risk of clogging on the membranes. These are mainly iron and aluminum hydroxides, and calcium phosphates. For these salts, precautions must be taken during the pre-treatment, so as to minimize their concentration.

In the absence of information on salt concentrations, the assumption of a recovery rate of 85% and an antiscalant dosage of 2 $mg \cdot L^{-1}$ can be a first step for conventional surface water sizing. It can then be verified and corrected based on field analyses. This simplified approach is not valid for very salty waters (sea water, brackish water), for which sulfate concentrations are very high because the efficiency limit of the antiscalant is reached, and it is often necessary to lower the recovery rate of the system.

A quick estimate of salt precipitation resulting from the concentration of the salt to its solubility limit is, in its simplest expression, a function of the permeate recovery rate:

For the equation: $A_aB_b \rightleftarrows aA^+ + bB^-$

$$Ksp = \frac{[A^{b+}]^a \, [B^{a-}]^b}{[1-Y]\,[1-Y]}$$

with:

– Ksp: solubility product;

– A^{b+}: concentration of cations, moles L^{-1};

– B^{a-}: concentration of anions, moles L^{-1};

– a, b: number of moles;

– Y: recovery rate.

18.5.3. *pH adjustment at the membrane inlet*

Given that there is good calcium rejection of the membranes (type NG 70, NF-90, low-pressure reverse osmosis [LPRO], RO), a pH of 7 at the membrane inlet generally keeps the calcium carbonate saturation index below 1.5, which is the efficiency limit currently accepted by the manufacturers of sequestering products and verified during the tests carried out by VERI-Veolia, as well as by feedback from the plants. In exceptional cases (in summer, with high hardness in the inlet water), the pH of the feed water may be lowered to 6.8. However, the choice of pH also depends on the pre-treatment (minimum aluminum residual at the filter outlet).

18.5.4. *Choice and application of the acid*

The feed water is acidified with sulfuric acid or hydrochloric acid.

Sulfuric acid is easier to use and less expensive than hydrochloric acid. Hydrochloric acid increases the passage of calcium (the coupling effect of sulfates with calcium tends to increase the rejection of calcium) and lowers the pH of the inlet water, without the risk of precipitation due to sulfates. However, it is delicate to implement and is, in most cases, substituted by sulfuric acid. Acidification is injected either at the pre-treatment level to improve its efficiency, or at the low-pressure pump discharge.

The pH of the feed water to the membranes must be maintained at a nominal value, with a maximum deviation of ± 0.1. Furthermore, the control system must be able to adapt quickly to flow variations with a pH stabilization time of less than 5 min, and must not exceed a pH of 7.5 during this adjustment phase.

18.5.5. *Influence of sulfates*

Sulfate ions are almost completely rejected by NF membranes. Given that there is a necessity to maintain the electroneutrality of the solution on both sides of the membrane, the sulfate content influences the rejection of other ions, in particular the hardness.

As sulfuric acid is generally used to acidify the water upstream of the membranes, a significant increase in sulfate content is therefore made. This parameter must be taken into account in the simulations.

18.6. Design and implementation

18.6.1. *Pre-treatment*

Pre-treatment is recognized as the key to the proper functioning of the process and the performance of the membranes. By pre-treatment, we mean any type of process upstream of the NF or RO stage, which is designed to feed the membranes with water that is as clogging-free as possible, thus allowing the membranes to function in the best possible conditions. This water quality must be as constant as possible.

The aim is to remove all particles that could damage the membranes, by producing water with a turbidity < 0.5 NTU at the outlet of the pre-treatment. As such, turbidity removal is done by coagulation–flocculation–settling–filtration, or by MF or UF membranes.

The advantage of "chemical" clarification is that it also removes some of the organics and reduces the feed to the high-pressure membranes, all while having an SDI < 5.

Pre-treatment processes typically involve the addition of acid, a sequestrant to prevent the precipitation of poorly soluble salts as the discharged ions become more concentrated, followed by 5–10 μm cartridge filtration to protect the NF-OI membranes from micro-particle clogging.

18.6.2. *Treatment processes*

The treatment process depends on the quality of the raw water and is designed based on the sizing parameters listed above. The treatment process prepares the feed water for the membranes in such a way that no parameter can disturb their operation or fail to meet the required guarantees. The treatment processes are diverse, varied and very different depending on whether surface water or groundwater is to be treated. Surface waters are classified in different categories, depending on the resource and the contaminants that constitute them. A surface water can be river water (with variable contaminants during the year), impounded water (ponds, dams, lakes), especially with the presence of algae, or estuary water contaminated with wastewater.

Figures 18.15–18.22 show some examples of the process according to raw water quality.

Case 1: Turbid surface water with the presence of Fe and Mn, and significant variation in organic matter, pesticides and micropollutants

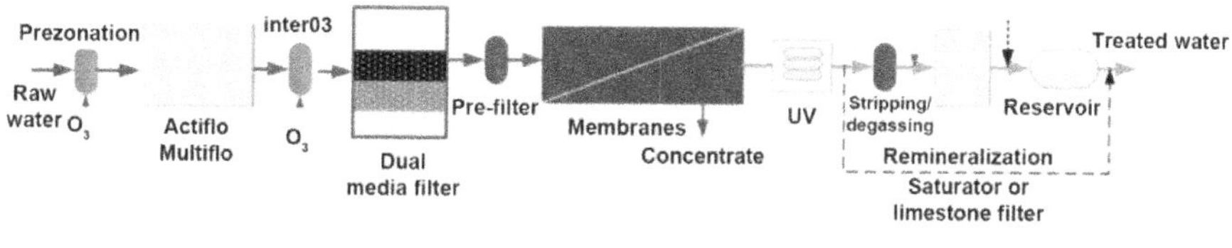

Figure 18.15. *Presence of Fe and Mn, and significant variation of organic matter, pesticides and micropollutants. For a color version of this figure, see www.iste.co.uk/gaid/watertreatment4.zip*

Case 2: Turbid water with organic matter, pesticides and micropollutants

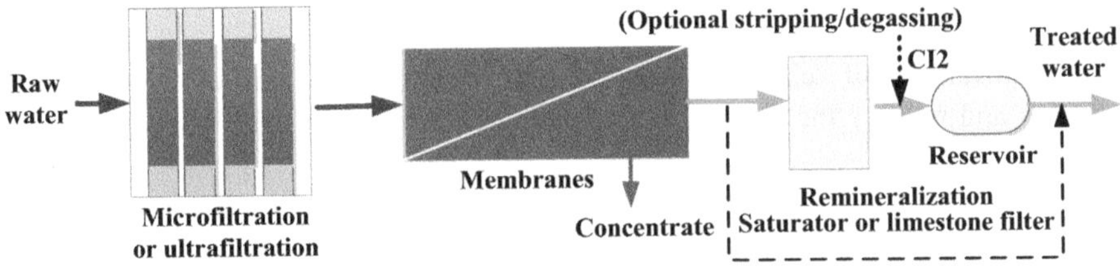

Figure 18.16. *Presence of organic matter, pesticides and other micropollutants. For a color version of this figure, see www.iste.co.uk/gaid/watertreatment4.zip*

Case 3: Turbid surface water with organic matter, pesticides and micropollutants, nitrates

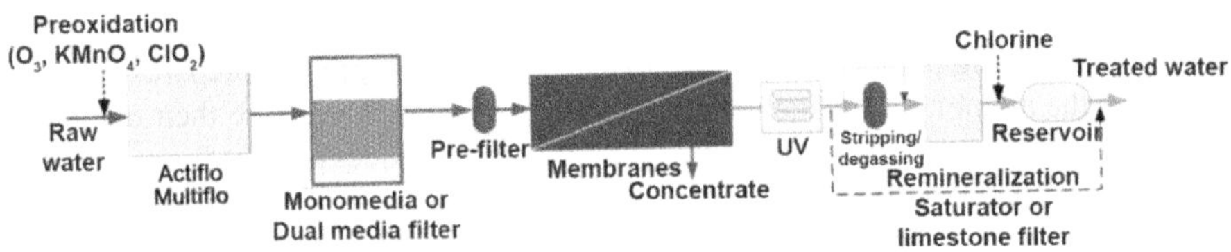

Figure 18.17. *Presence of organic matter, pesticides, other micropollutants and nitrates. For a color version of this figure, see www.iste.co.uk/gaid/watertreatment4.zip*

Case 4: Slightly turbid surface water with algae, toxins, organic matter, pesticides and micropollutants

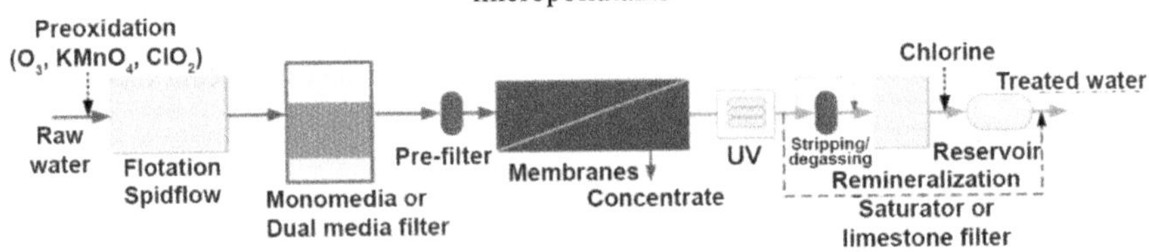

Figure 18.18. *Presence of organic matter, algae, pesticides and other micropollutants. For a color version of this figure, see www.iste.co.uk/gaid/watertreatment4.zip*

Case 5: Groundwater loaded with iron, manganese, calcium, magnesium, sulfates and carbonates

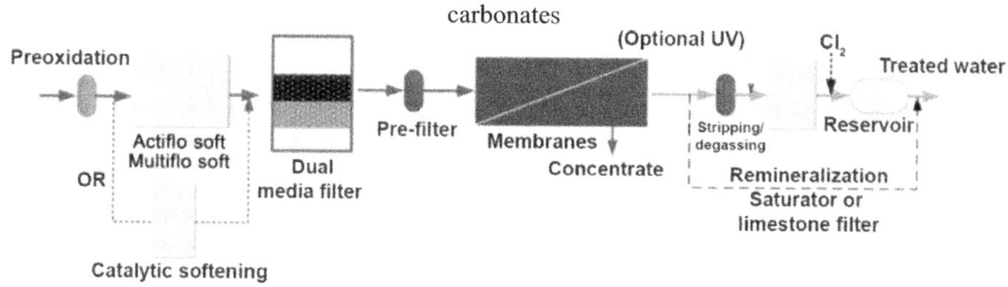

Figure 18.19. *Presence of iron, manganese, calcium, magnesium, sulfates and carbonates. For a color version of this figure, see www.iste.co.uk/gaid/watertreatment4.zip*

Case 6: Very low turbidity groundwater (<1.0 NTU) loaded with sulfates, pesticides and micropollutants

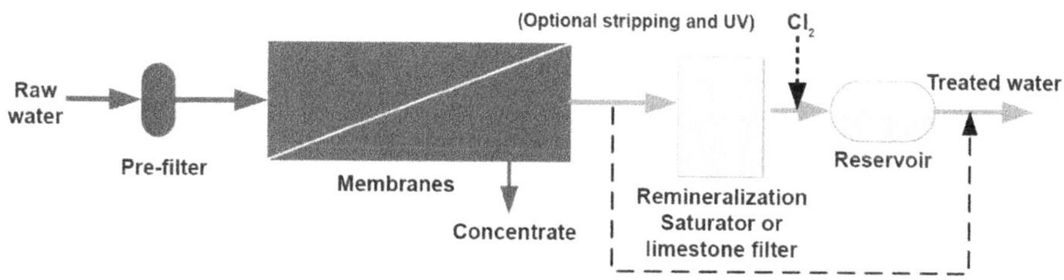

Figure 18.20. *Presence of sulfates, pesticides and other micropollutants. For a color version of this figure, see www.iste.co.uk/gaid/watertreatment4.zip*

Case 7: Very low turbidity groundwater (<1.0 NTU) loaded with heavy metals, perchlorates, PFAS and PFOS, and micropollutants

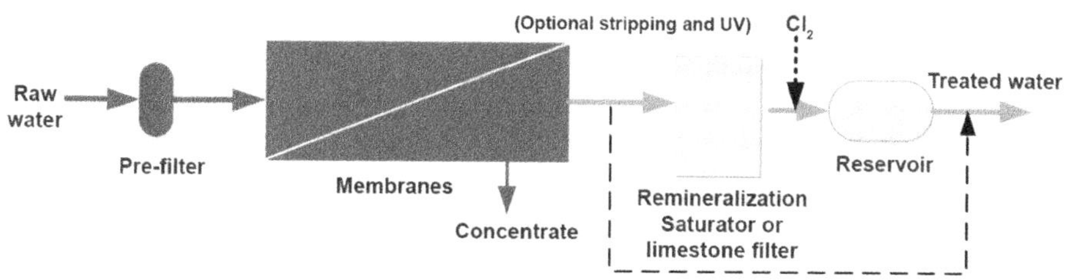

Figure 18.21. *Presence of heavy metals, fluorinated organic compounds and micropollutants. For a color version of this figure, see www.iste.co.uk/gaid/watertreatment4.zip*

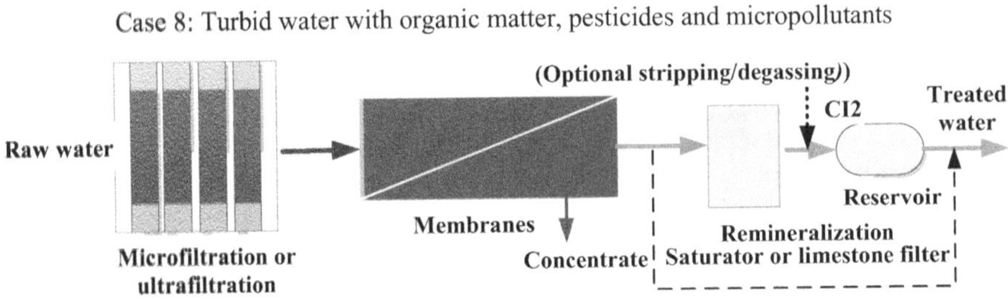

Figure 18.22. *Presence of turbidity, organic matter, pesticides and other micropollutants. For a color version of this figure, see www.iste.co.uk/gaid/watertreatment4.zip*

18.6.3. *Membrane station*

Membrane treatment of surface water and conventional groundwater includes high-pressure pumping and membrane trains. Seawater treatment is specifically detailed in Chapter 19.

The design of a membrane unit is simplified by the software offered by the membrane manufacturers, which selects the conversion rate, determines the drop in pressure and the quality of the permeate, and verifies the scaling potential of the concentrate.

In the absence of experience, NF bench tests are recommended, especially for surface water, because they make it possible to determine the degree of feasibility, to develop design criteria and to pre-qualify membranes. The pilot tests (continuous operation) allow the assessment of clogging risk and pre-treatment requirements and optimization of the cleaning conditions of the membranes.

A certain amount of equipment in addition to the membranes is necessary to ensure their proper functioning.

18.6.3.1. *Pre-filtration*

Pre-filtration is a safety filtration, because the membranes must be protected against the accidental and short-lived influx of clogging particles, especially in the case of pre-treatment failure. Several criteria control the choice of the system, such as the cut-off point, the operating mode and the operating costs.

The technology chosen, whether in pre-filtration bags, cartridges or in an automated system with continuous washing, must ensure a very good removal of micro-particles (>5 μm) and micro-algae.

The bags are disposable and manually replaceable. The cartridges are much more reliable and have a higher retention capacity than the bags and can be regenerated in place. The polypropylene cartridges have a good chemical tolerance for cleaning.

The cartridges are arranged in a set of filters operating in parallel and are installed just upstream of the high-pressure pumping systems, after the injection of the sequestrant, to act as the ultimate protection.

When the pH adjustment (6.5–7) is not made at the pre-treatment level, it is done upstream of the pre-filters, so as to control the precipitation of calcium carbonates in the concentrate.

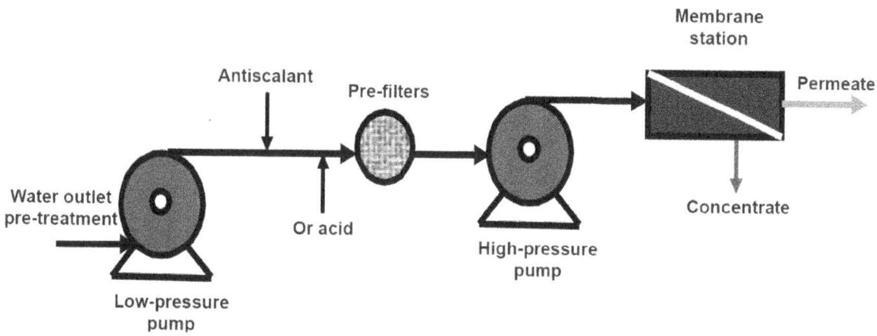

Figure 18.23. *Diagram of the pre-filtration station. For a color version of this figure, see www.iste.co.uk/gaid/watertreatment4.zip*

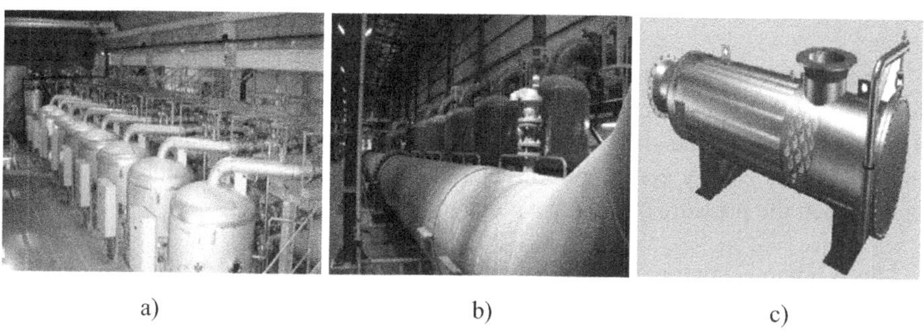

Figure 18.24. *Pre-filtration with (a) micro-filtration, (b) continuous operation and (c) manual cartridge replacement. For a color version of this figure, see www.iste.co.uk/gaid/watertreatment4.zip*

The pre-filtration is centralized, and therefore it can be managed independently of the membranes. The cleaning can be done automatically by injection of acid or basic solutions. The unit is sized in such a way that the entire flow rate can be passed during the cleaning process, so that the capacity of the installation is not reduced.

In the case of groundwater that does not have particulate matter and therefore does not require pre-treatment, this station can be the only pre-treatment. A low-pressure pump provides the necessary pressure to compensate for the pressure drop of the pre-filters (about 1–3 bars).

18.6.3.2. *Membrane station equipment*

After the pre-filtration station, the actual membrane station includes:

– a high-pressure pumping system;

– a line of membranes;

– a degassing tower (optional, depending on the case);

– a UV disinfection reactor (optional, depending on the case).

A cleaning station is added for the periodic cleaning of the membranes, as well as a neutralization station for the pre-filter and membrane cleaning waste.

18.6.3.3. *High-pressure pumping*

A high-pressure pump provides the necessary pressure for membrane operation, typically 5–20 bar. The purpose of this station is to provide the necessary pressure and flow rate to filter the water through the membrane and ensure the required conversion rate and production rate.

Each pump supplies a complete and independent line of membranes, as well as for degassing and disinfection. Thus, the pump only supplies the pressure needed for a dedicated line. As the clogging condition can be different from one line to another, this minimizes energy consumption (centralized arrangement of the pumps leads to operation at the pressure of the most clogged line).

Each pump is equipped with a variable frequency drive that maintains a constant membrane feed rate, and regulates itself according to a given set point when the HMT varies, whether that be due to variations in pre-filter pressure loss that is related to its degree of clogging, variation in water temperature, degree of membrane clogging or feed. The frequency converter helps to maintain very good energy efficiency over a wide operating range.

The HP pumps are each equipped with a bypass, which is used during the stop and start sequences of the membranes.

18.6.3.4. *Membrane configuration*

The number of membranes is determined by the surface flow to be used (flow rate compared to the total filtration surface). The choice of this flow is directly related to the quality of the water in question. Thus, surface waters require much lower specific flows, and therefore a larger number of membranes than groundwater, whose clogging load in organic matter and particles is much lower. In order to give flexibility to the installation, as well as to be able to distribute the water homogeneously over the membranes, it is necessary to group them in identical rows, which are themselves subdivided into blocks.

The choice of the number of lines and the size of the blocks must satisfy a number of criteria, namely:

– the desired degree of reliability and minimum percentage of total capacity when a line is stopped;

– the flexibility to adjust the production rate (In fact, each membrane line produces a constant flow rate. The adjustment of the total flow rate is therefore only done according to the number of lines in production);

– the homogeneous hydraulic flow distribution on all the membranes of a block;

– the number of membranes that can be cleaned simultaneously and associated with the size of the cleaning station in place.

The filtration mode operates in tangential mode, as opposed to frontal filtration used in most conventional filtration processes. The concentrate flows tangentially to the membrane, at a sweep rate that must be sufficient to carry away impurities and prevent their deposit on the membrane. The permeate collection tubes are connected together by connectors and seals, so that all the permeate from each module is collected at the end of the pressure tube.

There are six to seven membranes in a tube, depending on the water quality. Beyond that, the concentrate flow rate of the last module becomes too low to prevent accumulation of the membrane deposits. Like the membranes, the sizes of the pressure tubes are standardized.

In order to further increase the overall conversion rate, the pressure tubes themselves are arranged in parallel and in line to form stages (Figures 18.25 and 18.26). A membrane line is thus composed of two or three stages, depending on the quality of the water to be treated. The number of pressure tubes in parallel in a stage is a function of the flow rate to be treated (the feed flow rate of a tube being about 8 $m^3 \cdot h^{-1}$).

Since the conversion rate of a tube is about 50%, the inlet flow rate of the second stage (first stage concentrate) is half the inlet flow rate of the first stage. The number of tubes to be implemented is therefore divided by two from one stage to the next, and the configuration thus achieved is called the "Christmas tree configuration" (Figure 18.26).

So:

– 50% for 1 stage;

– 75–80% for the two stages together;

– 85% for all three stages.

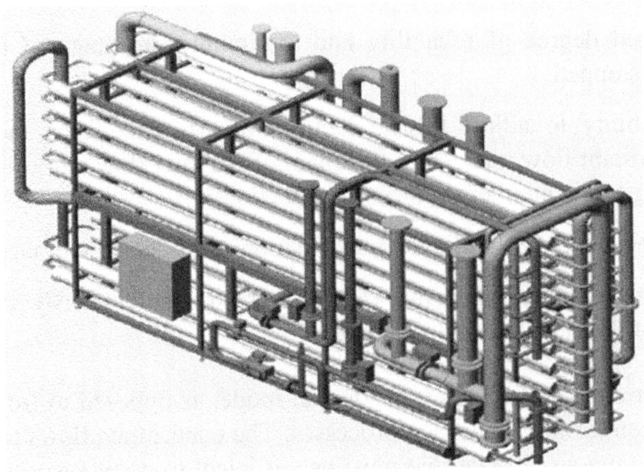

Figure 18.25. *High-pressure membrane train. For a color version of this figure, see www.iste.co.uk/gaid/watertreatment4.zip*

Several implementation diagrams are possible, and Figure 18.27 shows some examples such as:

– implementation on the entire flow rate: in order to avoid remineralization, a membrane with low salt rejection can be chosen;

– implementation with a fixed or variable bypass rate: NF can sometimes allow a much higher level of treatment than that required by the specifications. The use of a bypass reduces the investment cost. In this case, it must be checked that the bypass does not bring the feed water to a critical parameter level;

– mixing with water from an external source or from an existing line in the plant (without membrane treatment).

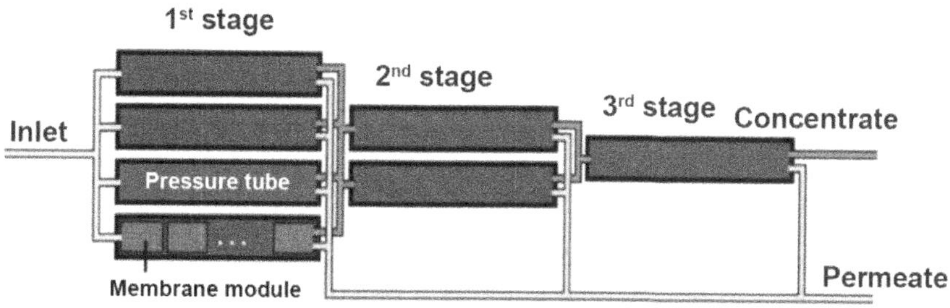

Figure 18.26. *Membrane installation configuration. For a color version of this figure, see www.iste.co.uk/gaid/watertreatment4.zip*

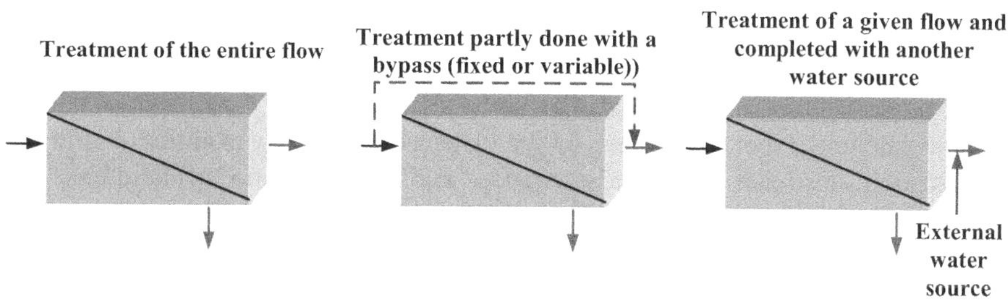

Figure 18.27. *Possible configurations of the membrane trains for the production of a global flow. For a color version of this figure, see www.iste.co.uk/gaid/watertreatment4.zip*

For the last two cases, the use of a membrane with high salt rejection has limited the final remineralization of the water brought by the bypass or by the external source.

In case of mixing, the mixing flow rate (QM) is the difference between the total required flow rate (Qt) and the flow rate produced by the permeate (Qp):

$$QM = Qt - Qp$$

The mixing water must not contain any organic contaminants, otherwise it will be polluted by the contaminants in the raw water. It can come from a treatment line parallel to the membrane plant, or from groundwater that only constrains salts. The tolerated concentrations in the raw water are used to determine the mitigation flow rate needed to meet the required final water quality objectives.

$$QM = Qp\frac{(Cs - Cp)}{(CM - Cs)}$$

with:

– Cs ($mg \cdot L^{-1}$): standard quality concentration of drinking water;

– Cp ($mg \cdot L^{-1}$): concentration of TDS or other elements in the permeate;

– CM ($mg \cdot L^{-1}$): concentration of the mixing water.

For example, for a permeate flow rate Qp of 100 $m^3 \cdot h^{-1}$, a TDS concentration of the mixing water CM of 2,500 $mg \cdot L^{-1}$, a TDS concentration Cp in the permeate of 100 $mg \cdot L^{-1}$ and a standard quality concentration Cs of 500 $mg \cdot L^{-1}$, the mixing rate is:

$$QM = 100 \frac{(500 - 100)}{(2{,}500 - 500)} = 20 \text{ m}^3 \cdot \text{h}^{-1}$$

18.6.4. *Post-treatment*

The calcium-carbonate balance of the water is profoundly modified due to the acidification upstream of the membranes, and the retention of hardness and alkalinity by the membrane. Post-treatment can include many of the same operations as conventional drinking water treatment, such as aeration, degassing, pH adjustment, addition of corrosion control chemicals and fluoridation. When the feed water is acidified, the aggressive concentrations of CO_2 in the treated water are substantial and require treatment to restore the calcium-carbonate balance.

Treatment usually consists of one of the following:

– an atmospheric degassing, followed by an injection of soda to give the water a slight scaling power;

– remineralization by injection of lime water (and if necessary CO_2);

– an injection of CO_2 or sulfuric acid on the limestone filter.

When carbonic acid alone is present, degassing a fraction of the permeate can be a valid solution. This fraction is then remixed with the fraction containing CO_2, so as to obtain the desired alkalinity by downstream neutralization (with soda, limewater or a limestone filter).

When CO_2 and H_2S are present, the total flow is degassed. Another solution is to perform a partial degassing and to neutralize the hydrogen sulfide (by chlorine oxidation) that is present in the bypassed fraction.

With some membranes, the removal of hardness is significant and further remineralization must be carried out by adding lime or by mixing with water of higher hardness.

The sizing details of the carbon-calcium balance and remineralization are detailed in section 20.

The permeate post-treatment methods (Table 18.10) are for membrane processes that do not include disinfection or corrosion inhibitor addition stages.

18.6.4.1. *UV disinfection downstream of the membranes*

A safety disinfection is often provided at the outlet of the membranes for the treatment of conventional surface water. This disinfection can be carried out by UV radiation, for example. In any case, the residual effect of chlorination in the network must be ensured.

Objectives	**Methods**
Removal of dissolved gases	Degassing
Re-mineralization	Mixture Addition of chemicals
Corrosion prevention	pH adjustment

Table 18.10. *Objectives and methods of post-treatment*

18.6.5. *Cleaning units in place*

A cleaning unit ensures the periodic cleaning of the membranes, as well as any device for neutralizing the pH of the cleaning solutions. The membranes are cleaned by circulating diluted chemical solutions. Cleaning with a basic detergent removes organic deposits. Acidic cleaning removes mineral deposits. The cleaning is carried out in a closed circuit (recirculation of the permeate and concentrate) for the time necessary for the effectiveness of the products. The membranes are rinsed abundantly after the passage of each product. The solutions are heated to a maximum temperature of 35°C to increase their efficiency.

The cleaning solutions are prepared from the returned water taken directly from the outlet manifold of the installation by means of a booster unit, common to all cleaning stations. Each cleaning unit is equipped with a frequency-controlled pump for mixing the solutions and distributing them to each membrane block, under controlled pressure and flow conditions. The flow rate applied in the different phases is regulated and kept constant, regardless of the system pressure losses, which can vary according to the state of clogging of the pre-filters and membranes.

The pumps are sized to ensure a minimum flow rate per tube of 7.5 $m^3 \cdot h^{-1}$, which is sufficient to evacuate the deposits from the membrane (tangential sweeping effect). Each station is also equipped with a cartridge filter in order to filter the solutions before sending them to the membranes and during recirculation.

The solutions are distributed in a looped network that can be connected successively to each membrane block. Each network is drained by gravity, because of the purposely made incline provided, and can operate in a closed circuit (return of the solutions to the tank) or an open circuit (rinsing sequences). The rinses are carried out with the returned water taken by the cleaning pumps, from the permeate outlet collector.

All waste produced during cleaning is directed to the neutralization tank. Thus, these discharges are well separated from the discharges generated in production mode.

The neutralization plant is designed to treat all the discharges generated by the membrane and pre-filter cleaning operations:

– the cleaning is carried out block by block, with a very low conversion rate and a high flow rate, which makes it possible to remove the polarization layer on the membrane and to take maximum advantage of the tangential sweep effect;

– the cleaning is triggered automatically, either when a pressure drop threshold is reached, or on a timer, based on the operator's judgment. The frequency of cleaning varies according to the quality of the water in question. When possible, a fully automated membrane cleaning unit is provided.

The cleaning plants have four identical units ensuring maximum safety:

– a unit for acidic cleaning;

– a unit for basic cleaning;

– a unit for rinsing;

– a backup unit or for a possible third reagent.

18.6.5.1. *Discharge of used cleaning solutions and rinsing of membranes, pre-filters and various circuits*

Neutralization is carried out by injection of soda or sulfuric acid, at a pH between 6.5 and 8.5. The reagents are dosed from the storage tanks provided for the cleaning of the membranes and pre-filters. Table 18.11 shows the different types of clogging observed on the high-pressure membranes, as well as the preventive and curative actions.

Different clogging can have the same impact on performance. To complete the diagnosis in case of severe problems, it is generally necessary to proceed with additional investigations: analysis of cleaning solutions and autopsy of the membranes.

Types of clogging	Possible causes	Impacts on performance	Preventive action	Curative action
SS, sand, particles	Poor pre-treatment	Increased pressure loss factors at the first stage and permeability drops	Better monitoring of pre-treatment	High-speed cleaning
Aluminum	Coagulant residual	Drop in permeability, especially at the third stage	pH adjustment Bonding	High pH cleaning
Phosphorus, $CaPO_4$ precipitate	Coagulant (HMPS, sequestrant)	Permeability drops at the third stage	Bonding	Cleaning base + acid
Precipitate of calcium carbonates	pH too high at the inlet of the membranes	Permeability drops at the third stage	Lowering of the pH Selection and dosage of sequestrate	Cleaning acid or base + acid
Bacterial clogging (biofilm)	Bacteria in feed water Water stagnation in the membranes, poor rinsing	Increase in pressure loss at the first stage Viscous film Permeability losses	Better pre-treatment (bonding) No large buffer storage upstream of the membranes	Disinfectant + detergent
$BaSO_4$, $SrSO_4$, $CaSO_4$, SiO_2	Content in feed water, recovery rate too high	Permeability drops at the third stage	Lowering of the conversion rate For $CaSO_4$, decarbonation Choice and dosage of sequestrant	More or less efficient cleaning
Mn, S, FeS precipitation on groundwater	Oxygenation and presence of iron	Permeability drops	Oxidation and filtration, or no oxygenation Choice of materials to avoid corrosion (Fe)	Cleaning base + acid

Table 18.11. *Types of clogging, and preventive and curative actions*

18.7. Functional and operating parameters

18.7.1. *Basic principles*

The association of the stages is known as a "series-rejection" association. This association consists of connecting the concentrates and the feeds in series. The concentrate from the first stage feeds the second stage, and the concentrate from the second stage feeds the third stage. The final concentrate leaving the third stage is the final discharge of the system. Some of the feed water passes through the membrane to the permeate collector, leaving the salts on the other side of the membrane. Thus, as the hydraulic flow advances through the tubes and stages, the salinity of the unfiltered water increases (this is the concentrate).

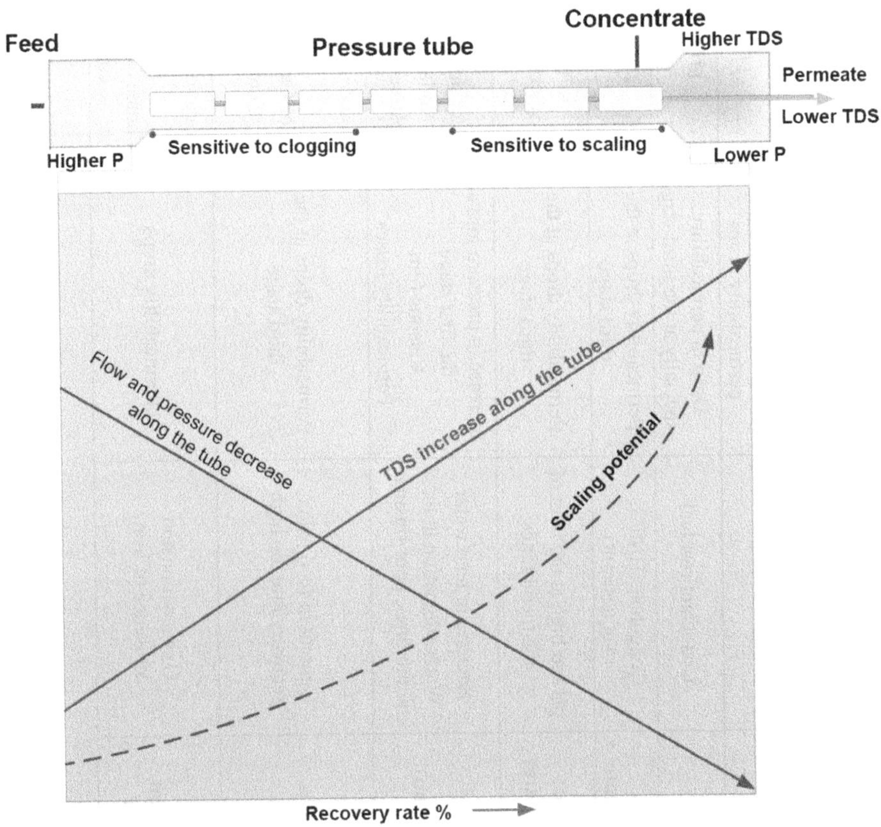

Figure 18.28. *Evolution of the different operating parameters along a pressure tube. For a color version of this figure, see www.iste.co.uk/gaid/watertreatment4.zip*

The evolution of the various parameters along a pressure tube (Figure 18.28) shows:

– a decrease in flow;

– a decrease in pressure;

– high pressure in the concentrate;

– a decrease in TDS in the permeate, but an increase in the concentrate.

The first membranes are sensitive to organic clogging. The last membranes are sensitive to scaling.

The kinetics and level of clogging of a membrane system depend on the permeate flow at each membrane module. The first elements in series of a pressure tube generally produce the highest permeate flows. This is the combined result of two factors. The first modules are fed with water that has a lower salt concentration and the highest feed pressure. The effect of decreasing pressure and increasing TDS affects the productivity of the membranes located in the second and third stages of the trains.

The quality of the filtered water depends on the quality of the feed water to the membranes. The concentration of salts in the permeate depends on the type of salt, as well as its rejection by the membrane.

On the other hand, the quality of the feed water to the membranes influences the conversion rate of the membranes. Thus, the efficiency of the first stage is higher than that of the second stage, which in turn is higher than that of the last stage, and the permeate flow rate simultaneously decreases from stage to stage.

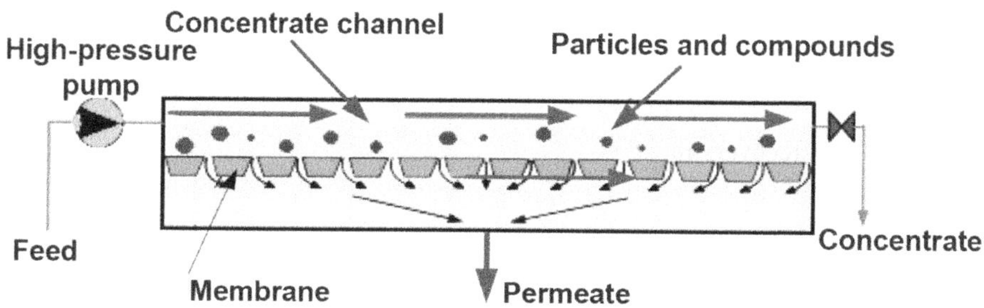

Figure 18.29. *Operating principle of tangential filtration. For a color version of this figure, see www.iste.co.uk/gaid/watertreatment4.zip*

The contribution of each stage to the total permeate flow is all the more significant as the hydraulic efficiency of the stage is high. For example, the treatment of water (500 $m^3 \cdot h^{-1}$) by NF with an overall conversion rate of 85%, a three-stage configuration and a flux rate of 25 $L \cdot m^{-2} \cdot h^{-1}$, is calculated as follows. The 85% recovery rate leads to a total permeate flow rate equal to:

total permeate flow rate: $500 \times 0.85 = 425\ m^3 \cdot h^{-1}$

The total surface of the membranes is:

$Smt = 425 \times 1{,}000/25 = 17{,}000\ m^2$

Based on a unit area of a membrane equal to 37 m^2, the number of membranes is then 17,000/37 = 459 membranes.

With a configuration of the related membranes with the following distribution 50% (first stage); 25% (second stage); 10% (third stage): the permeate flow rate per stage is therefore:

- first stage: $500 \times 0.50 = 250\ m^3 \cdot h^{-1}$;
- second stage: $500 \times 0.25 = 125\ m^3 \cdot h^{-1}$;
- third stage: $500 \times 0.10 = 50\ m^3 \cdot h^{-1}$.

The number of membranes per stage is deduced from:

$$Nmi = \frac{Qpi\ x\ Ntm}{Qin\ x\ Y}$$

with:

- Qpi: permeate flow rate per stage ($m^3 \cdot h^{-1}$);
- Ntm: total number of membranes;
- Qin: overall feed rate ($m^3 \cdot h^{-1}$);
- Y: overall recovery rate (%).

So:

- First stage: $Nm1 = \frac{250 \times 459}{425} =$ 270 membranes;
- second stage: $Nm2 = \frac{125 \times 459}{425} =$ 135 membranes;
- third stage: $Nm3 = \frac{50 \times 459}{425} =$ 54 membranes.

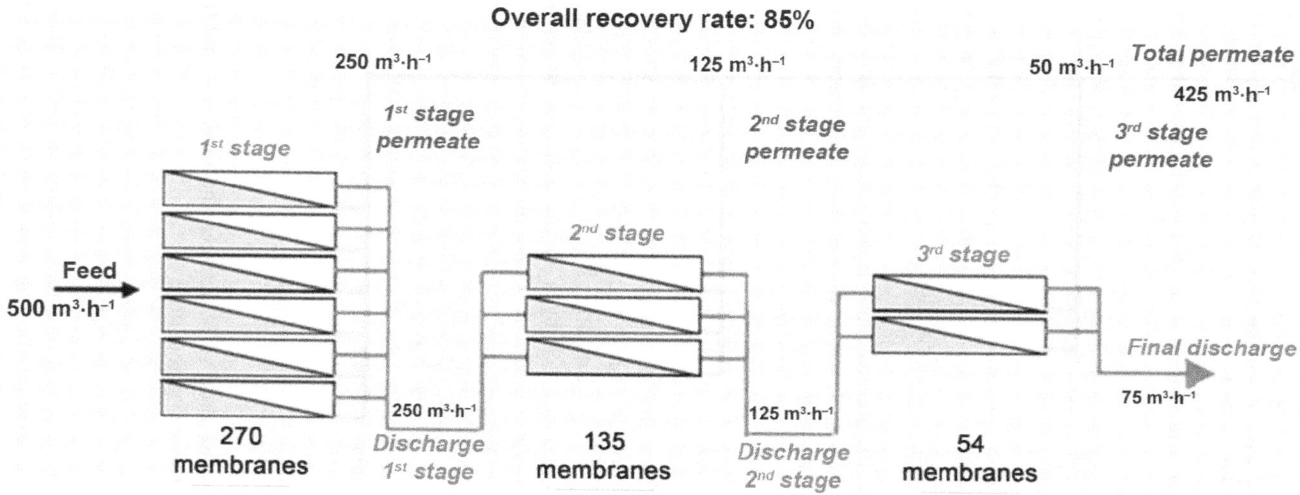

Figure 18.30. *Example of flow calculations, number of modules and pressure tubes. For a color version of this figure, see www.iste.co.uk/gaid/watertreatment4.zip*

The distribution of water flows per stage is as follows:

$$R(\%) = \frac{250}{425} + \frac{125}{425} + \frac{50}{425} = 58.8\% + 29.4\% + 11.8\% = 100\%$$

The total number of pressure tubes, with six modules per tube, is given as 459/6 = 77 tubes.

The number of tubes per stage is 270/6 = 45 at the first stage, then 23 at the second stage and 9 at the third stage.

This calculation is a first approximation. It must be confirmed by the suppliers' software and verified that the flux per module does not exceed 15%. In such a case, it will be necessary to refine these calculations.

This distribution is maintained and periodically verified by reading the instantaneous flux rates of each stage on the flow indicators installed on the permeate collectors of the three stages.

18.7.2. *Permeability (Lp)*

The permeability reflects the state of the membrane pore structure. A constant value of permeability, in other words, linear dependence of the feed water flow, indicates a stable state of the membrane. When there is no longer linearity, the permeability varies and indicates changes in the state of the membrane's pore structure. A decrease in Lp reflects that the membrane is clogged. It is then necessary to increase the feed pressure to maintain the permeate flow. Continuous monitoring of the permeability allows the operator to initiate chemical cleaning to restore the permeability to a higher level, in case of a progressive drop. A drop in Lp of more than 10% triggers an automatic chemical cleaning (CIP). Lp is therefore the main indicator of membrane clogging.

18.7.3. *Longitudinal pressure drop (ΔPfc)*

The longitudinal pressure drop ΔPfc between the feed side and the concentrate side is calculated by the following equation:

$$\Delta Pfc = Pin - Pc$$

It is standardized for flow rate and temperature at 25°C. An increase in the standardized longitudinal pressure drop is usually due to an accumulation of a deposit on both the feed spacer and the membrane. This deposit may contain mineral compounds, colloids, organic matter and micro-organisms. Very often, the increase in this parameter is due to a biofilm, made up of micro-organisms and the exo-polymers they secrete.

18.7.4. *Hydraulic resistance*

The hydraulic resistance of the membrane is given by the following relation:

$$Rm\ (m^{-1}) = \mu\ Lp$$

with:

– Lp: in $m^3{\cdot}m^{-2}{\cdot}h^{-1}{\cdot}Pa^{-1}$;

– μ: in $Ns{\cdot}m^{-2}$.

The total resistance is:

$$Total\ R\ (m^{-1}) = \frac{\Delta P - \Delta \Pi}{\mu\ Jw}$$

The resistance of the membrane due to clogging (Rmc) is therefore:

$$R\ mc\ = Total\ R - Rm$$

18.7.5. *Energy*

Decreasing the MWCO of an NF or LPRO membrane implies an increase in energy demand, due to higher TMP and crossflow operation. This energy demand comes from a combination of relatively high pressure and high-pressure pumping:

$$W\ (J{\cdot}m^{-3}) = \frac{\Delta Pf\ Q}{\eta\ Jw\ Amod}$$

with:

– Δ Pf: (Pa) pressure increase inside a module;

– Q: ($m^3{\cdot}s^{-1}$) flow rate of a module;

– Jw: ($m{\cdot}s^{-1}$) average flux per module;

– Amod: (m^2) membrane surface of a module;

– η: viscosity of the feed water.

The average pressure drop of 0.56 bar and flow rate of 0.003 $m^3 \cdot s^{-1}$ (cross flow of 0.5 $m \cdot s^{-1}$) results in an energy demand of 0.27 $kWh \cdot m^{-3}$.

18.7.6. *SDI*

To achieve an SDI of less than 4 on conventional surface water, the membrane feed water must contain less than 200 particles > 1.5 µm in size. This is possible by integrating an inter-ozonation stage after the settling stage and by including micro-coagulation upstream of the two-layer filtration that follows the ozonation stage. Nevertheless, SDI < 3 is less and less becoming the guarantee requested by membrane suppliers, and a value between 4 (80–90% of the time) and below 5 (100% of the time) is accepted. Therefore, inter-ozonation is no longer mandatory. On the other hand, micro-coagulation upstream of a dual media filtration is recommended. The operator monitors the SDI with a continuous SDI meter or with spot analyses.

18.7.7. *Chemical cleaning*

Since membranes have a natural tendency to clog, chemical cleaning is carried out regularly. The frequency of the cleaning depends on the quality of the pre-treated water and can be done every fortnight, every month or every quarter.

This operation requires the considered membrane unit to shut down for 1–2 days. This generally implies having a unit in stand-by to compensate the lack of water production during the chemical cleaning phases. The chemical solutions generally used are soda at a concentration of 0.3%, and/or a 2% citric acid solution. The volumes of the used cleaning solutions are small compared to the volumes of the concentrates.

The volume of solution spent for the monthly cleaning of a high-pressure membrane system does not exceed 0.51 m^3 of permeate.

When cleaning in two stages, the acid and alkaline solutions are mixed so that they neutralize each other.

18.7.8. *The fate of concentrates and used washing solutions*

The exact nature of the concentrate depends on the characteristics of the raw water, the pre-treatment, the membranes and the conversion rate. The discharge concentrates all the dissolved species of the raw water, as well as the chemicals necessary for the pre-treatment (acids and/or anti-clogging agents). This concentration depends on the retention and conversion rate. The concentration factor, for example, is 5 for a retention rate of 100% and a recovery rate of 80%.

The discharge can be considered industrial waste, and there are several solutions as to what to do with it: discharge into the river, discharge into the sewer, sprinkling, reinjection into the deep aquifer and evaporation in ponds.

River discharge is the most common, with salt and/or organic matter concentrations theoretically more than six times that of the river (for a recovery rate of 85%). The dilution effect assessed in the worst case depends on the low water flow of the river.

Discharge to the sewer is done on a case-by-case basis, depending on the characteristics of the substrate, the proximity of the wastewater treatment plant and the availability of hydraulic, mineral and organic loads in the plant.

These possibilities depend on the geographical location of the installation, its flow, the chemical composition of the concentrates, the proximity of a wastewater treatment plant, the climatic conditions which may allow the consideration of an evaporation solution by lagoon, the treatment cost, etc.

Parameters	Raw water	Permeate	Concentrate
TOC, mg·L^{-1}	1.33	0.23	2.87
Conductivity at 25°C, µS·cm^{-1}	2,830	230	6,500
Ca hardness (mg·L^{-1} $CaCO_3$)	1,094	26	2,887
Alkalinity (mg·L^{-1} $CaCO_3$)	396	220	589
Silica, mg·L^{-1}	8.97	1.87	18.80
Calcium, mg·L^{-1}	202	5	686
Magnesium, mg·L^{-1}	144	3	286
Sodium, mg·L^{-1}	310	60	720
Potassium, mg·L^{-1}	4	1	10
Total cations, meq·L^{-1}	35.46	3.15	89.30
Bicarbonates, mg·L^{-1}	484	27	719
Chlorides, mg·L^{-1}	85.4	40	274
Nitrates, mg·L^{-1}	9.1	6	7.5
Sulfates, mg·L^{-1}	1,200	70	3,345
Phosphates, mg·L^{-1}	<0.01	3.13	0.365
Total anions, meq·L^{-1}	35.47	3.161	89.28

Table 18.12. *Physicochemical composition of raw water, permeate and concentrate during the treatment of salt-loaded groundwater by "opened" nanofiltration*

Table 18.12 shows the physicochemical characteristics of concentrate from a drinking water plant using NF with a 70% recovery rate. For a first estimation, the punctual calculation of the concentration of the ions in the concentrate (Cc) can be done using the formula:

$$\mathrm{Cc} = \frac{1}{1-Y}\,\mathrm{Cin}$$

where:

– Y: recovery rate (%);

– Cin: concentration of the ion in the inlet water of the membrane system.

Table 8.12 presents average values of the raw water, permeate and concentrate, registered along years.

In the case of surface water with a membrane with high gold-organic rejection, recovery rate of 85% and medium (50–60%) salt rejection, the values in Table 18.13 are obtained. Table 8.13 presents average values of the raw water, permeate and concentrate, registered along years.

Parameters	Raw water	Permeate	Concentrate
Permanganate oxidizability, mg·L^{-1}	0.65	<0.1	8.1
Conductivity at 25°C, µS·cm^{-1}	600	265	19,400
Dry residue, mg·L^{-1}	411.9	175	1,672
Ca hardness (mg·L^{-1} $CaCO_3$)	317.5	110	1,378
Alkalinity (mg·L^{-1} $CaCO_3$)	200	92.5	762.5
Silica, mg·L^{-1}	11.6	9.5	21.3
Calcium, mg·L^{-1}	111.5	41.1	470
Magnesium, mg·L^{-1}	9.4	1.8	49.3
Sodium, mg·L^{-1}	19.1	15.3	36
Potassium, mg·L^{-1}	5.6	5.1	11.6
Total cations, meq·L^{-1}	7.324	2.99	29.43
Bicarbonates, mg·L^{-1}	244	112.9	930.6
Chlorides, mg·L^{-1}	34.4	25.7	81.5
Nitrates, mg·L^{-1}	22	20	26
Sulfates, mg·L^{-1}	76	<5	510
Phosphates, mg·L^{-1}	0.1	<0.1	0.5
Fluorides, mg·L^{-1}	0.2	<0.1	0.825
Total anions, meq·L^{-1}	6.921	2.90	28.65

Table 18.13. *Physicochemical composition of raw water, permeate and concentrate during "opened" nanofiltration treatment of surface water*

18.7.9. *Methods for assessing the impact of concentrate discharges in the natural environment*

The assessment of the environmental impact of discharges requires the use of microbiological analyses, physicochemical tests and biological tests. The microbiological and physicochemical approach to water quality characterizes individual micro-organisms or specific pollutants, but it assumes that the contaminants are identifiable and in relatively small numbers. This is rarely the case because during the construction of a drinking water plant, the concentrate is not available. On the other hand, an estimation of the physicochemical quality is possible from the analysis of the raw water and the rejection and conversion rates of the membranes. Simulations of the physicochemical composition of the concentrate provide information on the exposure of organisms to certain pollutants, and bioassays assess the hazards of the discharge to organisms.

Various complementary biological tools characterize the impact on the ecosystem, such as ecotoxicity tests, ecological bio-indicators and bio-markers. Bio-indicators carry out an inventory of the species present in the environment in order to assess its level of contamination. The principle of bio-markers is to measure the biochemical molecules produced or inhibited in the organism in the presence of toxic substances. Ecotoxicity tests are laboratory tests in which a population of aquatic organisms is exposed to an effluent, whose toxicity is to be estimated in order to assess the concentration levels causing effects such as mortality, reduced reproduction, reduced respiration, etc. These tests are carried out on different organisms of the trophic chain, so that the toxicity of a discharge is observed on the whole of the eco-system concerned. Model organisms are used and are chosen for their ease of growing, their representativeness of the entire trophic chain and their sensitivity to toxic substances.

Different types of tests can be used depending on the toxic effect to be highlighted, such as acute toxicity, chronic toxicity and genotoxicity tests.

Acute toxicity tests are short-term tests: effects must be seen within a short period of time (from a few hours to a few days, depending on the life cycle of the animal) after administration of a single dosage of substance. If no effects are observed, the substance is not acutely toxic (which does not mean that the substance is not chronically toxic). Responses are usually expressed as EC50 (effective concentration), LC (lethal concentration for mortality tests) or IC (inhibition concentration for activity tests). These values correspond to the dosage leading to 50% effects on organisms. Consequently, the higher the EC50, the less toxic the product is. This approach makes it possible to assess the overall ecotoxicity of the effluent, which is impossible to know from chemical analyses alone.

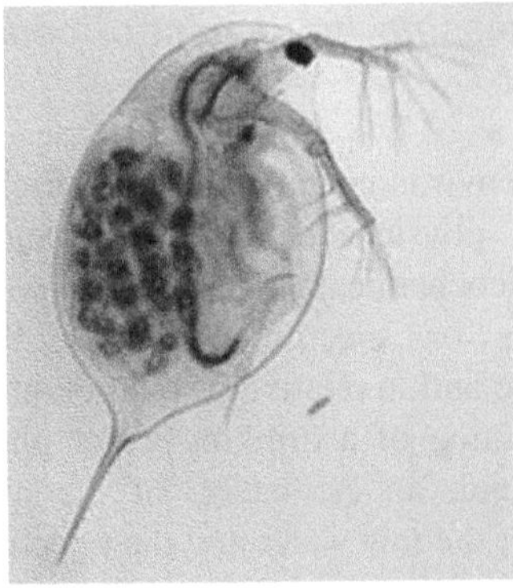

Figure 18.31. *The microcrustacean Daphnia magna Straus, used in ecotoxicity tests. For a color version of this figure, see www.iste.co.uk/gaid/watertreatment4.zip*

Chronic toxicity tests are long-term tests. The effects appear after prolonged exposure to the substance, but are imperceptible over a short time scale. The effects are observed over a much longer period than in acute tests (1/10 to 8/10 of the animal's life). They assess the latent onset of effects and their reversibility. They thus give access to an experimental concentration, below which no toxic effect is observed on the species studied under the test conditions: NOEC (no observed effect concentration). The LOEC (lowest observed effect concentration) is also used, which is the highest concentration causing no significant effect on growth compared to the control sample.

Genotoxicity tests highlight the changes in the genome induced by toxic substances. By altering the genes of an organism, pollutants can damage cells and organs, and disrupt physiological functions such as growth or reproduction.

The most commonly used aquatic ecotoxicity tests are given in Table 18.14. Among the species studied, invertebrates are the organisms that are most commonly used to study the effects of pollution on the natural environment. They have a relatively short lifecycle, relative immobility and a broad spectrum of sensitivity to organic pollution.

Acute toxicity test	Chronic toxicity test	Genotoxicity test
Inhibition of the mobility of *Daphnia magna* Microtox Fish test	Inhibition of *Ceriodaphnia dubia* reproduction Algae test *Brachionus caliciflorus*	Mutatox SOS chromotest

Table 18.14. *Most common ecotoxicity and genotoxicity tests*

The following categories (fish, daphnia and green algae tests) are as follows:

– very toxic: EC50 < 1 $mg \cdot L^{-1}$;

– toxic: 1 $mg \cdot L^{-1}$ < EC50 < 10 $mg \cdot L^{-1}$;

– harmful: 10 $mg \cdot L^{-1}$ < EC50 < 100 $mg \cdot L^{-1}$;

– non-toxic: EC50 > 100 $mg \cdot L^{-1}$.

To assess the impact of the release of a toxic substance into the natural environment, the risk can be evaluated by the ratio PEC-PNEC (predictive environmental concentration to predictive no-effect concentration). The PEC-PNEC ratio is generally used to assess the impact of a substance released into the aquatic environment:

– if PEC-PNEC < 1: the receiving environment is not directly threatened. The substances do not have adverse effects on living organisms in the natural environment;

– if PEC-PNEC > 1: there is a potential risk to the aquatic environment.

PECs, expressed in $\mu g \cdot L^{-1}$, are estimated from concentrations measured in the discharge and upon dilution of the effluent in the receiving environment:

$$\text{PEC } (\text{kg} \cdot \text{m}^{-3}) = \frac{\text{Quantity emitted } (\text{kg} \cdot \text{j}^{-1})}{(\text{Rejected flow } (\text{m}^3 \cdot \text{j}^{-1}) + \text{low river flow } (\text{m}^3 \cdot \text{j}^{-1}))}$$

The PNEC is derived by taking the lowest NOEC value and assigning a safety factor of 100, which takes into account differences in sensitivity between species and inter-species variations.

The river flow chosen as a reference corresponds to the 5-year low flow, which is the annual minimum flow averaged over 30 consecutive days, and over a 5-year frequency.

PNEC is the predicted no-effect concentration in the environment. A PNEC can be extrapolated from the results of ecotoxicity tests performed in the laboratory. The organisms studied must represent the three trophic levels of the aquatic environment: algae, invertebrates (most often daphnia) and fish.

The PNEC is determined from the lowest concentration of a substance having an effect on one of the organisms in the food chain, whether it is an EC50 for one organism or a NOEC for another. This value is then divided by an extrapolation factor, which depends on the amount and nature of information available for the ecosystem. The extrapolation factor is higher if little information is available. The

PNECs used for the most common pollutants are available on the INERIS website and can be used. The conditions for using the factors proposed by the European Community are described in Table 18.15.

Tests	Extrapolation factors
An EC50 from a short-term test for each of the three trophic levels (fish, invertebrates and algae)	1,000
A NOEC from a long-term test (fish or invertebrates)	100
Two NOECs from long-term tests for two species representing two trophic levels (fish and/or invertebrates and/or algae)	50
Three NOECs from long-term test for at least three species representing three trophic levels (fish, invertebrates, algae)	10

Table 18.15. *Use of scaling factors in PNEC assessment*

18.8. High-pressure membrane performance

Given their different cut-offs and compound-to-membrane surface interactions, the NF and RO membranes differ in their removal of monovalent ions, divalent ions and low MW organics and micropollutants (Figure 18.32). Chapter 7 (Volume 2) and Chapter 10 (Volume 3) present the performance of these membranes in detail.

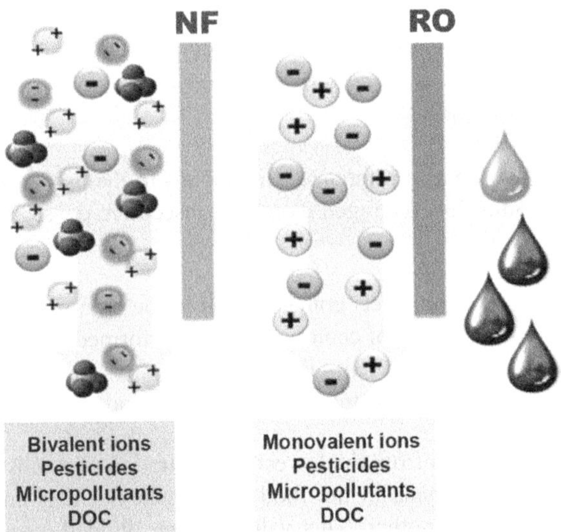

Figure 18.32. *Differences in the performance of NF and RO membranes. For a color version of this figure, see www.iste.co.uk/gaid/watertreatment4.zip*

18.8.1. *Organic matter*

Compared to UF, NF allows a much more thorough removal of organic matter, in particular of the biodegradable fraction at the root of bacterial regrowth phenomena in the networks. Section 7.9 (Volume 2) discusses the performance of high-pressure membranes in detail.

Figure 18.33 shows the removal percentage of DOC according to the membrane cut-off. An NF or LPRO membrane (MWCO 100–200 Da) removes more than 85% of the DOC.

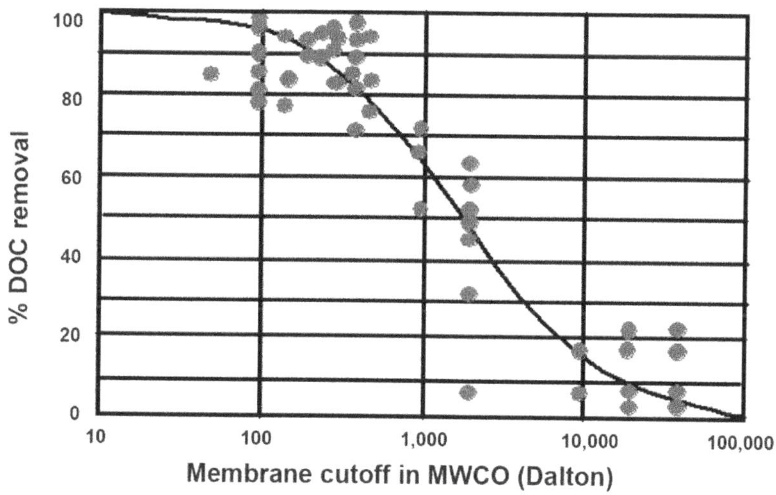

Figure 18.33. *DOC removal according to membrane MWCO*

Oxidizability to $KMnO_4$ (mg $O_2 \cdot L^{-1}$)	**Around 95%**
Dissolved organic carbon (mg $C \cdot L^{-1}$)	85–95%
Biodegradable dissolved organic carbon (mg $C \cdot L^{-1}$)	>85%
Reduction in chlorine demand (mg $Cl_2 \cdot L^{-1}$)	>90%
Reduction of trihalomethane formation precursors (µg$\cdot L^{-1}$)	>90%

Table 18.16. *Nanofiltration membrane performance with regard to DOC*

18.8.2. *Pesticides, drug residues, endocrine disruptors and industrial residues*

Section 10.4.5 (Volume 3) discusses in detail the performance of high-pressure membranes with regard to these organic compounds. Performance varies with membrane characteristics, water temperature (removal is best in cold water) and inlet concentration. Figure 18.34 shows the influence of the membrane cut-off on micropollutant removal. LPRO, which has the lowest cut-off (MWCO), leads to the best removal of these substances.

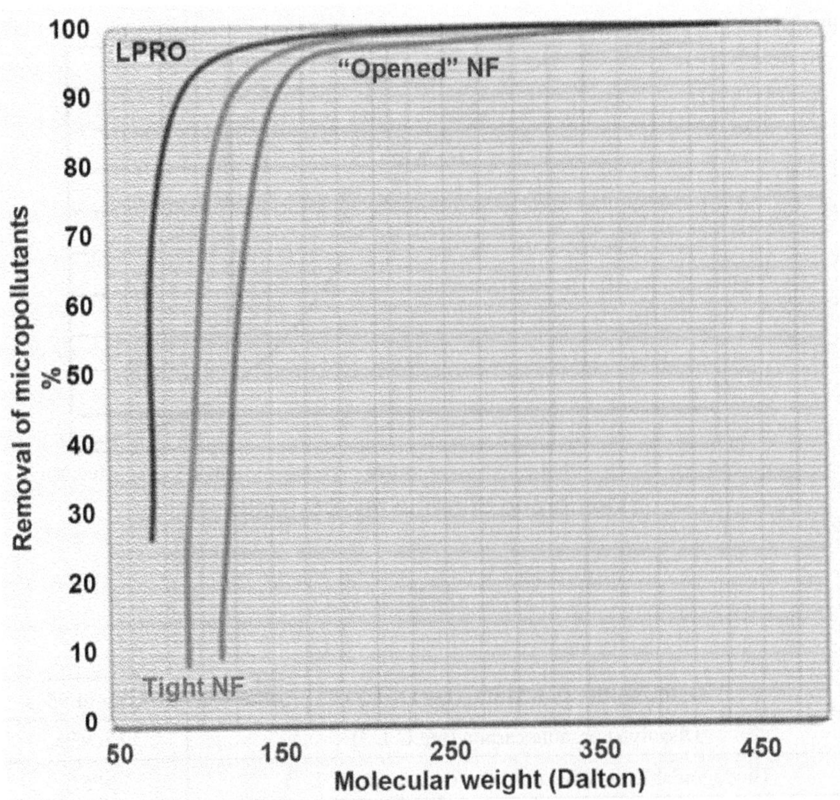

Figure 18.34. *Removal of micropollutants according to their molecular weight and the MWCO of NF-OIBP membranes. For a color version of this figure, see www.iste.co.uk/gaid/watertreatment4.zip*

18.8.3. *Various toxic and undesirable substances*

Section 16.2.2.4 (Volume 3) discusses in detail the performance of high-pressure membranes with regard to metals. Table 18.17 lists the performance with regard to some metals.

Compounds	Removal
Arsenic (As)	90–98%
Cadmium (Cd)	>90%
Lead (Pb)	>90%
Mercury (Hg)	75–95%
Aluminum (Al)	>90%

Table 18.17. *Performance of high-pressure membranes with regard to some heavy metals*

18.8.4. *Salts*

Salt rejection is very variable depending on the type of membrane, the ionic composition of the water (in particular the sulfate content) and the temperature:

– tight NF membrane and LPRO membranes: high salt rejection (>90%). The water produced requires remineralization or half-steeping, with a good quality external source and sufficient Ca^{2+};

– "Opened" NF membrane: lower rejection (about 50%). Re-balancing the water is more limited in this case.

Table 18.18 gives some values obtained with a tight NF membrane (at 14°C) and a "Opened" NF membrane (at 10°C) on salts.

In all cases, selectivity varies greatly from one salt to another. The ranking of the best retained salts to the least well retained is:

$$SO_4 > Mg \geq Ca > HCO_3 > Cl > Na > K > NO_3$$

Therefore, bivalent species are better retained than monovalent ions. With loose membranes such as NF-200 or NF-270, monovalent ions in the form of Na^+ (hydrated radius = 3.32 Å), K^+ (3.31 Å) and Cl^- (3.58 Å) are moderately rejected (rejection < 30%) due to their small hydrated radius. Nitrates are even less rejected (rejection < 10%), not just because of a steric effect caused by their hydrated radius (3.35 Å), which is close to that of Na^+, K^+ and Cl^- ions, but because of their high mobility (absolute ionic mobilities relative to K^+ at 25°C: NO_3^- 0.972, Cl^- 0.682 and Na^+ 1.039), as well as the conditions of electroneutrality imposed by the membrane surfaces. For example, when a mobile co-ion such as NO_3^- is introduced at a much lower concentration than a mobile counter-ion present in the feed water, such as Na^+, a small percentage of these counter-ions can penetrate and, therefore, force a large percentage of the less concentrated mobile co-ions to penetrate, maintaining electroneutrality.

	Tightly packed NF or LPRO membranes		**"Opened" NF membranes**	
	Inlet	**Permeate**	**Inlet**	**Permeate**
Ca ($mg \cdot L^{-1}$)	115	7.8	115	43
Mg ($mg \cdot L^{-1}$)	11	0.6	10.5	2.4
Na ($mg \cdot L^{-1}$)	27	9.2	19	15.3
K ($mg \cdot L^{-1}$)	7	3.1	5.3	4.8
SO_4 ($mg \cdot L^{-1}$)	153	<5	80	5,0
HCO_3 ($mg \cdot L^{-1}$)	198	29.7	271	121.0
Cl ($mg \cdot L^{-1}$)	45	13.5	34	25.0
NO_3 ($mg \cdot L^{-1}$)	23	10.0	23	18.0
SiO_2 ($mg \cdot L^{-1}$)	14	7.8	12.4	10.0
TDS ($mg \cdot L^{-1}$)	579	79	558	239
Ca Hardness ($mg \cdot L^{-1} CaCO_3$)	333	22	288	118
Alkalinity ($mg \cdot L^{-1} CaCO_3$)	162	24	222	109
pH	6.5	5.7	7	6.76

Table 18.18. *Salt rejection with two types of nanofiltration membranes*

18.8.5. *Micro-organisms*

NF and RO membranes provide an excellent physical barrier to all micro-organisms. They increase the reliability of disinfection for the following reasons:

– the removal achieved on industrial installations is very good: 6 to 8 logs of removal for all germs, especially for parasites such as *Giardia* and *Cryptosporidium*;

– the water produced contains no SS or turbidity. With NF, organic matter, which can react with chlorine to the detriment of bactericidal action, is greatly reduced. Chlorine can be applied at a very low dosage downstream of the NF, so as to ensure continuous control and residual effect in the network, without the inconveniences associated with its massive use (taste, odors, by-products, etc.).

However, this barrier cannot be considered absolute, due to the extremely low but not zero risk of leakage, as well as the limited means of continuously monitoring the integrity of the membranes.

For spirally wound modules (NF), leakage can mainly occur at the joints and interconnections between the permeate tubes. Leakage can occur during assembly and disassembly, or during hydraulic jerks that cause displacement of the membranes in the tubes. Various means of control exist on the market. Veolia has a patent for monitoring sulfates on a membrane placed in the parameter of a pressure tube, which makes it possible to continuously monitor the evolution of this parameter.

18.8.6. *Overall performance*

The performance of NF membranes (tightly packed NF-90 or equivalent membranes) can be summarized as follows:

– significant reduction of pesticides and micropollutants (drug residues, endocrine disruptors);

– reduction of the nitrate content by a factor of about 2;

– reduction of the final chlorination requirements compared to a conventional process;

– reduction of the dissolved organic matter content of the water, in particular of the biodegradable fraction at the root of bacterial regrowth phenomena in the networks;

– reduced risk of THM formation (linked to low OM content and low requirement for final chlorination).

18.9. Lifetime warranties

Most manufacturers offer 5-year warranties. However, feedback from the field shows that a properly sized and operated treatment system can increase the life of membranes to 8–10 years. If the life of a membrane is short, it is due to inadequate pre-treatment or an accident. Many membranes are destroyed by chlorine when the de-chlorination system fails.

18.10. Parameters affecting the performance of NF membranes

The most significant operating parameters affecting the performance of NF and RO membranes are the following:

– the pressure difference is the driving force behind the NF and RO process. A decrease in this parameter affects the performance and permeability of the membranes;

– the increase in temperature increases the flow, due to the decrease in viscosity. Salt rejection decreases;

– increasing the longitudinal flow velocity leads to an increase in the average flow, due to the removal of the deposit layer that is left on the surface of the membrane. However, the mechanical resistance of the membrane determines the maximum longitudinal flow velocity that can be applied. Operating an NF or RO membrane at too high a longitudinal flow velocity can, on the one hand, lead to membrane and module failure and, on the other hand, it promotes pressure drop.

An increase in the conversion rate leads to a decrease in rejection. This is because the diffusion of compounds across the membrane increases when the conversion rate is higher. This results in an accentuation of the concentration-polarization mechanism, of the membrane-compound interactions and even of the adsorption mechanisms on the membrane surface. These phenomena have an adverse effect on membrane performance.

pH affects the performance of NF and RO membranes in several ways.

Charged sites (carboxyl and sulfonic groups) on the membrane surface are negatively charged at neutral or higher pH, but lose their charge at acidic pH. Many NF and RO membranes have lower rejections at low pH, or after an acid rinse. Given that different membrane manufacturers use different chemistries to produce their thin film composite layer, the pH dependence of a membrane must be assessed for each type of membrane.

In addition, too low a pH can lead to variations in the solubility of the ions, as well as in the dissociation state, which can then induce different rejection rates.

The rejection of monovalent ions decreases as their concentration in the feed water increases. The rejection of divalent ions is much less affected:

1) rejection of NaCl and $MgSO_4$ at 2,000 ppm, ΔP: 4.9 bar, T: 25°C, Y: 15%;

2) rejection of $CaCl_2$ (500 ppm) and $MgSO_4$ at 2,000 ppm, ΔP: 4.9 bar, T: 25°C, Y: 15%;

3) rejection of NaCl and $MgSO_4$ at 2,000 ppm, ΔP: 5.2 bar, T: 25°C, Y: 15%;

4) rejection of $CaCl_2$ (500 ppm), ΔP: 5.1 bar, T: 25°C, Y: 15%;

5) rejection of NaCl at 2000 ppm, ΔP: 10 bar, T: 25°C, Y: 15%;

6) rejection of NaCl at 2,000 ppm, ΔP: 15.5 bar, T: 25°C, Y: 15%.

Membranes	Delta P (bar)	pH range	Max T (°C)	NaCl rejection (%)	$MgSO_4$ rejection (%)
NF-90	41	3–10	45	85–95.1	97
NF-200	41	3–10	45	35–50.2	97
NF-270	41	3–10	45	40–60.2	97
NF-70	41	3–10	45	60–70.3	95
ESNA 1 LF	41	3–10	45	75–92.4	–
ESPA 3	41	3–10	45	–	99
RE BLF	41	3–10	45	99.2	99
BW 30	41	2–11	45	99.5	99
BW 30 LE	41	2–11	45	99.6	99
SW 30	55	2–11	45	99.5	99.5

Table 18.19. *Characteristics of some NF, LPRO and RO membranes*

18.10.1. *Taking clogging into account*

Clogging causes deposits to adhere to the surface of the membrane, reducing its permeability to water. This leads to an increase of the inlet pressure. The different types of NF and LPRO membrane clogging are as follows:

– initial clogging: this is an irreversible phenomenon that occurs at the beginning of the membrane's life and can last from a few days to 3 months, depending on the type of membrane;

– irreversible clogging: this is the loss of permeability not recovered by cleaning, measured in relation to the state of the membrane after initial clogging;

– reversible clogging: this is the additional drop that is allowed between two cleanings. It is calculated in relation to the permeability of the cleaned membrane.

On some membranes, after several years of operation, a rise in permeability is observed and is often accompanied by a decrease in rejection. This can be due to a wear and tear phenomenon of the selective skin of the membrane. If this is altered, holes are produced which increase the passage of water and solutes. However, this phenomenon has not been observed during the first 3 years of operation and should not be taken into account in the design, as it is unlikely that higher permeabilities will be achieved than with the new membrane.

Types of clogging	Decrease in permeability to be considered for the sizing
Initial clogging	NF-70: 30% (3 months) Other membranes: 10% (2 weeks)
Irreversible clogging	5% per year starting in the second year (the first year's clogging is included in the initial clogging)
Reversible clogging (cleaning trigger)	15%

Table 18.20. *Indicative values of the permeability decrease according to the clogging*

In addition, clogging can also manifest itself on the increase of the longitudinal loss of the membrane, in addition to the permeability. This phenomenon is not restricting for the sizing of high-pressure pumps, but must be considered for the sizing of cleaning pumps.

18.11. Monitoring and control parameters: standardization of raw data

Membrane performance is highly dependent on operating parameters such as temperature, water quality and flow rate. In order to follow the evolution and the clogging state of the membranes, it is essential to be able to constantly compare the current performances with the initial performances measured at the time of commissioning.

This long-term monitoring is crucial for the management of the installation. For this purpose, the operating parameters must be standardized, that is to say, expressed in relation to reference conditions.

Four criteria are therefore considered for performance monitoring:

– standardized permeability at 25°C: it characterizes the "transverse" clogging of the membrane by very fine particles or by salts;

– standardized pressure drop factor: it characterizes the "longitudinal" clogging of the membrane, obtained in case of serious failure of the pre-treatment or the pre-filters (algae, filter piercing, large particles, etc.);

– the specific surface flux: it provides the distribution of the clogging load between the stages;

– standardized salt passage at 25°C: it allows detection of a modification of the membrane's cut-off power.

The continuous process control points are defined in Table 18.21.

Types of water	Measured parameters	Means of control and monitoring
Membrane station feed water	SDI Al, Fe, Mn Turbidity Conductivity DOC Oxidizing residual pH	Instrumentation to be checked regularly
Water upstream of the pre-filters (after addition of sequestrant) **Water downstream of the pre-filters (high pressure suction)**	pH Turbidity Particles Conductivity	Measurements carried out by sampling upstream and downstream of the pre-filters with the same apparatus
Water downstream of the pre-filters	Oxidizing residual Temperature	Instrumentation to be checked
Permeate from each block	Conductivity	Possibility to measure manually, tube by tube. Can also measure on the general permeate of the line by scanning
Degassed water	UV transmission	Sensor integrated in the UV reactors
Water produced	pH DOC Alkalinity, hardness Conductivity	Instrumentation to be checked
General concentration	Conductivity pH	Instrumentation to be checked
Cleaning stations	pH Turbidity Temperature	To be checked for each cleaning station
Neutralization	pH Turbidity	pH measurement in each tank

Table 18.21. *Instructions for monitoring water quality at different points of the membrane station*

18.12. Veolia's experience: examples of treatment processes

Some examples of treatment systems for various types of water are presented, involving different pre-treatments.

18.12.1. *Surface water No. 1*

The Méry-sur-Oise drinking water treatment plant, operated by Veolia Water on behalf of the Syndicat des Eaux d'Ile de France (SEDIF), supplies drinking water to approximately 800,000 people living in the northern suburbs of Paris. The plant treats surface water using two parallel processes, a conventional channel (coagulation – flocculation – settling – filtration – sand-inter-ozonation – filtration – activated carbon – disinfection) and a channel including an NF membrane. The NF file (overall capacity: 140,000 $m^3 \cdot d^{-1}$) was installed in 1999, bringing the overall capacity of the plant to 340,000 $m^3 \cdot d^{-1}$, after mixing the waters. Its purpose is to improve the quality of the water by reducing its hardness, its concentration of natural organic matter and micropollutants.

The raw water is a surface water (close to the river Oise) that presents a physicochemical composition, which regularly degrades with an increase in the concentration of natural organic matter (6.7 mg $DOC \cdot L^{-1}$) during periods of heavy rainfall, and which the conventional process is unable to treat adequately to obtain a DOC residual of less than 2 mg $C \cdot L^{-1}$.

Other contaminants, such as micropollutants, are regularly found such as atrazine (1 $\mu g \cdot L^{-1}$) and its metabolites (desethyl atrazine and deisopropyl atrazine). Ultimately, the relatively high hardness (200–312 mg $CaCO_3 \cdot L^{-1}$) requires a sufficient reduction in order to satisfy the consumers' requirements in terms of taste and comfort, as well as to homogenize the water hardness in the distribution network.

The new treatment plant (Figure 18.35), which runs parallel to the existing plant, consists of a pre-treatment including pH control at 6.9, ballasted flocculant coagulation-settling (Actiflo) followed by inter-ozonation and two-layer filtration (anthracite-sand). A micro-coagulation is injected upstream of the two-layer filtration. The membrane station includes a 6 μm absolute pre-filtration, upstream of which the sequestrant is injected. The NF unit is composed of eight identical trains, each consisting of three successive stages, and contains 9,120 spirally wound NF-200B (Dow Chemical Company) 8-inch elements. The NF membrane was specifically developed to reject pesticides and DOC, while allowing partial passage of calcium.

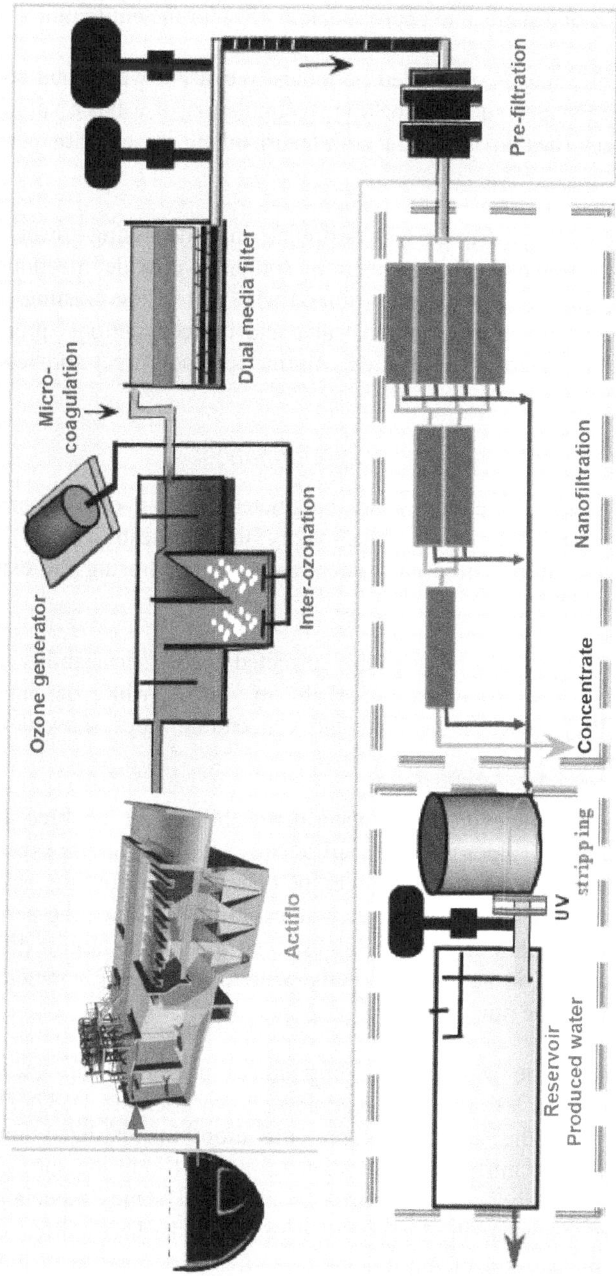

Figure 18.35. *Diagram of the treatment process with nanofiltration (source: SEDIF and Veolia). For a color version of this figure, see www.iste.co.uk/gaid/watertreatment4.zip*

Each stage is designed to filter 50% of the feed flow. The permeate recovery rate is therefore 50% for one stage, and approximately 85% for the entire unit.

CIP chemical cleaning is conducted periodically every 6 weeks and consists of different successive stages such as soaking, recirculation of a basic solution, then citric acid-detergent (heated to 35°C) with intermediate rinses with permeate at 30°C.

The post-treatment consists of CO_2 stripping in a degassing tower, medium pressure UV disinfection and pH adjustment by sodium hydroxide injection. Finally, after chlorination, the water produced is mixed with that of the existing plant at a ratio of 80/20, between the nanofiltered water and the conventional process. The water is then dechlorinated to the desired chlorine residual target and fed into the distribution network.

18.12.1.1. *Monitoring of operation and performance*

During the 10 years of operation, many parameters of the hydraulic performance of the membranes was monitored, but mainly the permeability (Lp) and the longitudinal pressure drop. These parameters are calculated using the expressions presented in section 18.4.

After 10 years, Lp (osmotic pressure-corrected water permeability at 25°C) decreased by only 25%, from 6 to 4.5 ($L \cdot h^{-1} \cdot m^{-2} \cdot bar^{-1}$), while the longitudinal pressure drop (ΔPfc) between the feed and concentrate sides showed relatively moderate increases.

Figure 18.36 shows the evolution of water permeabilities for the three stages of a membrane train. The Lp values remained sufficiently high and the life expectancy of the membrane was doubled because of extreme controlling of fouling, achieved by both efficient pre-treatment and effective chemical cleaning. Thus, the permeability has always remained within acceptable values, even though the largest drops in Lp have generally occurred in autumn and early winter, which are characterized by a higher organic load in the raw water.

The evolution of the standardized longitudinal pressure drop (ΔPfc) for a membrane train over the 10 years of operation is shown in Figure 18.37. The values shown are relative to the longitudinal pressure drop. Standardized longitudinal pressure drops remained fairly stable between 1.2 and 1.4 until the spring of 2002, and then increased periodically, primarily in the third stage, especially in the summer. The third stage has the largest increase in standardized ΔPfc, reaching 3 (200% increase) by the end of the membrane's life. The increase in longitudinal pressure drop is clearly related to the season, as it is generated by the biofilm, whose biological growth is activated at favorable temperatures (spring and summer).

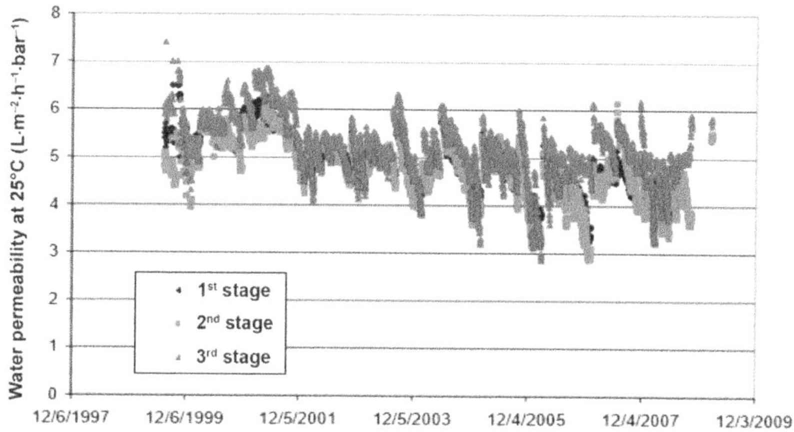

Figure 18.36. *Evolution of the permeability (25°C) during the 10 years of operation. For a color version of this figure, see www.iste.co.uk/gaid/watertreatment4.zip*

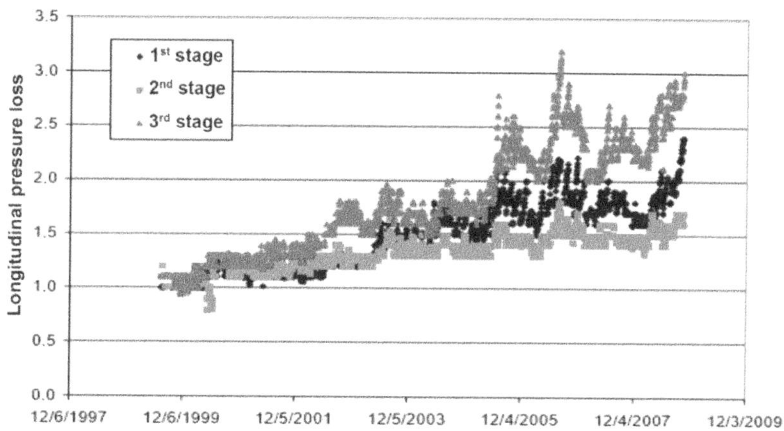

Figure 18.37. *Evolution of the longitudinal pressure drop during the 10 years of operation. For a color version of this figure, see www.iste.co.uk/gaid/watertreatment4.zip*

18.12.1.2. *Characterization of the deposits*

Autopsy campaigns showed that all membrane elements are covered with a sticky, smelly, green-brown material. Microscopic observations show that the deposit is a few micrometers thick, highly porous and predominantly organic.

Further chemical characterizations indicate that the organic matrix contains a large number of biopolymers, particularly proteins and exo-polysaccharides (EPS), which are features of biofilm (Figure 18.38).

During the 10 years of operation, the membranes maintained their selectivity relative to the water quality objectives, in other words, a TOC reduction of up to 95%, removal of micropollutants and reduction of calcium (rejection between 20 and 55%, depending on temperature). The NF permeate met the requirements of TOC < 0.5 mgC·L^{-1} and BODC (biodegradable DOC) less than 0.2 mgC·L^{-1}. This has resulted in a decrease in chlorine demand, a sharp decrease in the formation of disinfection by-products and a decrease in bacterial regrowth in the system. The average hardness value is between 150 and 170 ppm $CaCO_3$.

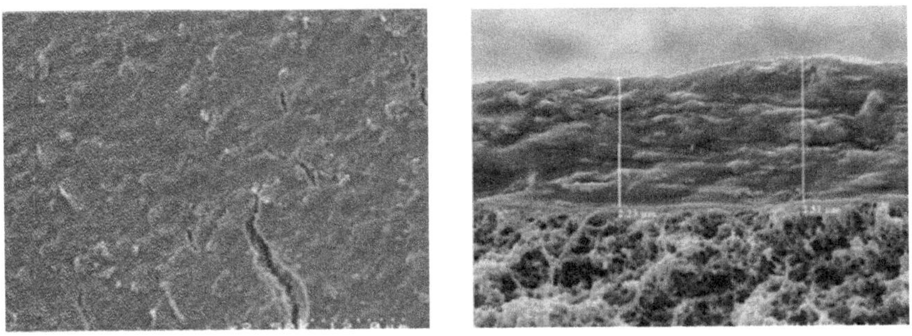

a) b)

Figure 18.38. *Electron microscope observations of a (a) clean and (b) fouled NF-200 membrane*

The preventive (pre-treatment) and corrective (CIP) actions taken to control fouling have maintained hydraulic performance. This has made it possible to extend the life of the membrane from 5 to 10 years. Indeed, the membranes were initially designed for a 5-year life span but, as their performance was stable, they were renewed after 10 years of operation.

18.12.2. *Surface water No. 2*

The Magenta plant (St. Brieuc, France) has an NF unit designed to reduce the concentration of natural organic matter (6-13 mg TOC·L^{-1}), nitrates (40–60 mg NO_3·L^{-1}) and pesticides (>1 µg·L^{-1}). The treatment process (Figure 18.39) includes coagulation–flocculation–settling (Multiflo) followed by sand filtration, pre-filtration (5 µm), membrane station, balance and disinfection. Due to the presence of nitrates, tight NF (NF-90 400, DOW chemical) is selected.

The key findings are:

– Acidification to pH = 6.0–6.5 improves coagulation in Aqualenc. Dosages vary between 60 and 120 $g \cdot m^{-3}$. The concentration of dissolved aluminum is regularly controlled so that it is less than 50 $\mu g \cdot L^{-1}$ in order not to damage the NF membranes.

– The turbidity of the settled water is lower than 1 NTU, that of the sand filtered water is lower than 0.5 NTU and that of the pre-filtered water is lower than 0.1 NTU.

– The membrane station has three stages, with a total of 28 modules. The total filtration area is 2,898.48 m^2. The NF membranes are fed at a flow of 22.5 $L \cdot m^{-2} \cdot h^{-1}$. The conversion rate is 85%. The SDI of the water fed to the membranes is between 3 and 3.5.

– After 9 years, the water permeability (at 25°C, corrected for osmotic pressure) remained between 4 and 5 ($L \cdot h^{-1} \cdot m^{-2} \cdot bar^{-1}$). Figure 18.40 shows the evolution of water permeabilities for the three stages of a membrane train. The Lp values remained sufficiently high and the life expectancy of the membrane was doubled, due to very thorough fouling control, which was achieved by both effective pre-treatment and effective chemical cleanings.

The water quality produced remained stable during the 9 years of operation, with a TOC output < 1.5 $mg \cdot L^{-1}$ and a nitrate concentration <20 mg $NO_3 \cdot L^{-1}$ (Figures 18.41 and 18.42).

Remineralization of the water, with CO_2–lime water, delivers water in calco-carbonic balance with an alkalinity of 90–100 $mg \cdot L^{-1}$.

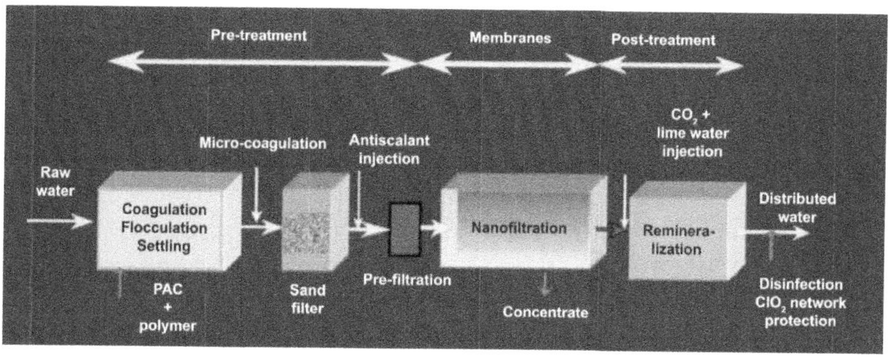

Figure 18.39. *Treatment process at the Magenta plant. For a color version of this figure, see www.iste.co.uk/gaid/watertreatment4.zip*

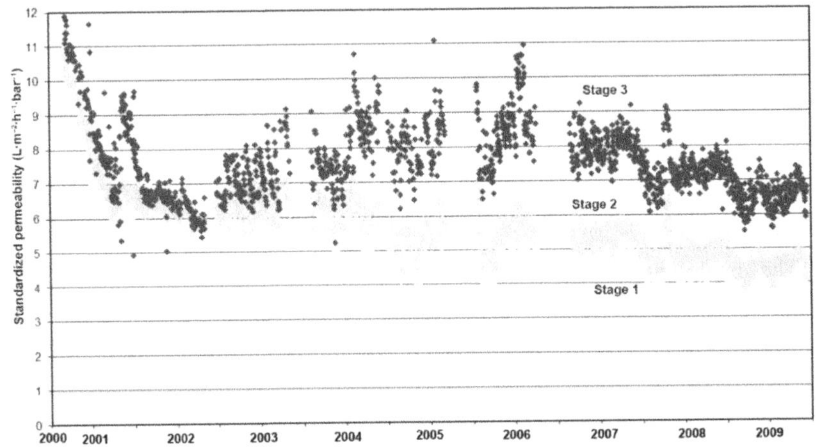

Figure 18.40. *Evolution of normalized permeability during 9 years of operation. For a color version of this figure, see www.iste.co.uk/gaid/watertreatment4.zip*

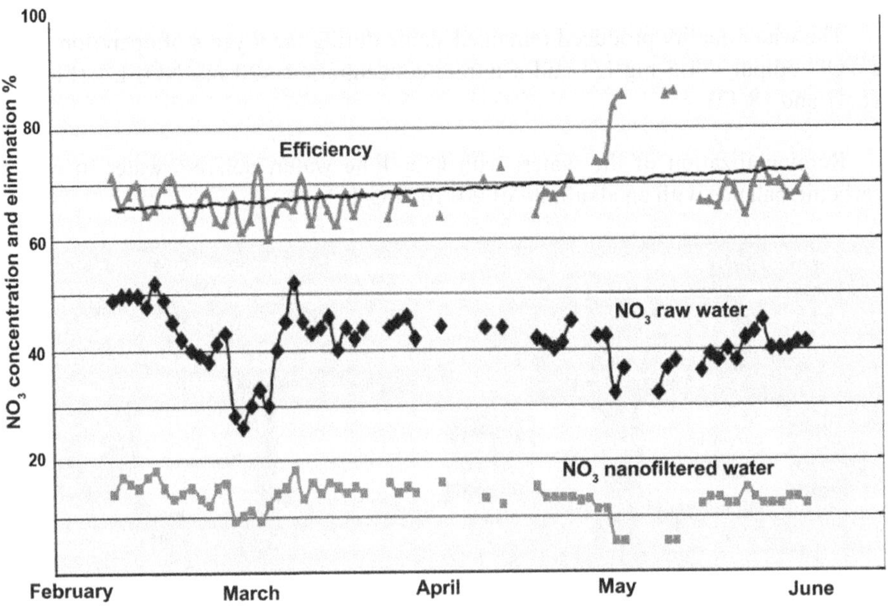

Figure 18.41. *Nitrate removal. For a color version of this figure, see www.iste.co.uk/gaid/watertreatment4.zip*

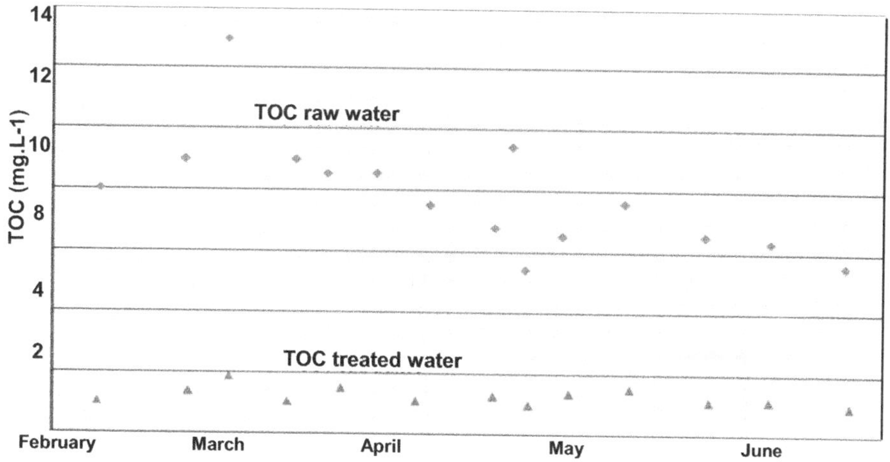

Figure 18.42. *Removal of total organic carbon. For a color version of this figure, see www.iste.co.uk/gaid/watertreatment4.zip*

Figure 18.43. *Membrane station in the Magenta plant. For a color version of this figure, see www.iste.co.uk/gaid/watertreatment4.zip*

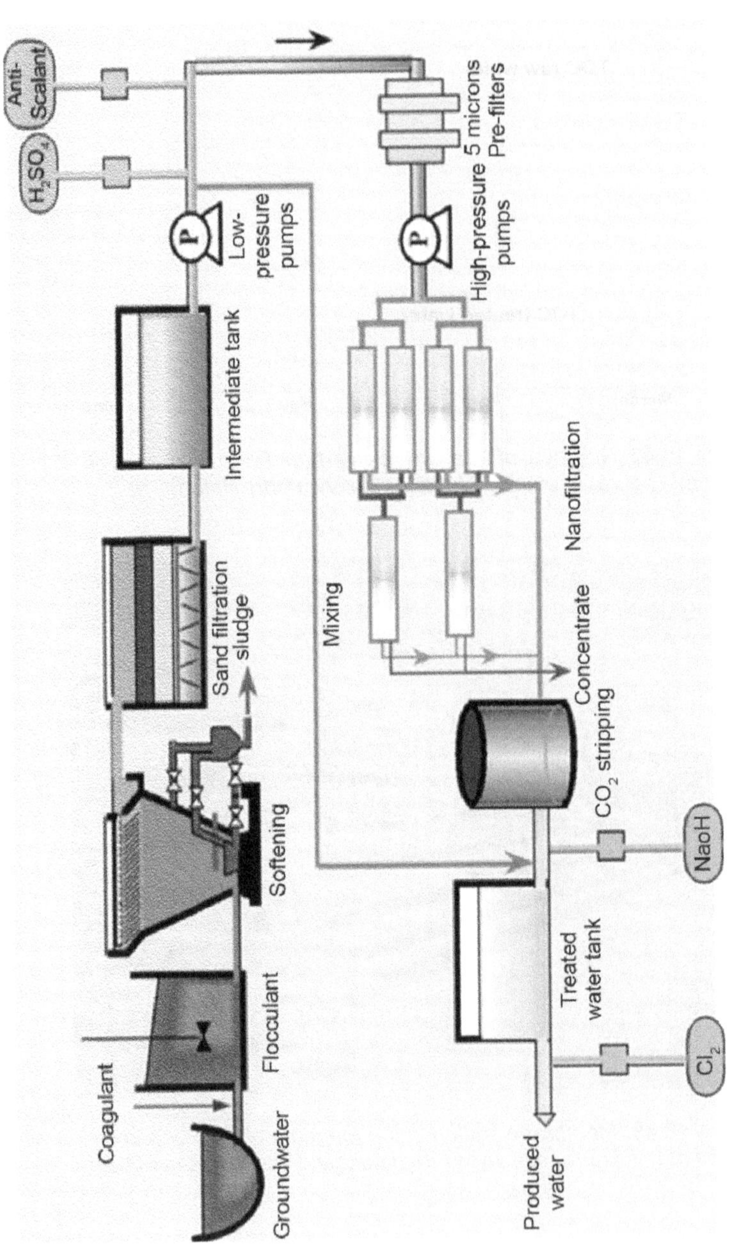

Figure 18.44. *Diagram of the nanofiltration installation in Jarny. For a color version of this figure, see www.iste.co.uk/gaid/watertreatment4.zip*

18.12.3. *Groundwater No. 1*

Two of these types of installations have been built in the Meurthe-et-Moselle department: one in Jarny and the other in Moineville. The NF plant in Jarny is made up of two autonomous lines, treating a flow of 2,500 $m^3 \cdot d^{-1}$ with a 75% conversion rate. The treatment process includes coagulation–flocculation–settling, followed by sand filtration. After acidification and a sequestrant dosage, the water is pre-filtered on bags with a 5 µm cut-off. The HP pumps then raise the pressure to 10 bar (depending on the water temperature and the state of clogging of the membranes) in order to produce a constant flow rate (Figure 18.44).

The membrane station consists of NF-70 membranes (DOW chemical) distributed over two stages. The conversion rate is 75%. The post-treatment includes a partial CO_2 degassing tower. After mixing, the CO_2 in the mixture is neutralized with soda to form sodium carbonate. A final chlorination is performed to obtain a chlorine residue of 0.1 $mg \cdot L^{-1}$.

Parameters	Feed water	Permeate	Distributed water (average count)
pH	6.83	6.18	7.80
Ca hardness ($mg \cdot L^{-1}$ $CaCO_3$)	1,307	41	196
Alkalinity ($mg \cdot L^{-1}$ $CaCO_3$)	266	35	77.4
Conductivity ($\mu S \cdot cm^{-1}$)	3,020	301	526
Calcium ($mg \cdot L^{-1}$)	272	6.0	38.2
Magnesium ($mg \cdot L^{-1}$)	153	6.0	24.8
Sodium ($mg \cdot L^{-1}$)	320	50.0	73.5
Potassium ($mg \cdot L^{-1}$)	6.0	1.0	1.6
Bicarbonates ($mg \cdot L^{-1}$)	325	43	93.4
Chlorides ($mg \cdot L^{-1}$)	0.0	48.0	34.0
Sulfates ($mg \cdot L^{-1}$)	1,675	48.0	214

Table 18.22. *Chemical characteristics of feed water, permeate and distributed water after mixing*

The results obtained show the effectiveness of the NF membranes in terms of the quality of the water produced (Table 18.22). An average hardness removal of 96.8% is noted, while sulfate removal exceeds 97%. The high sulfate concentrations and the very high conductivity justify the use of NF, because simulations with other techniques such as ion exchange had shown that the exchange cycles were very short and generated restrictive operating costs.

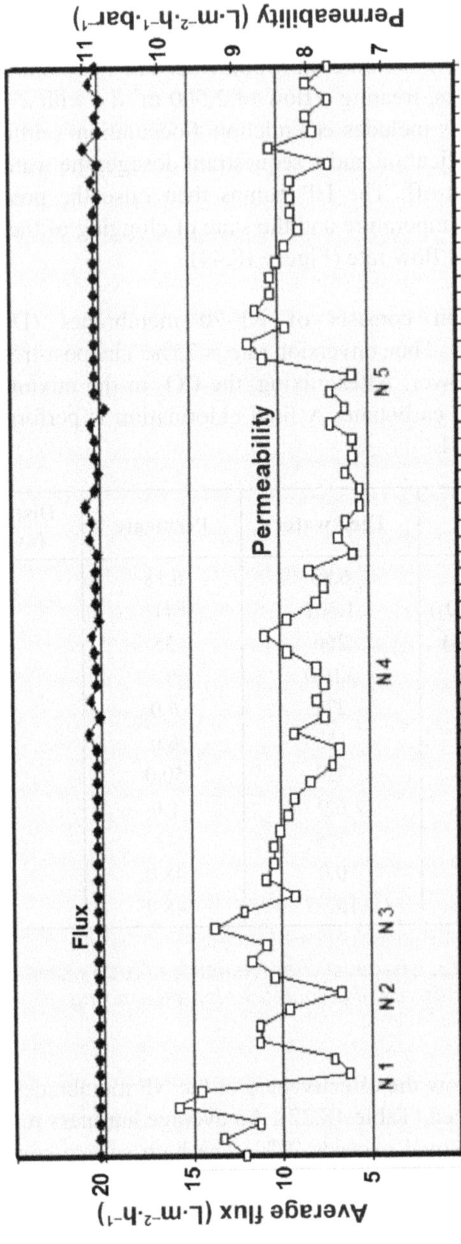

Figure 18.45. *Evolution of flow (Jw) and permeability as a function of time (N represents chemical cleanings 1, 2, etc.). For a color version of this figure, see www.iste.co.uk/gaid/watertreatment4.zip*

As far as the operating parameters are concerned, the regularity of the product flux (Jw) and the evolution of the permeability should be noted, whose sequential drops determine the cleaning frequencies (Figure 18.45).

Figure 18.46. *Nanofiltration membrane train in Jarny. For a color version of this figure, see www.iste.co.uk/gaid/watertreatment4.zip*

The post-treatment includes a partial CO_2 degassing tower. After mixing, the CO_2 in the mixture is neutralized with soda to form sodium carbonate. A final chlorination is performed to obtain a chlorine residue of 0.1 $mg \cdot L^{-1}$.

18.12.4. *Groundwater No. 2*

The construction of the drinking water treatment plant aims to increase and secure the production of drinking water for the municipality of Saint-Bon Courchevel, especially during the winter season when consumption can reach levels of around 15,000 $m^3 \cdot d^{-1}$.

The raw water comes from the Ariondaz reservoir (maximum altitude of 2,185 mNGF), with a branch to the Biollay reservoir. The sulfate content is around 700 $mg \cdot L^{-1}$. The minimum temperature is 2°C.

The Saint–Bon–Courchevel drinking water treatment plant includes the following stages:

– a raw water supply;

– a flow and pressure regulating device;

– a pre-treatment station consisting of a sand filtration (two metal pressure filters installed);

– a mixing bypass;

– an NF membrane station (two NF lines) with sequestrant injection, including pre-filtration and high-pressure pumps;

– post-treatment with carbon dioxide (CO_2) injection, limestone filtration (two pressure filters installed) and soda pH adjustment;

– chlorine disinfection and network protection;

– an existing drinking water tank;

– a filter washing unit and a chemical cleaning unit (CIP), a rinsing unit, as well as the implementation of NF skids for preservation.

Sand filtration allows to retain the particles sucked during the pumping. It is followed by a pre-filtration station placed before the membranes, so as to stop accidental particles making their way through in case the pre-treatment fails. This station is made up of filter cartridges with a cut-off of 10 µm.

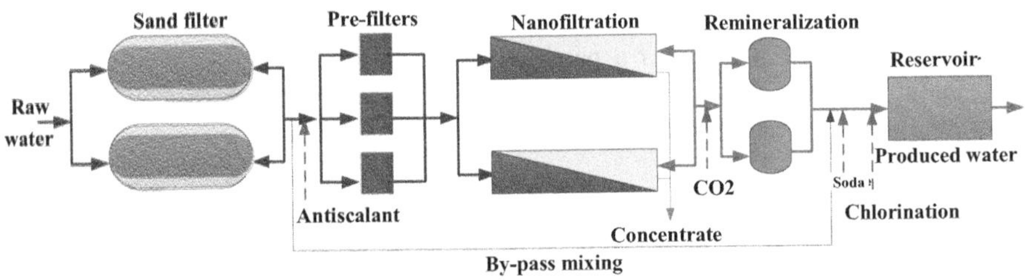

Figure 18.47. *Treatment process of the Saint–Bon–Courchevel plant. For a color version of this figure, see www.iste.co.uk/gaid/watertreatment4.zip*

Upstream of the pre-filters, antiscalant (2 $mg \cdot L^{-1}$) is injected to prevent precipitation of salts present in the concentrate onto the membrane.

A high-pressure pump provides the necessary pressure for the operation of the membranes. It is equipped with a frequency converter, whose purpose is to vary the pump flow rate and to adjust it to the set point of the NF feed flow rate. The NF feed flow rate is measured continuously (autonomously) by adding the permeate and concentrate flow rate measurements for each line.

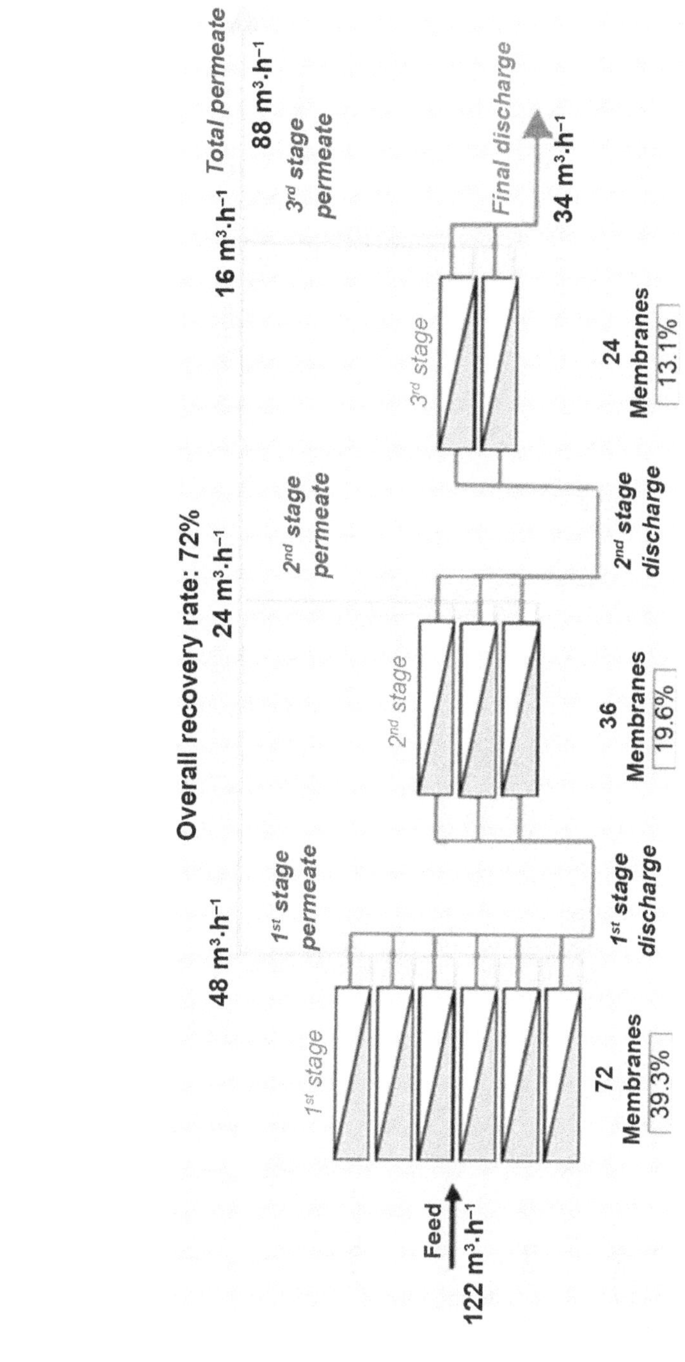

Figure 18.48. *Configuration of the three stages of the membrane station. For a color version of this figure, see www.iste.co.uk/gaid/watertreatment4.zip*

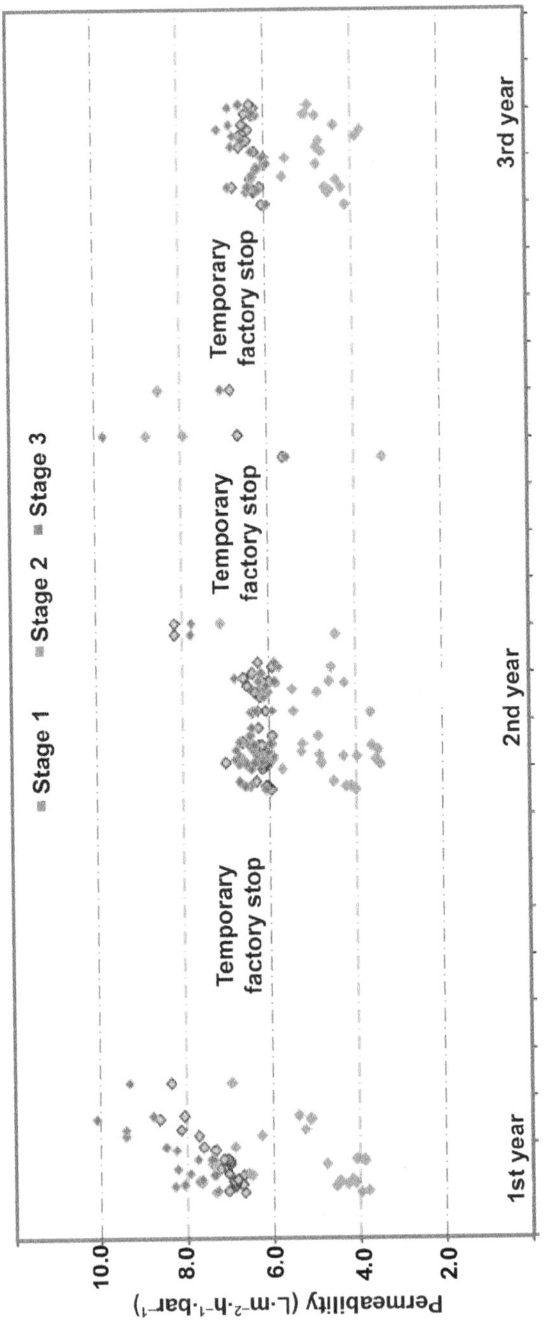

Figure 18.49. *Evolution of permeability during 3 years of operation. For a color version of this figure, see www.iste.co.uk/gaid/watertreatment4.zip*

Two lines of membranes are identically equipped and are designed to operate alone or in parallel. The NF-200 B400 NF membranes are housed in a series of six inside-port pressure tubes. The NF frame is designed to accommodate a maximum of 12 pressure tubes, divided into six tubes for the first stage, three tubes for the second stage and two tubes for the third stage. A spare space is left available. The total number of membranes is 132. The recovery rate is 72%, in other words, a permeate flow rate of 88 $m^3 \cdot h^{-1}$ for a 122 $m^3 \cdot h^{-1}$ feed. The applied flux is 24.9 $L \cdot m^{-2} \cdot h^{-1}$.

Due to the acidification at the inlet of the membranes, the permeate is aggressive and a degassing and/or an addition of soda is carried out in order to bring the water to a calco-carbonic balance. The latter is obtained by passing through neutrality filters and by mixing with very hard water. This station is optimized, so as to have balanced water at the plant outlet with a pH of 8–8.5 and an alkalinity of 80 $mg \cdot L^{-1}$ $CaCO_3$. Lastly, a safety disinfection by chlorine injection is planned. The residual effect by chlorination in the network is ensured.

In addition, a cleaning station (CIP) is used for rinsing during the stop sequence as well as during chemical cleaning.

Figure 18.50. *Nanofiltration membrane trains at the Courchevel plant. For a color version of this figure, see www.iste.co.uk/gaid/watertreatment4.zip*

The monitoring of the membrane permeability during the operation periods shows that it remains stable and conforms to the expectations for NF-200 B 400. It is between 6 and 8 $L \cdot m^{-2} \cdot h^{-1} \cdot bar^{-1}$ for the first two stages and between 4 and 6 $L \cdot m^{-2} \cdot h^{-1} \cdot bar^{-1}$ for the third stage. The sulfate concentration is about 50–70 $mg \cdot L^{-1}$ at the plant outlet.

The conditioning of the membranes during the shutdown of the plant, outside of the operating period, proved to be particularly delicate, and the restart involved modifications to the positioning of the modules within the pressure tubes. The final solution was to supply the plant with water of similar quality to that of the permeate produced in summer, which was achieved by producing water through the NF unit for a few hours a week during the off-season at a low flow configuration of the plant (44 $m^3 \cdot h^{-1}$).

18.13. References

Abouzaid, A., Mouzdahir, A., Rumeau, M. (2003). Étude de la rétention des sels monovalents et bivalents par nanofiltration. *Comp. Rend. Chim.*, 6, 431–436.

Afonso, M., Hagmeyer, G., Gimbel, R. (2001). Streaming potential measurements to assess the variation of nanofiltration membranes surface charge with the concentration of salt solutions. *Sep. Purif. Technol.*, 22, 529–541.

Ahmad, A.L., Ooi, B.S., Mohammad, A.W., Choudhoury, J.P. (2004). Development of a highly hydrophilic nanofiltration membrane for desalination and water treatment. *Desalination*, 168, 215–221.

Al-Zoubi, H. (2006). Development of novel approach to the prediction of nanofiltration membrane performance using advanced atomic force microscopy. PhD Thesis, University of Nottingham.

Azais, A., Mendret, J., Gassara, S., Petit, E. (2014). Nanofiltration for wastewater reuse: Counteractive effects of fouling and matrice of the rejection of pharmaceutical active compounds. *Sep. Purif. Technol.*, 133, 313–327.

Balannec, B., Vourch, M., Rabiller-Baudry, M., Chaufer, B. (2005). Comparative study of different nanofiltration and reverse osmosis membranes for dairy effluent treatment by dead-end filtration. *Sep. Purif. Technol.*, 42, 195–200.

Bandini, S. (2005). Modelling the mechanism of charge formation in NF membranes: Theory and application. *J. Memb. Sci.*, 264, 75.

Bannoud, A.H. (2001). Élimination de la dureté et des sulfates contenus dans les eaux par nanofiltration. *Desalination*, 137, 133–139.

Bellona, C., Drewes, J.E., Xu, P., Amy, G. (2004). Factors affecting the rejection of organic solutes during NF-RO treatment – A literature review. *Water Res.*, 38, 2795–2809

Ben-David, A., Bason, S., Jopp, J., Oren, Y., Freger, V. (2006). Partitioning of organic solutes between water and polyamide layer of RO and NF membranes: Correlation to rejection. *J. Membr. Sci.*, 281, 480–490.

Berg, P., Hagmeyer, G., Gimbel, R. (1997). Removal of pesticides and other micropollutants by nanofiltration. *Desalination*, 113, 205–208.

Bertrand, S., Lemaitre, I., Wittmann, E. (1997). Performance of a nanofiltration plant on hard and highly sulphated water during two years of operation. *Desalination*, 113, 277–281.

Bohdziewicz, J., Bodzek, M., Wasik, E. (1999). The application of reverse osmosis and nanofiltration to the removal of nitrates from groundwater. *Desalination*, 121, 139–147.

Bouchard, C., Kouadio, P., Ellis, D., Rahni, M., Lebrun, R.E. (2000). Les procédés à membranes et leurs applications en production d'eau potable. *Vecteur Environ.*, 33(4), 28–38.

Boussahel, R., Bouland, S., Moussaoui, K.M., Montiel, A. (2000). Removal of pesticide residues in water using the nanofiltration process. *Desalination*, 132, 205–209.

Boussu, K., Zhang, Y., Cocquyt, J., Van der Meeren, P., Volodin, A., Van Haesendonck, C., Martens, J.A., Van der Bruggen, B. (2006). Characterization of polymeric nanofiltration membranes for systematic analysis of membrane performance. *J. Memb. Sci.*, 278, 418–427.

Bowen, W.R. and Welfoot, J.S. (2002). Modelling the performance of membrane nanofiltration – Critical assessment and model development. *Chem. Eng. Sci.*, 57, 1121–1137.

Brandt, D.C., Leitner, G.F., Leitner, W.E. (1993). Reverse osmosis membranes state of the art. In *Reverse Osmosis: Membrane Technology, Water Chemistry, and Industrial Applications*, Amjad, Z. (ed.). Van Nostrand Reinhold, New York.

Buisson, H., Lebeau, T., Lelievre, C., Herremans, L. (1998). Les membranes : point sur les évolutions d'un outil incontournable en production d'eau potable. *L'eau l'industrie les nuisances*, 135, 42–47.

Côté, P.L., Randon, G., Mercier, M. (1993). Nanofiltration for surface water treatment. In *Proceedings of the 2nd International Conference on Advances in Water and Effluent Treatment*, Cumbria.

Cyna, B., Chagneau, G., Bablon, G., Tanghe, N. (2002). Two years of nanofiltration at the Méry-sur-oise plant. *Desalination*, 147, 69–75.

Erikson, P., Kyburz, M., Pergande, W. (2005). NF membrane characteristics and evaluation for sea water processing applications. *Desalination*, 184, 281–294.

Escobar, I.C., Hong, S., Randall, A.A. (2000). Removal of assimilable organic carbon and biodegradable dissolved organic carbon by reverse osmosis and nanofiltration membranes. *J. Membr. Sci.*, 175, 1–17.

Draževič, E., Košutić, K., Freger, V. (2014). Permeability and selectivity of reverse osmosis membranes: Correlation to swelling revisited. *Water Res.*, 49, 444–452.

Gaid, K., Bertrand, S., Binot, P., de Larminat, G. (1996). La nanofiltration : une eau potable de qualité. *Conference Aquatech*, 22 November, Limoges.

Gaid, K., Bablon, G., Turner, G., Franchet, J., Protais, J.C. (1998). Performances of 3 years's operation of a nanofiltration plant. *Conference Water Treatment*, Amsterdam.

Hilal, N., Al-Zoubi, H., Mohammad, A. (2005). Characterization of nanofiltration membranes using atomic force microscopy. *Desalination*, 177, 187–199.

Hofman, J.A.M.H., Noij, T.H.M., Schippers, J.C. (1993). Removal of pesticides and other organic micropollutants with membrane filtration. *Water Supply*, 11, 101–111.

Hwan Kim, J., DiGiano, F.A., Geens, J., Vandecasteele, C. (2004). Influence of MF pretreatment on NF performance for aqueous solutions containing particles and an organic foulant. *Sep. Purif. Technol.*, 36, 203–213.

Kecili, K. (2006). Étude et caractérisation de membranes synthétiques organiques immergées de microfiltration (MF) et d'ultrafiltration (UF) employées pour la production d'eau destinée à la consommation humaine : intensification des opérations de déconditionnement et de nettoyage. PhD Thesis, Université Pierre et Marie Curie, Paris.

Kecili, K., Pontié, M., Thekkedath, A., Habarou, H., Suty, H., Croué, J.P. (2007). Membrane autopsy as a sustainable management of fouling phenomena occurring in MF, UF and NF processes. *Desalination*, 204(1), 155–169.

Kecili, K., Djafer, M., Démocrate, C., Heim, V. (2009). Review of nanofiltration membrane autopsies after nine years of operation in a drinking water plant. In *Proceedings of AWWA Membrane Technology Conference*, Long Beach.

Lee, K.P., Arnot, T.C., Mattia, D. (2011). A review of reverse osmosis membrane materials for desalination – Development to date and future potential. *J. Membr. Sci.*, 370, 1–22.

Lin, L., Feng, C., Lopez, R., Coronell, O. (2016). Identifying facile and accurate methods to measure the thickness of the active layers of thin-film composite membranes – A comparison of seven characterization techniques. *J. Membr. Sci.*, 498, 167–179.

Marconnet, C., Djafer, M., Coriton, G., Di Martino, P., Heim, V. (2008). Prevention of membrane biofouling by UV irradiation. In *Proceedings of AWWA Water Quality Technology Conference*, Cincinatti.

Mohsen, M.S., Jaber, J.O., Afonso, M. (2003). Desalination of brackish water by nanofiltration and reverse osmosis. *Desalination*, 157(1–3), 157–167.

Peltier, S., Cotte, M., Gatel, D., Herremans, H., Cavard, J. (2003). Nanofiltration: Improvements of water quality in a large distribution system. *Water Sci. Technol. Water Supply*, 3(1/2), 193–200.

Petersen, R.J. (1993). Composite reverse osmosis and nanofiltration membranes. *J. Membr. Sci.*, 83, 81–150.

Pontié, M., Dach, H., Leparc, J., Hafsi, M., Lhassani, A. (2008). Novel approach combining physico-chemical characterizations and mass transfer modelling of nanofiltration and low pressure reverse osmosis membranes for brackish water desalination intensification. *Desalination*, 221, 174–191.

Schaep, J. and Vandecasteele, C. (2001). Evaluating the charge of nanofiltration membranes. *J. Membr. Sci.*, 188, 129–136.

Schaep, J., Van der Brugen, B., Uyetherhoven, S., Croux, R., Vandecasteele, C., Wilms, D., Van Houtte, E., Vanlerberghe, F. (1998). Removal of hardness from groundwater by nanofiltration. *Desalination*, 119, 295–301.

Teixeira, M., Rosa, M., Nyström, M. (2005). The role of membrane charge on nanofiltration performance. *J. Membr. Sci.*, 265, 160–166.

Van der Bruggen, B., Schaep, J., Wilms, D., Vandecasteele, C. (1999). Influence of molecular size, polarity and charge on the retention of organic molecules by nanofiltration. *J. Membr. Sci.*, 156, 29–41.

Van der Bruggen, B., Everaert, K., Wilms, D., Vandecasteele, C. (2001). Application of nanofiltration for removal of pesticides, nitrate and hardness from ground water: Rejection properties and economic evaluation. *J. Membr. Sci.*, 193, 239–248.

Ventresque, C. and Bablon, G. (1997). The integrated nanofiltration system of the Mery-sur-Oise surface water treatment plant (37 mgd). *Desalination*, 113, 263–266.

Ventresque, C., Gisclon, V., Bablon, G., Chagneau, G. (2000). An outstanding feat of modem technology: The Mery-sur-Oise nanofiltration treatment plant (340,000 m'/d). *Desalination*, 131, 1–16.

Wijmans, J.G. and Baker, R.W. (1995). The solution-diffusion model: A review. *J. Membr. Sci.*, 107, 1–21.

Zhang, X., Cahill, D.G., Coronell, O., Mariñas, B.J. (2009). Absorption of water in the active layer of reverse osmosis membranes. *J. Membr. Sci.*, 331, 143–151.

19

Desalination by Reverse Osmosis

Reverse osmosis desalination (SWRO) is an important alternative for coastal regions with a shortage of drinking water. Seawater desalination is a globally efficient and economical way to produce water for human consumption, as well as for agricultural and industrial needs. Reverse osmosis (RO) plants for the conversion of seawater to drinking water and other uses have been in operation in many parts of the world for over four decades. The design and operation of RO plants depends on the quality of the pre-treated water, and the choice of the best pre-treatment technology depends on the quality of the seawater and its variations.

Numerous large-scale experiments have shown that pre-treatment is the key to this application. Therefore, in recent years, a major effort has been made to identify and characterize the various organic and inorganic components present in seawater in order to optimize pre-treatment.

Advanced analytical methods have been developed to refine water characterization and clogging indicators, to create predictive tools adapted both to seawater variations and to anticipate the consequences for RO membranes.

19.1. Characterization of the water to be treated

The global approach to characterizing seawater samples is presented through analytical tools that give access to many parameters, such as the type and nature of organic substances and inorganic mineral elements, as well as the micro-organism and phytoplankton count. These tools help to anticipate the effect of each of these parameters on the fouling of RO membranes, and identify the different pre-treatments needed to reduce or remove these elements or substances.

The physicochemical characterization of the water is essential for the definition and sizing of the pre-treatment stage. It includes parameters that are traditionally used in drinking water such as suspended solids (SS), turbidity, total organic carbon (TOC) and chlorophyll a, counting and identification of algae species, a complete ionic balance, water clogging capacity and analysis of various organic and mineral compounds that can damage the membranes (hydrocarbons, industrial residues, metals and specific elements).

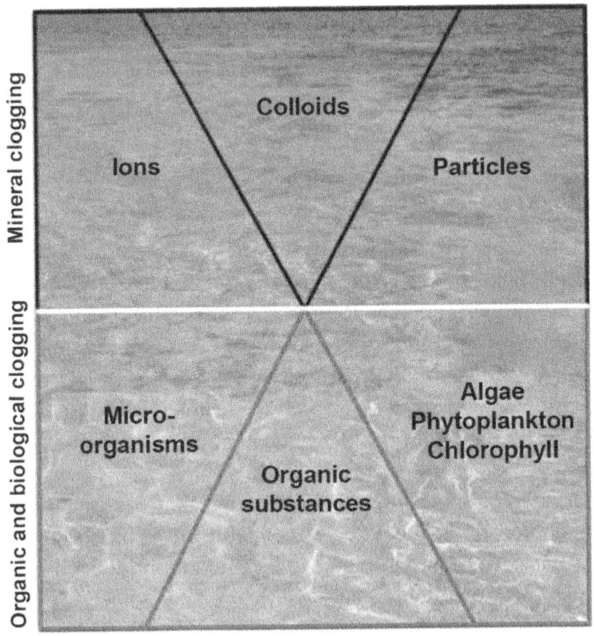

Figure 19.1. *Substances and elements involved in RO membrane clogging. For a color version of this figure, see www.iste.co.uk/gaid/watertreatment4.zip*

Seawater is a mixture of various salts, organic substances, algae, bacteria and microparticles. Figure 19.1 and Table 19.1 summarize the most commonly encountered substances and elements that contribute to membrane clogging:

– Inorganic and particulate fouling: due to the accumulation of particles on the surface of the membranes that have not been removed from the raw water during pre-treatment. Indicators of sufficient SS and particulate reduction are turbidity values below 0.5 NTU, and particle density index (PDI) values below 5. Scaling of the membrane surface is caused by precipitation of salts if the appropriate chemical reagents are not injected at the pre-treatment level or upstream of the membranes.

– Colloidal fouling: this is caused by the deposition of metal oxides, proteins, silicates, organic matter and particles (such as clay), creating a colloidal layer on the membrane surface. Colloidal fouling tends to cause harsh clogging of the membrane system rather than fouling of the membrane surface.

– Biological fouling: this is due to bacterial growth on the surface of the membranes. A biofilm is created, which is a gelatinous material consisting of exo-polysaccharides secreted by the bacteria. This biofilm not only causes a loss of flow, but can also physically degrade certain types of membranes.

– Organic fouling: this is the result of the adsorption of organic matter, in particular humic acid (HA) and fulvic acid (FA). Organic fouling is complex in nature, and has been identified as a cause of the decrease in membrane performance.

The physicochemical and bacteriological analyses make it possible to determine the most suitable pre-treatment for raw water and its variations during the year.

Inorganic	**Organic**	**Biological**
Silica Quartz Carbonates/sulfates	Lipids Proteins Polysaccharides	Algae Microplankton Unicellular micro-organisms

Table 19.1. *Seawater constituents involved in RO membrane clogging*

19.1.1. *Physical characteristics*

19.1.1.1. *Density of seawater*

Seawater density generally increases with decreasing temperature, increasing salinity and increasing ocean depth. The density of seawater at the ocean surface ranges from 1,020 to 1,029 $kg \cdot m^{-3}$. In the deepest parts of the oceans, seawater density can reach 1,050 $kg \cdot m^{-3}$.

Other physical characteristics that are assessed are SS, temperature, turbidity and the PDI.

19.1.1.2. *Suspended solids*

The SS concentration is assessed to estimate the pre-treatment level. This parameter is very variable depending on the oceans, seas or marine estuaries. SS concentrations can vary between 1 and 500 $mg \cdot L^{-1}$ (observed during typhoons and tornados).

19.1.1.3. *Temperature*

Temperature is a dimensional parameter of the membranes. The evolution of the temperature during the year is an important element to be considered, because it varies during the seasons. Guarantees in terms of salt rejection must be consistent, and the membrane performance assessment must take into account the maximum and minimum temperatures of the seawater.

19.1.1.4. *Turbidity*

This parameter assesses the concentration of microparticles present in the water. Turbidity is measured in nephelometric turbidity units (NTU).

19.1.1.5. *Clogging indicators*

Clogging is the major issue that needs to be managed when using membranes. Several parameters are proposed to measure clogging potential for use as a predictive tool, so as to assess pre-treatment performance. The silt density index (SDI) and the modified fouling index (MFI) are currently the only standard methods, although they do not reflect the actual fouling potential because particles smaller than 0.45 µm, which cause fouling, are not considered.

19.1.1.5.1. The SDI

The clogging index most often used by the water industry and membrane manufacturers is called the SDI. The SDI is therefore representative of a particle density that makes it possible to qualify the clogging nature of a water. It is calculated from the time required to filter a given volume of water through a flat membrane with a pore diameter of 0.45 µm (microfiltration [MF]) and a filtering surface of 1.73×10^{-4} m^2, at a constant transmembrane pressure of 2.07 bar. The membrane has a frontal flow. The principle of the measurement is to record the time required (t_1) to filter 500 mL of water through a new test membrane, and then after a filtration time between 1 and 60 min, to once again record the time required (t_2) to filter another 500 mL.

The SDI is expressed in %/minute. It is calculated by:

$$SDI\ (tf) = \frac{100}{tf}\left(1 - \frac{t_1}{t_2}\right)$$

with:

– t_1: time needed to collect the first 500 mL of the sample (min);

– t_2: time needed to collect the last 500 mL of the sample (min);

– tf: total filtration time (min).

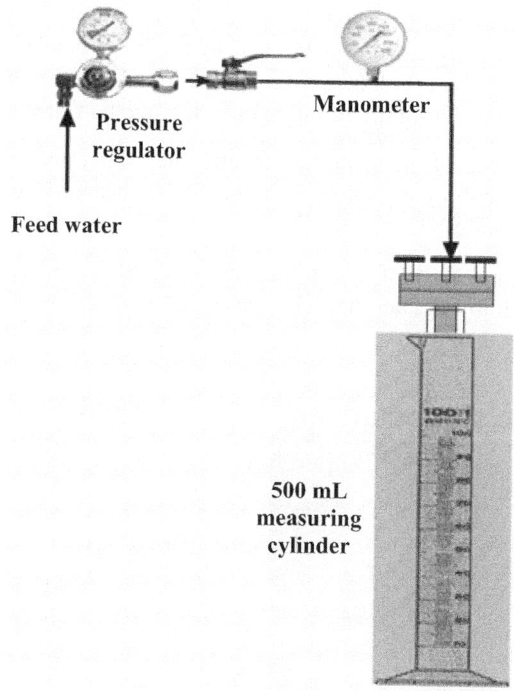

Figure 19.2. *SDI measuring equipment. For a color version of this figure, see www.iste.co.uk/gaid/watertreatment4.zip*

This index only provides an indication of particulate clogging and a static measure of filtration resistance. The more the water clogs, the shorter the overall filtration time because the membrane clogs quicker.

The ranges of validity of the measurements are therefore as follows: SDI_{15}: "0" to 6.6; SDI_{10}: "0" to 10; SDI_5: "0" to 20; SDI_3: "0" to 33.3.

The reference time of 15 min for tf is often the reference for treated water.

The advantage of SDI is that its determination is relatively easy. The objective of the pre-treatment is to feed the membranes with water characterized by an $SDI_{15} \leq 5$. Factors interfering with SDI measurements are:

– the pH influence, which can lead to an increase in SDI values from 4 to 6 when the pH rises from 7 to 8, mainly due to the presence of elements such as Ca^{2+}, which precipitates with increasing pH;

– the type of membrane, which is a determining factor for SDI values. A comparative study between hydrophobic and hydrophilic membranes (Figure 19.4) shows that higher results are obtained for the hydrophobic membrane than for the hydrophilic membrane during the filtration on a dual-media filter;

– at the beginning of the filtration cycle, the SDI_{15} measured with PVDF membranes (hydrophobic) is higher than the SDI_{15} measured with nitrocellulose membranes (hydrophilic);

– during the main part of the filtration cycle, SDI_{15} measured with PVDF membranes is much lower than the SDI_{15} measured with nitrocellulose membranes, and is always lower than 3.

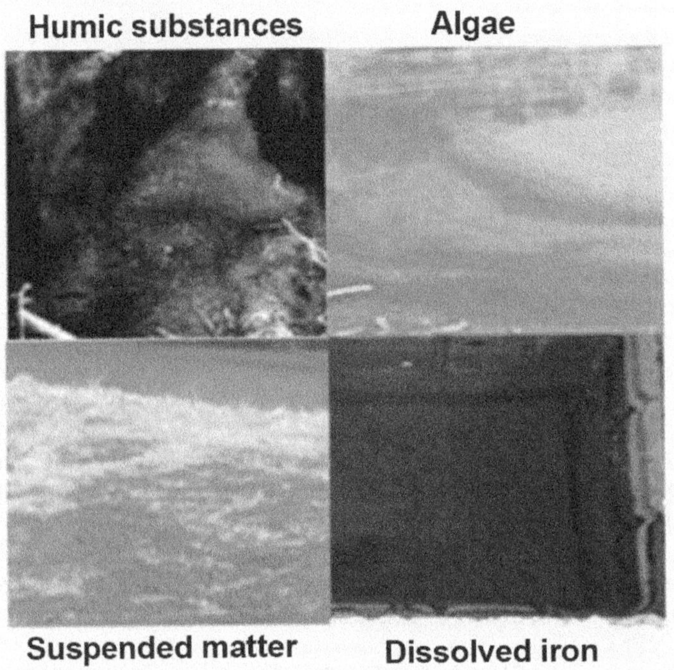

Figure 19.3. *Origin of the substances and elements responsible for clogging, assessed globally by the SDI measurement. For a color version of this figure, see www.iste.co.uk/gaid/watertreatment4.zip*

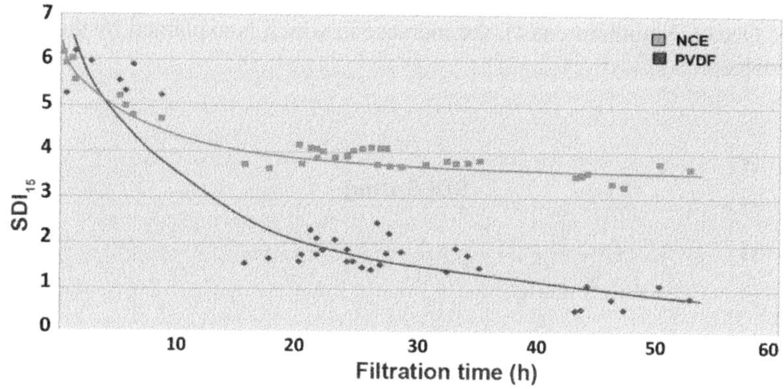

Figure 19.4. *SDI_{15} comparison of nitrocellulose (NCE) versus PVDF membrane. For a color version of this figure, see www.iste.co.uk/gaid/watertreatment4.zip*

Figures 19.5 and 19.6 show the evolution of SDI in spring and fall, at a site located in the Mediterranean Sea. This evolution can influence the quality of the pre-treated water if the selected technology and design are not based on the worst-case values. The seasonal evolution of this parameter shows that the SDI is quite stable in the fall, while in spring, there are frequent variations.

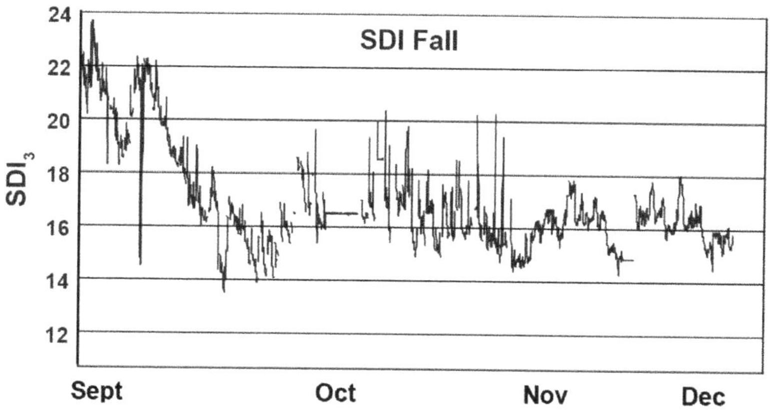

Figure 19.5. *Evolution of the SDI_3 of the Mediterranean Sea between September and December*

Figure 19.7 shows the variation of SDI according to seawater temperature (South Atlantic Ocean, Namibian coast), the increase in which is explained by the presence of microalgae.

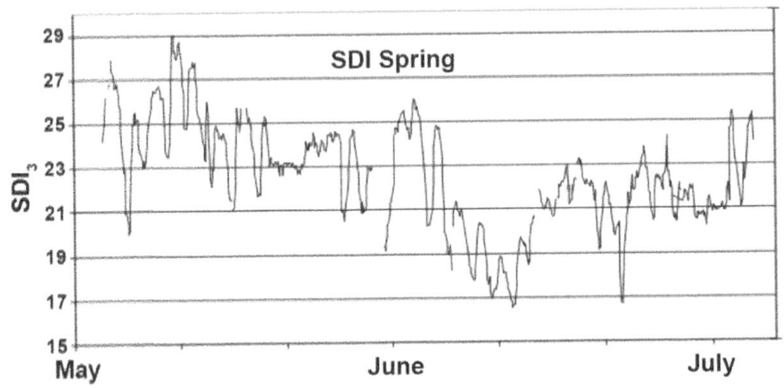

Figure 19.6. *Evolution of the SDI_3 of the Mediterranean Sea between May and July*

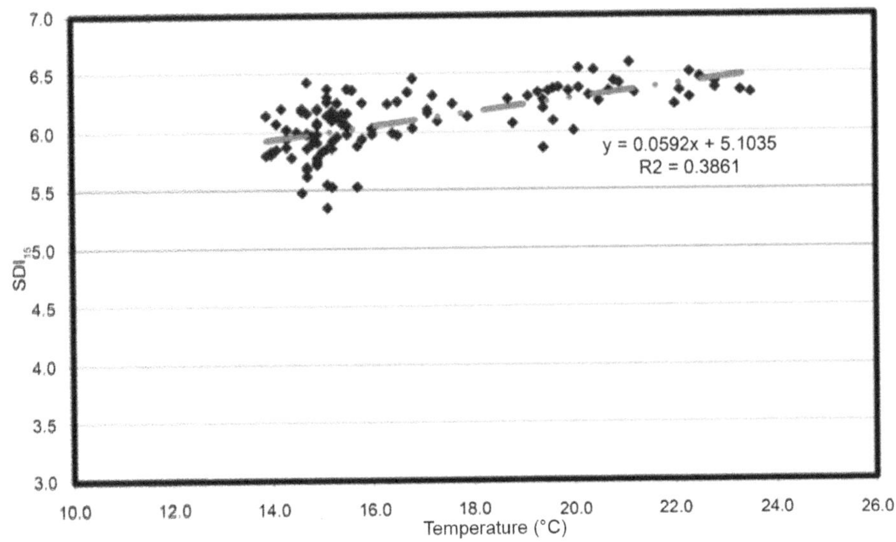

Figure 19.7. *Evolution of the SDI with temperature (South Atlantic Ocean-Namibian coast). For a color version of this figure, see www.iste.co.uk/gaid/watertreatment4.zip*

19.1.1.5.2. The modified fouling index (MFI)

Compared to the SDI, the MFI physically represents the resistance of one unit of deposit mass per unit area of filter membrane. The clogging of a membrane in frontal filtration with a constant pressure gradient takes place in three stages: pore blockage, formation of an incompressible cake and formation of a compressible cake. The denser the deposit formed, the more difficult it is for the fluid to be filtered. The MFI is therefore high.

When an incompressible cake is formed, a linear relationship exists between t/V and V, as shown in Figure 19.8. The MFI indicator specifically corresponds to the clogging by formation of a cake on the surface of the membrane, and is represented by the slope of the linear part of the curve t/V = f(V).

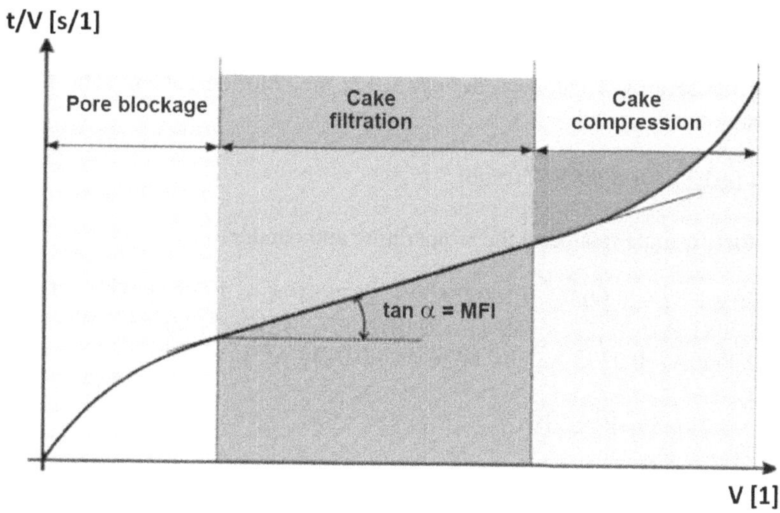

Figure 19.8. *Experimental determination of the MFI. For a color version of this figure, see www.iste.co.uk/gaid/watertreatment4.zip*

The experimental determination of the MFI is:

$$MFI(0.45) = \frac{\mu 20}{\mu T}\left(\frac{\Delta P}{210}\right) tan\ \alpha$$

with V, t, ΔP and μ are filtrate volume, filtration time, applied transmembrane pressure and dynamic viscosity of water, respectively.

The MFI value is considered at a temperature of 20°C and a pressure of 210 kPa.

In practice, the determination of the MFI is similar to that of the SDI. The same experimental setup is used. The tests consist of a series of time measurements and cumulative permeate volumes that allow t/V to be plotted against V.

19.1.2. *Chemical composition: ionic content*

19.1.2.1. *Total dissolved salts*

The vast majority of salts analyzed in seawater have a continental origin. Six ions make up about 99% of marine salts: chlorides (Cl^-), sodium (Na^+), sulfates (SO_4^{2-}), magnesium (Mg^{2+}), calcium (Ca^{2+}) and potassium (K^+).

Given that salinity is directly proportional to the amount of chlorides present in seawater, and the fact that they are accurately measured by simple chemical analysis, the salinity S has been defined using the chloride content. The following relationship is often used:

$$S\ (g \cdot L^{-1}) = 1.80655\ Cl\ (o/oo)$$

Another formula related to the temperature and conductivity is given by:

$$S\ (g \cdot L^{-1}) = 0.008 - 0.1692\ A^{0.5} + 25.3851\ A + 14.0941\ A^{1.5} - 7.026A^2 + 2.7081\ A^{2.5} + \{(T - 15)/\ (1 + 0.0162\ (T - 15))\} * (0.0005 - 0.0056\ A^{0.5} - 0.0066A - 0.0375\ A^{1.5} + 0.0636\ A^2 - 0.0144\ A^{2.5})$$

$$A = \sigma_{s,t,p}/(42.914 * B)$$

$$B = 0.6766097 + 0.0200564T + 0.00011043\ T^2 - 6.9698 \times 10^{-7}\ T^3 + 1.003 \times 10^{-9}\ T^4$$

T: Temperature °C

$\sigma_{s,t,p}$: conductivity, $ms \cdot cm^{-1}$

*From Aminot and Kérouel (2004).

For specific sites, we propose an adaptation of this formula as follows:

$$S\ (g \cdot L^{-1}) = 0.008 - 0.1692\ A^{0.5} + 26.6851\ A + 15.435\ A^{1.5} - 7.026A^2 + 2.7081\ A^{2.5} + \{(T - 15)/(1 + 0.0162\ (T - 15))\} * (0.0005 - 0.0056\ A^{0.5} - 0.0066A - 0.0375\ A^{1.5} + 0.0636\ A^2 - 0.0144\ A^{2.5})$$

The other parameters are the same.

The concentration of total dissolved salts (TDS) in seawater (25–45 $g \cdot L^{-1}$) consists of all the dissolved salts listed in Table 19.2. Their proportions are often the same. This can be explained by the fact that differences in salinity are caused by the evaporation of fresh water or the addition of river water. The main ions that make up 99.9% of the total dissolved solids are presented in Table 19.2.

Ions	Valences	Concentrations $g \cdot kg^{-1}$ $mmol \cdot kg^{-1}$		% in salinity	Atomic weights
Chloride Cl	–1	19,345	546	55.03	35.453
Sodium Na	+1	10,752	468	30.59	22.990
Sulfate SO_4	–2	2,701	28.1	7.68	96.062
Magnesium Mg	+2	1,295	53.3	3.68	24.305
Calcium Ca	+2	416	10.4	1.18	40.078
Potassium K	+1	390	9.97	1.11	39.098
Bicarbonate HCO_3	–1	145	2.34	0.41	61.016
Bromide Br	–1	66	0.83	0.19	79.904
Borate BO_3	–3	27	0.46	0.08	58.808
Strontium Sr	+2	13	0.091	0.04	87.620
Fluoride F	–1	1	0.068	0.003	18.998

Table 19.2. *Main ions in seawater*

Landlocked seas like the Black Sea and the Baltic Sea have different concentrations. This world map shows a variation in ocean TDS from 32 to 40 $g \cdot L^{-1}$. Low TDS is analyzed in cold seas, especially during the ice melt period.

High salinity (Figure 19.9) is observed in oceans that coincide with continental deserts. As a result of the cool, dry air that descends and warms up, these desert areas have very little precipitation and high evaporation. The Red Sea, which is located in the desert but is almost completely enclosed, shows the highest salinity of all (42–45 $g \cdot L^{-1}$), while the Mediterranean Sea is in the middle range (35–38 $g \cdot L^{-1}$). The lowest salinity is found in the upper reaches of the Baltic Sea (5 $g \cdot L^{-1}$). The Dead Sea has a salinity of about 240 $g \cdot L^{-1}$, mostly containing magnesium chloride $MgCl_2$. The estuaries contain 1–5 $g \cdot L^{-1}$.

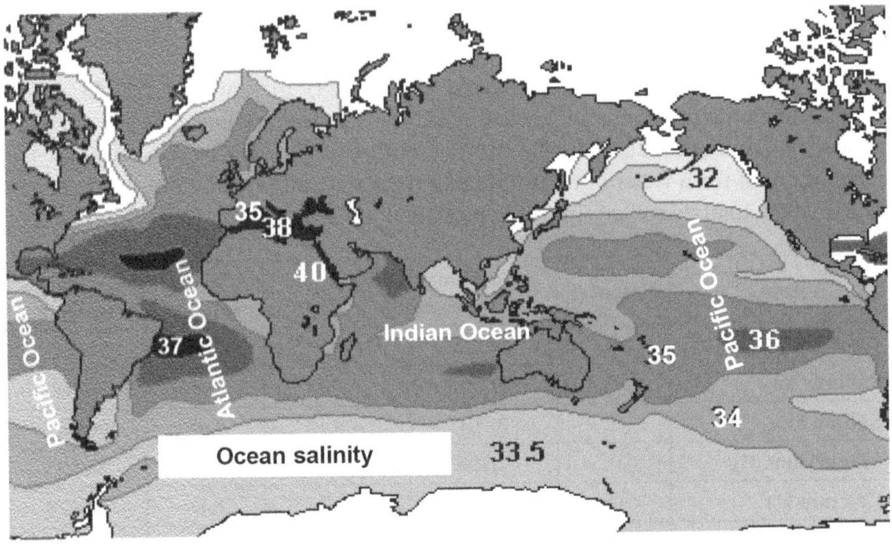

Figure 19.9. *Salinity in different oceans and seas of the world. For a color version of this figure, see www.iste.co.uk/gaid/watertreatment4.zip*

The overall ion content of the Persian Gulf is higher compared to the oceans and the Mediterranean Sea. Overall, these differences in salinity do not impact the choice of pre-treatment, but do impact the sizing of the membrane systems.

Table 19.3 shows the main characteristics of some seas and oceans.

However, while most of the ions analyzed in the seawater do not impact the pre-treatment design, iron and manganese must be removed before the water is fed to the RO membrane.

Table 19.4 shows the variation of salinity (35 $g \cdot L^{-1}$) with temperature and at atmospheric pressure.

This gives us:

$$\rho_T = \rho_{To} - \alpha\, \Delta T \text{ and } \sigma_T = \sigma_{To} + \beta\, \Delta T$$

With:

– β: fraction of variation in conductivity per unit change in temperature (°C);

– α: fraction of variation in density per unit change in temperature (°C).

Seas or oceans	Sampling sites	Temperatures (°C)	Salinity ($g \cdot kg^{-1}$)	Conductivity ($mS \cdot cm^{-1}$ to 20 °C)	TDS ($g \cdot L^{-1}$)	Bore ($mg \cdot L^{-1}$)	Clogging power (3 min)
Atlantic Ocean	Morocco–Laayoune OCP	15–25	35–37	48–50	36–39	4.4	15–30
Pacific Ocean	Australia–Perth	14–26	37–38	49.5–52	38–40	5.2	18–22
North Sea	Terneuzen–Holland	12–20	25–27	32.5–37.7	25–29	–	25–30
Mediterranean Sea	Israel–Ashkelon	16–30	33–34	45.5–49.4	35–38	4–5	15–25
Persian Gulf	Emirates–Abu Dhabi	18–40	39–42	52–56	40–43	5.8	17–32
Red Sea	Jeddah	25–34	37–40	51–54.5	39–42	4.5	25 to >33.3
Arabian Sea	Kuwait	28–36	39–42	52–58.5	40–45	<5	25 to >33.3

Table 19.3. *Characteristics of some seas and oceans*

Temperatures (°C)	ρ density ($g{\cdot}cm^{-3}$)	β, $°C^{-1}$ (dσ/dT)	10^7 α·$°C^{-1}$ (dρ/dT)	Conductivity σ ($\mu S{\cdot}cm^{-1}$)	Viscosity η (mPas)
0	1.028106	–	–	29,048	1.892
5	1.027675	884	862	33,468	1.610
10	1.026952	905.5	1,154	38,103	1.388
15	1.025973	925.6	1,422	42,933	1.221
20	1.024763	944.3	1,671	47,934	1.085
25	1.023343	961.6	1,905	53,088	0.966
30	1.021729	1,007.4	2,125	59,271	0.871
35	1.019934	1,060.7	2,334	66,174	0.785
40	1.017973	1,120	2,533	73,881	–

Table 19.4. *Variation of salinity parameters according to temperature and for P = 1 atm*

19.1.2.2. *Specific ions: iron and manganese*

Iron oxides, as well as manganese, play an important role in the removal of trace elements from seawater. In sediments, iron and manganese oxides, which are carried by sedimented particles, are reduced to ferrous and manganese ions during the oxidation of organic matter. The ferrous and manganese ions diffuse upward through the pore water, and are again transformed into iron and manganese oxides at the sediment–water interface.

19.1.2.2.1. Case of iron

Iron's behavior is very different from that of manganese. The hypothesis that dissolved iron concentration is a key variable controlling phytoplankton processes in ocean surface waters is now proven. Iron is an essential micro-nutrient for phytoplankton growth, as a component of biochemical processes such as photosynthetic and respiratory electron transport, nitrate and nitrite reduction, chlorophyll synthesis and a number of other biosynthetic reactions.

The oxidation rate constant of Fe^{2+} increases with increasing pH and temperature, and decreases with increasing salinity (ionic strength). The relationship is given by:

$$Log\ k = 21.56 - \frac{1.545}{T} - 3.29\ I^{\frac{1}{2}} + 1.52\ I$$

With:

– k: oxidation rate constant;

– T: absolute temperature;

– I: ionic strength.

Fe^{3+} speciation in seawater is dominated by its hydrolysis and tendency to form particulate Fe^{3+} hydroxides. In general, in oxic seawater at pH 8, iron is primarily present as particulate iron oxyhydroxide (FeOOH), which is highly insoluble and thermodynamically stable.

The solubility of Fe^{3+} in seawater is controlled by complexation with organic substances, which then regulates dissolved iron concentrations.

Dissolved Fe profiles generally show low surface concentrations (0.2–50 $\mu g \cdot L^{-1}$), with values that oscillate between 10 and 150 $\mu g \cdot L^{-1}$ at depths of 500–1000 m. The vertical profiles are similar to those of Fe^{3+} solubility, suggesting that dissolved Fe concentrations in deep ocean waters are primarily controlled by the binding of Fe^{3+} to naturally occurring organic ligands that have been released during the decomposition of biogenic organic matter at mid-depth and deeper. In oxic seawater, phytoplankton growth is controlled by Fe^{3+} solubilities and iron dissolution rates.

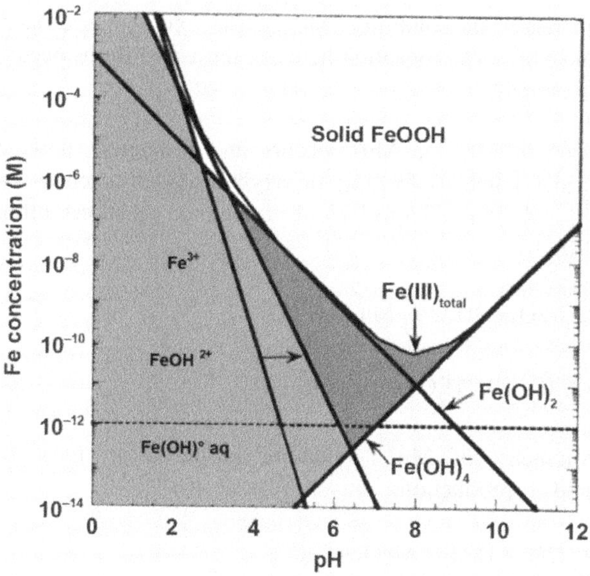

Figure 19.10. *Solubility of Fe according to pH and its amorphous form FeOOH. For a color version of this figure, see www.iste.co.uk/gaid/watertreatment4.zip*

The exchange reaction between the organic-Fe^{3+} complex and the high concentrations of the main alkaline earth metals (such as Ca^{2+} and Mg^{2+}) in seawater is one of the processes that lead to the dissociation of the organic-Fe^{3+} complexes and the precipitation of Fe^{3+}. In estuarine waters and coastal areas where rivers flow, organic–Fe^{3+} complexes (such as fulvic substance–Fe^{3+}) play an important role in the biological iron supply due to its dissociation during the mixing with seawater.

The limit recommended by the manufacturers is 50 $\mu g \cdot L^{-1}$ because of its possible oxidation on the membrane, which can cause irreversible damage to the membrane surface (Figure 19.11).

19.1.2.2.2. The case of manganese

Manganese chemistry in seawater is complex and is largely governed by pH and redox conditions. Mn^{2+} is present at low pH and redox potential, with an increasing proportion of colloidal manganese oxyhydroxides above pH 5.5 in non-dystrophic waters.

Given that Mn^{2+} is more soluble than Mn^{4+}, it tends to be more bio-available with a decrease in pH and redox potential, especially since the presence of chlorides and sulfates increases its solubility. Manganese exists in seawater in two main forms: Mn^{2+} and Mn^{4+}. The transition between these two forms occurs via oxidation and reduction reactions.

The oxidation of Mn^{2+} to Mn^{4+} occurs in two stages, in which solid-phase Mn-bearing oxides (such as Mn_3O_4) or oxyhydroxides (such as ß-MnOOH) are initially formed and then undergo lenient protonation reactions, ultimately forming Mn^{4+} oxides (MnO_2).

$$3Mn^{2+} + 1/2O_2 + 3H_2O \rightarrow Mn_3O_4 + 6H^+$$

$$Mn^{2+} + 1/4O_2 + 3/2H_2O \rightarrow MnOOH + 2H^+$$

The stoichiometry of Mn^{2+} oxidation is based on measurements of O_2 consumption and H^+ production:

$$Mn^{2+} + 1/2O_2 + H_2O \rightarrow MnO_2 + 2H^+$$

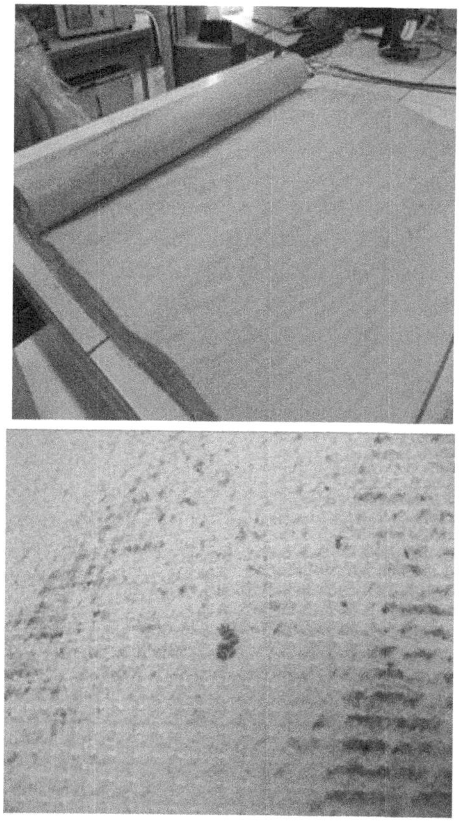

Figure 19.11. *Iron deposited on reverse osmosis membranes. For a color version of this figure, see www.iste.co.uk/gaid/watertreatment4.zip*

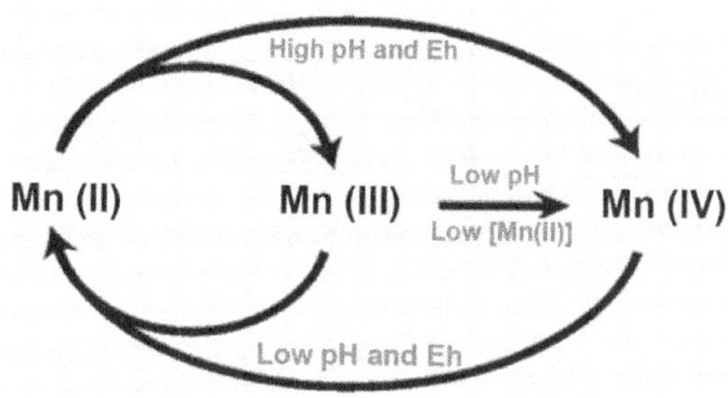

Figure 19.12. *Mn cycle and oxidation states*

In oxygenated waters, Mn^{2+} is thermodynamically unstable toward oxidation to insoluble manganese oxides. However, due to the relatively slow kinetics of Mn^{2+} oxidation in seawater, low equilibrium concentrations are rarely reached. The oceanic distribution of the metal appears to be dominated by external input sources that lead to maximums in surface waters.

Manganese concentrations in open seas vary from 0.4 to 10 $\mu g \cdot L^{-1}$. In the Atlantic Ocean, the English Channel and Indian Oceans, manganese levels range from 2 to 230 $\mu g \cdot L^{-1}$. The levels found in the coastal waters of the Irish Sea and the North Sea off the coast of the United Kingdom vary from 2 to 25.5 $\mu g \cdot L^{-1}$. The concentration of dissolved Mn^{2+} in anoxic waters is likely limited by its solubility compared to $MnCO_3$.

Manganese deficiency limits the growth rate of marine phytoplankton, especially when manganese-poor deep-sea water rises to the surface. At low manganese concentrations (<0.04 $mg \cdot L^{-1}$), green and blue-green algal flora may predominate, while at higher manganese levels, diatoms dominate. Streams with high iron and manganese concentrations may promote the development of blooms of ferruginous and manganiferous bacteria on areas of metal oxides.

Seawater	**Manganese concentrations ($\mu g \cdot L^{-1}$)**	**Iron concentrations ($\mu g \cdot L^{-1}$)**
North Sea	<5–4	5–4
Indian Ocean	2–180	2–220
Northeast Atlantic Ocean	5–80	25–230
UK Channel	3–4	3–4
Irish Sea	2–25	2–25
Arabian Sea	<5	<10
Red Sea	5–10	5–100
Mediterranean Sea	5–10	<15
Oman Sea	<5	<10

Table 19.5. *Fe and Mn concentrations in some seas and oceans*

The limitation recommended by membrane suppliers for manganese is 20 $\mu g \cdot L^{-1}$, because of its potential oxidation on the membrane which will irreversibly damage the membrane surface. Figures 19.13 and 19.14 show the damage observed on the RO membrane due to the oxidation of Mn^{2+} to Mn^{4+}.

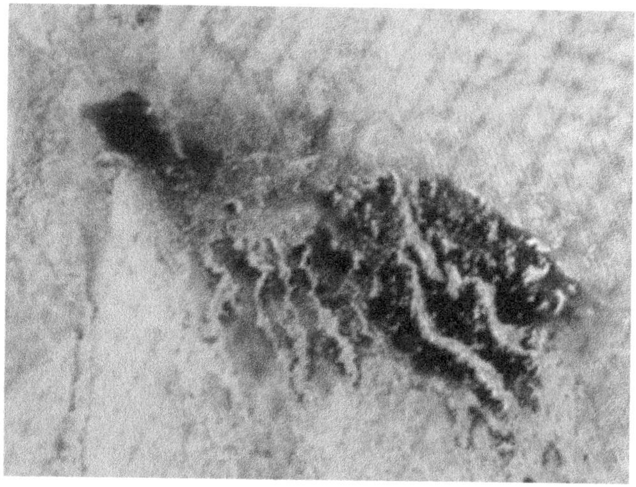

Figure 19.13. *MnO_2 deposit on an RO membrane. For a color version of this figure, see www.iste.co.uk/gaid/watertreatment4.zip*

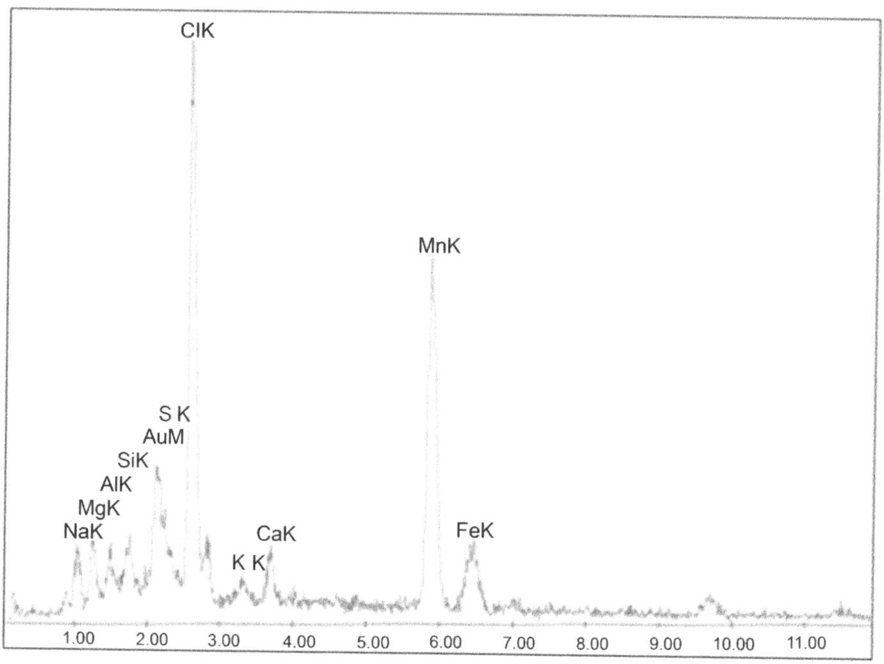

Figure 19.14. *Abnormal presence of Mn on RO membrane*

Scanning electron microscopy-energy dispersive X-ray analysis (SEM-EDXA) is often used to identify the mineral constituents of a membrane that is clogged.

The determination of ionic elements by the SEM-EDXA combination is based on the analysis of X-rays produced by electron beam excitation of a sample. This technique allows analysis in selective areas. The limited depth of analysis (usually a few microns), as well as the possibility of selecting the area of interest under the electron beam, allow local analysis to reveal compositional differences. The samples are coated with gold (Au) for analysis.

Figure 19.15 shows the surface of a membrane covered by a deposit, made up of granulated particles. EDX analysis of the particles shows the presence of chloride (Cl), sodium (Na), manganese (Mn), sulfur (S), magnesium (Mg), iron (Fe), calcium (Ca) and potassium (K).

An analysis of the Mn concentration in the feed water before the RO membrane is recommended.

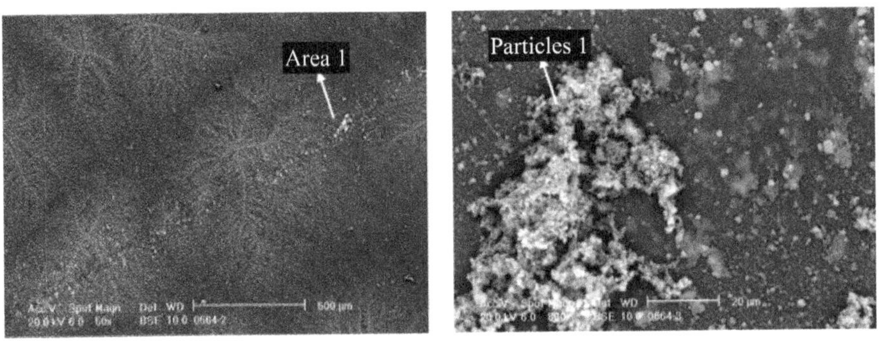

a) b)

Figure 19.15. *General view of (a) a membrane and (b) its surface deposits*

19.1.3. *Chemical composition: organic substances*

Clogging of RO membranes causes a progressive decrease in membrane flow, thus decreasing permeate production, but also leading to an increase in PMT, which results in a higher energy demand. Membrane degradation can be observed.

Therefore, organic substance removal is achieved to its maximum possible extent during the pre-treatment.

The dissolved organic matter is a mixture of aliphatic and aromatic compounds. More than 90% of the dissolved fraction consists of humic substances (HS). The two main types of organic matter are natural allochthonous organic matter (NOM), which is dominated by HS, and autochthonous or algal organic matter, which mainly consists of macromolecular and extracellular debris.

Organic materials generally have functional groups such as carboxyl (-COOH) and phenolic (-OH) groups that play an important role in organic fouling due to interactions with organic materials that vary according to water quality.

HS mainly consist of HA and FA. HA and FA have a negative charge density, very high molecular weights and rather voluminous forms.

Natural organic materials have relatively high specific UV absorbance values and contain relatively large amounts of aromatic carbon. The rate and extent of organic fouling is accelerated with decreasing pH, increasing ionic strength and increasing divalent cations. Electrostatic repulsion is highest at low pH, almost completely deprotonating carboxylic and phenolic groups. The electrostatic repulsion is reduced at high ionic strength, due to the high concentrations of divalent cations that act on the electric double layer around the charged organic materials.

There are several protocols for measuring NOM in seawater, such as:

– TOC, which represents the total concentration of dissolved and particulate organic substances;

– dissolved organic carbon (DOC), which represents the concentration of dissolved organic substances;

– UVA absorbance at 254 nm, which reflects the aromatic nature of NOM;

– SUVA (UVA254/COD ratio), which differentiates the fraction of HS from non-HS.

19.1.3.1. *Liquid chromatography*

DOC, or UV254 absorbance, pH, turbidity and color are common water quality parameters assessed in drinking water plants.

However, they do not provide information on the nature of the NOM, such as molar mass (MM) or hydrophobicity.

The liquid chromatography-organic carbon detection (LC-OCD) technique has been developed for the characterization of natural organic matter in surface waters. The organic fractions described below are usually found in surface waters.

The qualitative determination is based on the separation of hydrophilic molecules according to their hydrodynamic molecular size, performed by the size exclusion chromatography (SEC) column. High molecular weight molecules, which do not penetrate the pores of the column material, appear first on the chromatogram, while smaller molecules follow afterwards.

Quantification is based on the determination of TOC in the Gräntzel reactor. Section 7.2 (Volume 2) details the operating protocols and expected results with this analytical technique.

DOC reactions	Molecular weight (Da)	Compositions
Biopolymers	>20,000	Polysaccharides (such as TEP) and proteins
Humic substances	~1,000	Humic and fulvic acids
Building blocks	300–500	HS oxidation products
Organic acids with low MW	<350	All aliphatic organic acids with low PM
Neutral to low MW	<350	Alcohols, aldehydes, ketones and amino acids

Table 19.6. *Typical fractions of DOC obtained by LC-OCD*

Open seawaters generally have fairly low and stable TOC levels, ranging from 0.8 to 1.5 $mg \cdot L^{-1}$. In contrast, coastal wells have lower TOC concentrations. Polysaccharides are removed during slow filtration through the sand layer. The coastal wells are of very good quality (SDI and TOC), and little organic and biological clogging can be observed.

In surface water, the characterization of NOM by LC-OCD chromatography shows that their concentration varies during the seasons and allows the operator to anticipate the operating and maintenance conditions of the membranes.

In particular, Figures 19.16 and 19.17 show that samples collected in the fall have lower polysaccharide levels than samples collected in the summer. These low polysaccharide levels analyzed in the cold season also correspond to lower SDI3 values and bacterial counts. However, in warmer weather, biological activity (bacteria and algae) becomes a major cause in the fouling potential of membranes.

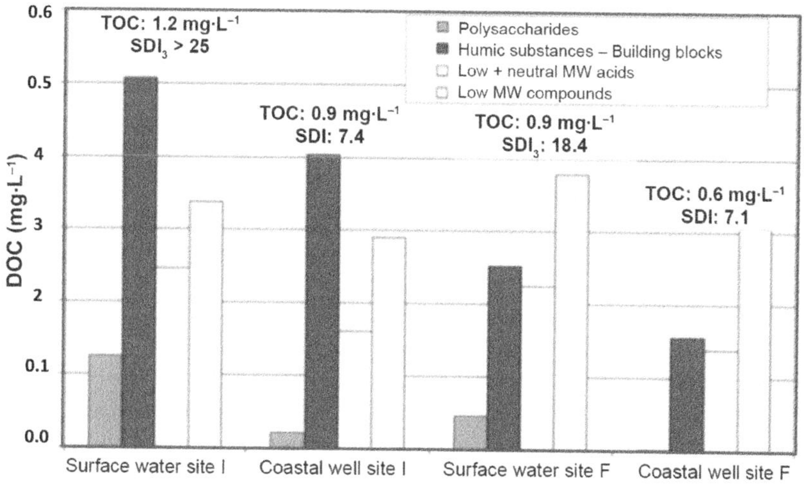

Figure 19.16. *TOC and SDI at two sites, with seawater and coastal well comparison. For a color version of this figure, see www.iste.co.uk/gaid/watertreatment4.zip*

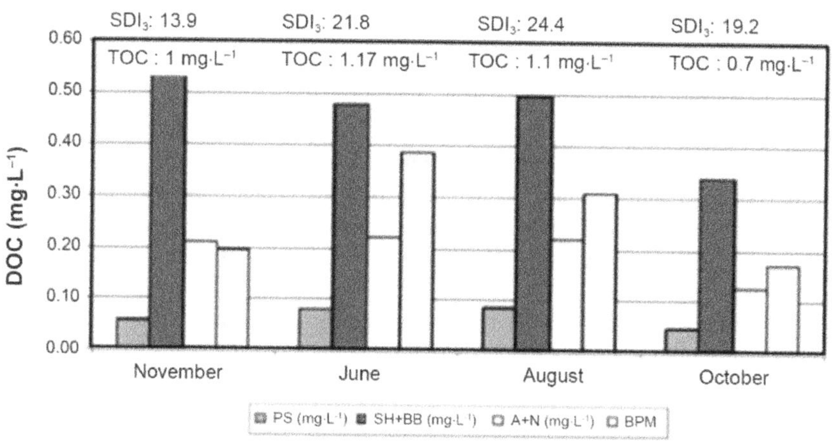

Figure 19.17. *Seasonal variations of SDI and TOC in the Mediterranean Sea. For a color version of this figure, see www.iste.co.uk/gaid/watertreatment4.zip*

19.1.3.1.1. Algae and bacteria

Algae

Micro-algae are photosynthetic eukaryotic micro-organisms, living in a humid environment and grouping several thousands of species. Generally speaking, autotrophic in nature, these unicellular species fix the surrounding inorganic carbon to transform it into organic matter (conversion of solar energy into chemical energy) in the euphotic zone; in other words, the layer of water on the surface through which the sun's light passes. This zone is very variable in depth, depending on the turbidity (from a few centimeters to more than 100 m deep).

The dominant groups of marine phytoplankton are the diatoms (*Bacillariophyceae* or *Diatomophyceae*, nearly 1,500 species) and the dinoflagellates (nearly 2,000 species). Less than 80 recorded species are toxin producers, and 90% belong to the dinoflagellate group. The most represented groups of toxic dinoflagellates belong to the genera *Alexandrium*, *Dinophysis*, *Gymnodinium* and *Prorocentrum*.

Algae, dinoflagellates and cyanobacteria represent a large group of photosynthetic organisms found in the seas and oceans. Algal and cyanobacterial cells contain chlorophyll and other photosynthetic pigments. They exist in various forms ranging from single cells and/or chains of cells to complex multicellular algae. Although most algae are microscopic (ranging from 2 to 100 µm), some are macroscopic and can grow to over 3 m in length. Planktonic organisms are classified by size, from femtoplankton (less than 0.2 µm) and picoplankton (0.2–2 µm) to mega-plankton (0.2–2 mm). Phytoplankton consists of organisms ranging from bacteria to diatoms and large dinoflagellates (such as the sea sparkler, *Noctiluca scintillans*). Their biomass can be estimated by measuring their chlorophyll (green pigment) using light-based methods. However, other pigments (brown, red) are also common and the amount of chlorophyll is only a small part of the biomass.

The best-known algae are the red, brown and green algae, which are part of a group of multicellular algae called macroalgae. However, the majority of algae and cyanobacteria are single-celled species that float freely in the water column; they are invisible to the naked eye and collectively form a group called phytoplankton. Excessive phytoplankton growth can occur in coastal seawater and estuaries, generally giving the water a red or brown color (Figure 19.18), due to the density and number of algae.

In the case of microalgae, and more particularly non-toxic dinoflagellates (e.g. *Heterocapsa triquetra*), red tides exist because of the pigments they contain. Out of more than 5,000 species of microalgae identified, 300 would be likely to form these red tides and 70 would produce toxins.

Red tide is a common name for a phenomenon where certain species of phytoplankton contain reddish pigments and "bloom", making the water appear red. A red tide is a complex phenomenon, involving many species with different characteristics and covering large areas. These species disrupt ecosystems, causing large-scale environmental damage. Most red tides cause large-scale fish deaths that end up on the shoreline, inducing bad odors on beaches due to decomposition.

Figure 19.18. *Algal bloom in different seawater. For a color version of this figure, see www.iste.co.uk/gaid/watertreatment4.zip*

Red tide events often occur during the spring-summer period and result in increased algae content (turbidity in the intake can reach up to 15 NTU), increased organic matter concentration (TOC increases to 4–5 $mg \cdot L^{-1}$) as well as apparent color. One or more red tide events may occur per year and each period may last 6–8 weeks. Red tide is a common name for harmful algal blooms (HABs). Algal blooms have become common in the Persian Gulf, but their intensity varies seasonally. The recent red tide in 2006 and 2008 in the Persian Gulf caused significant environmental damage and economic losses in the Gulf countries as well as in Iran, Iraq and Pakistan.

According to some forecasts, similar red tides could occur frequently in the future in the Persian Gulf. It is therefore a serious problem and a threat to public health, coastal fishing, industry, aquaculture and tourism. It is not caused by a single organism, although some are more common than others. Many of these species are regional and are quite adapted to local conditions. The two main toxic species are represented by certain phytoplankton that mainly produce chemical toxins that are harmful to fisheries, and a group of dinoflagellates that mainly produce neurotoxins that are harmful to humans and marine mammals. Most of the HABs from 1988 to 2008 in Oman were caused by some type of dinoflagellates.

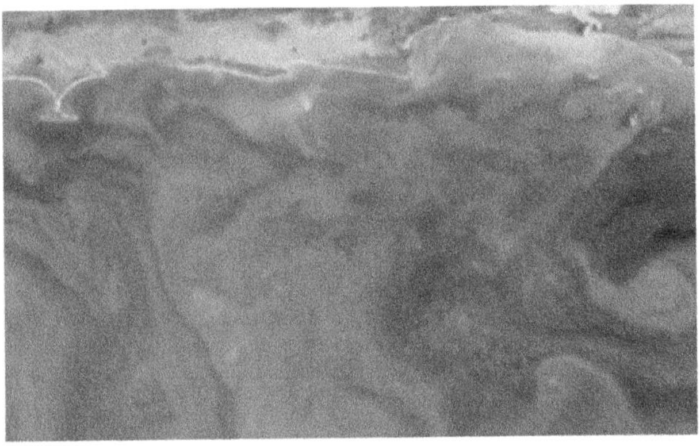

Figure 19.19. *Algal bloom on the coast of the Red Sea. For a color version of this figure, see www.iste.co.uk/gaid/watertreatment4.zip*

Many algal blooms are the result of excess nutrients (especially phosphorus and nitrogen) in the water. As they die, the organic matter becomes food for bacteria that grow using the dissolved oxygen in the water. When this oxygen decreases, many fish and aquatic insects cannot survive. This is the phenomenon of eutrophication.

Some species of algae produce neurotoxins that cause severe biological impacts on wildlife. Algal blooms known to naturally produce biotoxins are often referred to as "harmful algal blooms".

Seawater desalination plants, power plants and other plants that use seawater for cooling purposes were forced to shut down during the last red tide in the Persian Gulf region in order to avoid pre-treatment operating constraints as well as membrane fouling problems.

Results obtained (VERI-Veolia) on seawater samples have shown that bacterial and phytoplankton counts are related to the concentration of polysaccharides, organic clogging compounds and also with the SDI of the seawater before and after pre-treatment.

Other parameters such as turbidity do not show any correlation with the clogging potential of the pre-treated water. Biological clogging due to the attachment and growth of bacteria on the membranes is one of the main reasons for the malfunctioning of the membrane plants, especially since the presence of polysaccharides in the pre-treated water increases the risk of clogging, because these organic compounds are very likely to be adsorbed on the membrane surfaces and then used as nutrients by the bacteria.

Parameters	Analytical methods
Picophytoplankton	Cytometry
Bacteria	Cytometry
Algae pigments: chlorophyll and pheophytin	Fluorimetry

Table 19.7. *Analytical methods of some parameters of algae and bacteria*

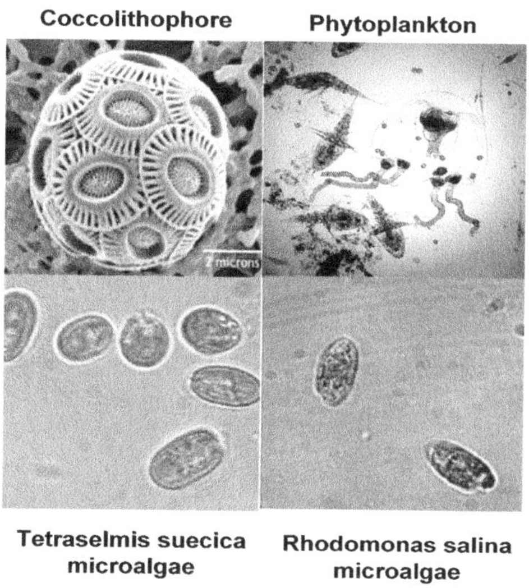

Figure 19.20. *Types of marine algae. For a color version of this figure, see www.iste.co.uk/gaid/watertreatment4.zip*

Picophytoplankton species correspond to very small phytoplankton species that should be monitored, because they are not simple to remove by conventional pre-treatments (Figure 19.20), whereas phytoplankton species larger than 100 μm are well removed by pre-treatments.

Figure 19.21 shows the concentrations of phytoplankton species at various desalination sites (Veolia). It appears that:

– the seawater of the Persian Gulf has much higher algal activity than that of the Mediterranean Sea;

– the positive impact of the coastal well is once again demonstrated, as the concentrations of phytoplankton species are very low or even non-existent.

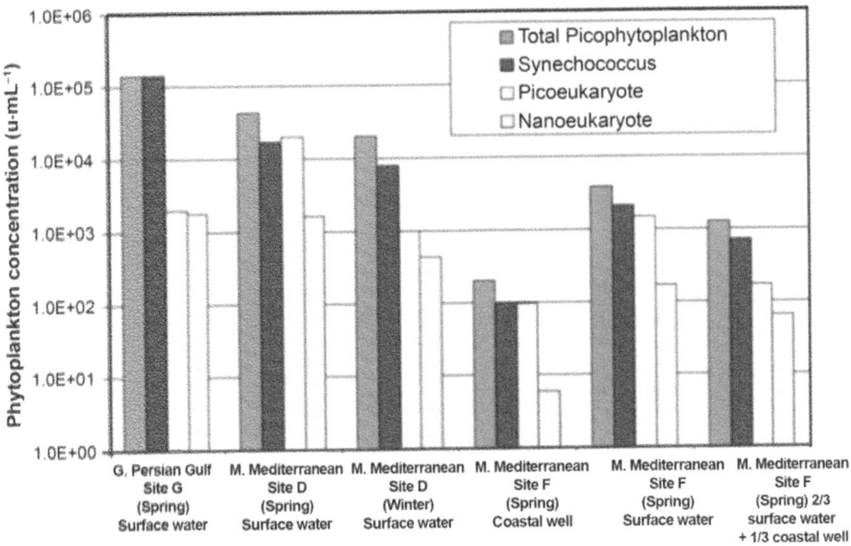

Figure 19.21. *Phytoplankton concentrations in different seawater. For a color version of this figure, see www.iste.co.uk/gaid/watertreatment4.zip*

Bacteria

Figure 19.22 shows a profile of bacterial counts, taken over several months. The bacteria content of the raw seawater in the summer (June and August samples) is higher than that of the sample taken in the fall (October sample). However, beyond the simple comparison of the total number of bacteria, the variation in distribution between active and dormant bacteria is very interesting. Indeed, the ratio of active to dormant bacteria seems to be reversed between the summer and fall samples, with the bacteria being more active in summer. This observation may have a large impact on membrane clogging kinetics. Of course, this observation may be related to higher sun exposure and higher water temperature in the summer (about 30°C in the summer compared to 15°C in the fall).

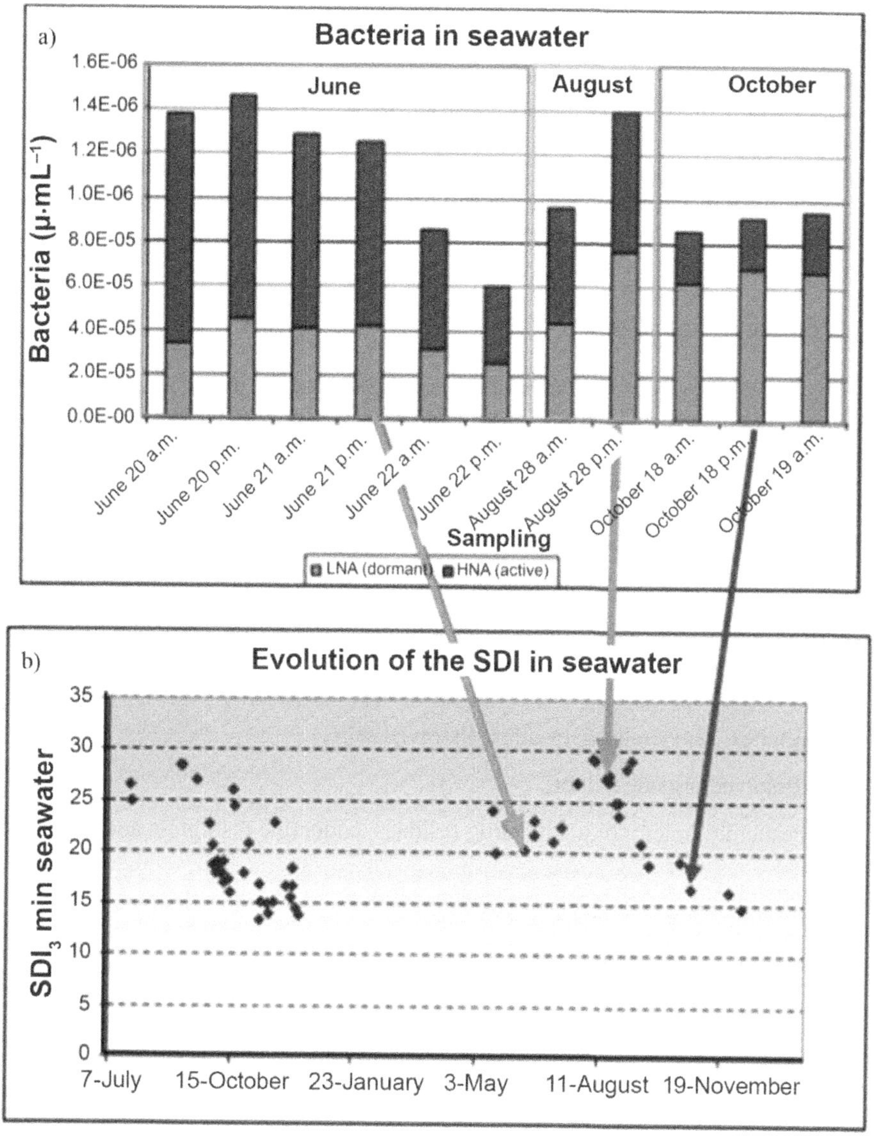

Figure 19.22. *a) Bacterial count and b) SDI evolution. For a color version of this figure, see www.iste.co.uk/gaid/watertreatment4.zip*

19.1.3.2. *Hydrocarbons and industrial chemicals*

Concentrations of hydrocarbons and industrial chemicals are often detected in seawater with concentration ranges up to 10 mg·L^{-1}. These substances can lead to irreversible clogging of the membranes, or even their deterioration. Flotation appears to be the most effective treatment for hydrocarbons and refining with activated carbon is sometimes used to maintain acceptable levels (when concentrations are too high), and thus limit the risks.

19.2. Fields of application

Desalination by RO is a technology applicable for the removal of salts from seawater (TDS between 15 and 45 g·L^{-1}) and brackish water (TDS between 2 and 15 g·L^{-1}). It is applied to fresh waters (TDS $<$ 2 g·L^{-1}) for the removal of organic substances, micropollutants, heavy metals, hardness, nitrates, etc.

The treatment objectives sought are as follows:

– the reduction of water salinity and other dissolved salts;

– the reduction of certain elements such as:

- boron (mainly for seawater or iron ore mine drainage water),

- nitrates, heavy metals, or other dissolved salts,

- dissolved organic matter,

- pesticides, micropollutants, drug residues, endocrine disruptors and industrial products.

It is a refining process that must be supplied with water of good physical quality, which usually requires a pre-treatment stage.

19.3. Operating principle of RO

When two solutions, one diluted and the other concentrated, are separated by a membrane that is permeable to water and ion, a homogeneous mixture is naturally created on both sides of the membrane.

When these two solutions are separated with a semi-permeable membrane, that is to say, one that is permeable to water but not to ions (as is the case with membranes used in RO), there is a natural flow of water from the less concentrated compartment to the more concentrated one.

There is therefore a dilution effect within the concentrated compartment, with an increased water level in the latter.

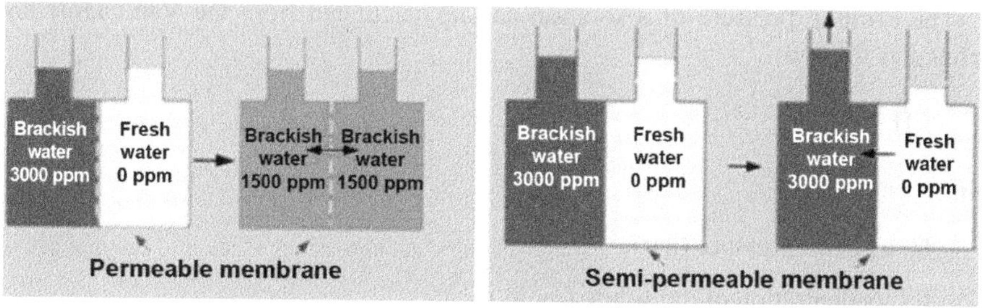

Figure 19.23. *Principle of reverse osmosis 1 and 2. For a color version of this figure, see www.iste.co.uk/gaid/watertreatment4.zip*

This flow will continue until it reaches a height of elevation H, corresponding to the value of the osmotic pressure of the concentrated compartment (Figure 19.23).

The principle of RO consists of reversing the direction of the natural flow, by applying pressure on the side of the most concentrated compartment (Figure 19.24).

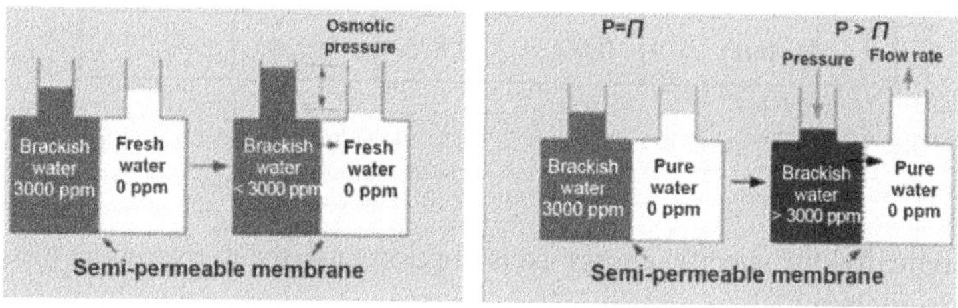

Figures 19.24. *Principle of reverse osmosis 3 and 4. For a color version of this figure, see www.iste.co.uk/gaid/watertreatment4.zip*

To make this reversal possible, the applied pressure Pa must be:

$$Pa > \Pi + dP + dH$$

with:

– Π: osmotic pressure;

– dP: pressure drop across the membrane;

– dH: difference of level between the two compartments corresponding to the required water return.

The osmotic pressure of a solution can be calculated from the Van't Hoff law defined as follows:

$$\Pi = CRT$$

with:

– Π: osmotic pressure (bar);

– C: concentration of the solute ($mole \cdot m^{-3}$);

– R: perfect gas constant ($8.314\ J \cdot mol^{-1} \cdot K^{-1}$);

– T: temperature of the solute–solvent mixture (K).

Examples of osmotic pressure values for NaCl at 35°C:

$$\Pi = 2CRT/100$$

Concentrations of NaCl solution ($mg \cdot L^{-1}$)	100	1,000	35,000	50,000
Π (bar)	0.0087	0.875	30.64	43.77

Table 19.8. *Osmotic pressure and concentration of the saline solution*

Theoretically, the specific energy consumption required to overcome the osmotic pressure is given by:

For 35°C and 35,000 $mg \cdot L^{-1}$, $\Pi = 30.64$ bars

So:

$$3{,}064\ kPa \text{ or } 3{,}064\ kN \cdot m^{-2} \times 1\ m^3 = 3{,}064\ kN \cdot m^{-1} = 3{,}064\ kJ$$

Or:

$$3{,}064 \times 2.778 \times 10^{-4}\ kWh \cdot m^{-3} = 0.85\ kWh \cdot m^{-3}$$

The osmotic pressure achieved with the total concentration of dissolved salts (TDS in mg·L^{-1}) can be obtained with one of the proposed relations, which is the following:

$$\Pi(bars) = 0.00264 \text{ TDS } (T + 273)/(1{,}000 - \frac{TDS}{1{,}000})$$

where T is the temperature in °C.

Thus, at 35°C and a TDS concentration of 37,000 mg·L^{-1}, the osmotic pressure is equal to 31.24 bars.

19.4. The membranes used in desalination

The membranes used in desalination (seawater and brackish water) are membranes with a porous structure and an MWCO of 50–90 kDa. They are often made of polyamide with a polysulfone support. Some made of cellulose acetate still exist, proposed by suppliers like Toyobo (Japan).

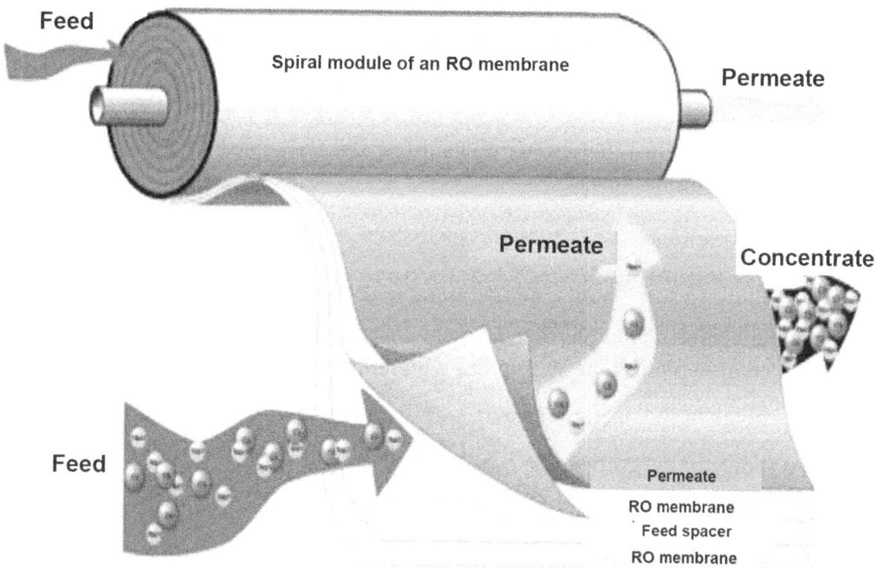

Figure 19.25. *Separation mode in RO membrane. For a color version of this figure, see www.iste.co.uk/gaid/watertreatment4.zip*

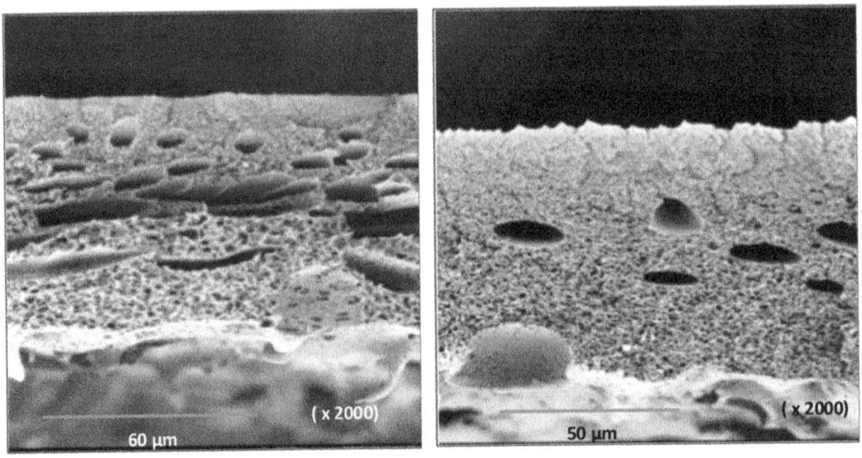

a) b)

Figure 19.26. *Electron microscope view of a membrane section a) SW30-XLE (DOW Filmtec) and b) BW 30 (DOW)*

The manufacturing and separation process in a RO membrane is similar to that described for a nanofiltration or low-pressure RO membrane (see Volume 4, section 18.1).

Figure 19.26 shows the porous polysulfone structure used as a support and the polyamide layer on the surface.

19.5. Sizing parameters

The knowledge of the physicochemical characteristics of the water to be treated is fundamental in SWRO, because it has a direct impact on:

– the choice of the operating flow;

– the choice of the conversion rate;

– the operating pressure range of the osmosis unit;

– the requirement to dose an antiscalant;

– the range of water quality produced by the osmosis unit.

The quality of the water to be produced defines the type of membrane to be used, as well as the configuration of the osmosis unit (one or two passes). Particular attention should be paid to the Boron parameter (see section 19.11.1).

The material transfer model that is applicable in RO is that of solubilization-diffusion.

Two equations govern the transfer of solvent (water) and solutes (ions and molecules).

19.5.1. *Flow*

As the salt concentration in the feed water increases, the required pressure also increases. If this pressure is kept constant, an increase in the salt concentration leads to a decrease in the flow. The increase in osmotic pressure, due to the increase in salt concentration, compensates for the required pressure.

The higher the flow ($L \cdot m^{-2} \cdot h^{-1}$) is, the higher the power consumption. The choice of the flow value has a significance with regard to the amount of investments, but also on the quality of the permeate, as a low flow corresponds to a low feeding pressure, and thus a higher salt passage. Moreover, the increase in the flow has a direct effect on the life span of the membranes and on the frequency of chemical cleaning:

$$Jw = Kw\,(TMP - \Delta\Pi)$$

And:

$$TMP = \frac{(Pa+Pc)}{2} - Pp$$

with:

– Jw: solvent flow ($L \cdot m^{-2} \cdot h^{-1}$);

– Kw: membrane permeability to solvent ($L \cdot m^{-2} \cdot h^{-1} \cdot bar^{-1}$);

– TMP: transmembrane pressure (bar);

– $\Delta\Pi$: osmotic pressure differential across the membrane;

– Pa: feed pressure (bar);

– Pc: concentrate pressure (bar);

– Pp: permeate pressure (bar).

The term (TMP -ΔΠ) is also referred to as "Net Driving Pressure".

Solute transfer:

$$Ji = Ks\ \Delta C$$

with:

– Ji: solute flow ($mg \cdot m^{-2} \cdot h^{-1}$);

– ΔC: solute concentration differential across the membrane ($g \cdot m^{-3}$);

– Ks: membrane permeability coefficient toward salts.

19.5.2. *Concentration polarization*

Due to the existence of a laminar hydraulic regime in the vicinity of the membrane (boundary layer theory), the solute concentration in the vicinity of this membrane is higher than the average solute concentration in the circuit. The polarization factor is then defined as follows:

$$\beta = \frac{Cm}{Co}$$

with:

– β: polarization factor;

– Cm: solute concentration near the membrane;

– Co: average solute concentration in the circuit.

The concentration polarization factor directly impacts the solute flow through the membrane, and thus its rejection.

The salinity of the permeate is given by: Cp ($mg \cdot L^{-1}$) = Ji/Jw.

19.5.3. *Conversion rate*

The higher the conversion rate, the lower the supply rate. However, this decrease does not compensate for the increase in feed pressure, and thus power consumption. This is because it leads to an increase in the salt concentration rate, and thus in the osmotic pressure.

The conversion rate affects the salt passage and the permeate water flow. Its increase also results in a higher concentration of salts in the concentrate.

Economically speaking, the conversion rate must be as high as possible, although it is technically limited by four main constraints:

– the maximum operating pressure of the RO modules;

– the risk of salt precipitation;

– the minimum sweep rate of the modules in the concentrate circuit;

– the maximum concentration bias value (β factor).

The conversion rate of a module is typically between 10 and 20%, depending on its position in the pressure tube.

The conversion rate (Y) of a module is its hydraulic efficiency. It is defined as follows:

$$Y(\%) = 100\frac{Qp}{Qin}$$

with:

– Y: conversion rate (%);

– Qp: permeate flow rate ($m^3 \cdot h^{-1}$);

– Qin: feed flow rate ($m^3 \cdot h^{-1}$).

In order to prevent salts from precipitating, the filtration is always tangential. Therefore, there are three water circuits:

– the raw water circuit known as the feed;

– the treated water circuit known as the permeate;

– the discharge circuit known as the concentrate.

19.5.4. *Passage rate and rejection rate in salts*

The salt passage rate is defined as the ratio of the salt concentration in the permeate to the (average) concentration in the membrane feed water:

SP = 100 × (Cp/Cfeed)

with:

– SP: salt passage rate (%);

– Cp and Cfeed: salt concentrations in the permeate and feed water of the membranes, respectively.

The rejection rate is then obtained by:

$$SR = 100 - SP \text{ or even } SP = 100 \times (Cfeed - Cp)/Cfeed$$

19.5.5. *Influence of the temperature*

The permeate flow rate increases with increasing water temperature, because the viscosity decreases and the diffusion coefficient of the water through the membrane increases.

However, the increase in water temperature leads to a lower rejection of salts.

The variation of the permeate flow rate with temperature is expressed using a temperature correction factor (TCF):

$$TCF = \text{Exp}\left(K \times \left(\left(1/(273 + T) - 1/298\right)\right)\right)$$

With:

– TCF: temperature correction factor;

– K: characteristic linked to the membrane material;

– T: water temperature (°C).

Taking a temperature of 25°C as a reference, which gives TCF = 1, we obtain a variation of the permeate flow rate of 3% per degree.

Alternatively, the can also be expressed as follows:

$$TCF = a^{(T-25)}$$

where a is a coefficient between 1.024 and 1.030.

The flow rate of a membrane with a nominal flow rate Q_{nom} at 25°C is equal to a temperature T (°C):

$$Q(T) = TCF(T) \times Q_{nom}$$

The same correction is made for the flow. Table 19.9 shows the expected increase in flow with an increase in temperature using 25°C as a reference (with a = 1.030).

Temperature (°C)	TCF	Increasing of flux (%)
30	1.160	16
35	1.344	34.4
40	1.550	55

Table 19.9. *Increase of the flow with temperature*

19.5.6. *Determining the number of modules and pressure tubes*

The average hydraulic flow for the sizing of the osmosis units depends on the quality of the feed water to the membranes.

The number of modules to be implemented is defined as follows:

$$N\ modules = 1{,}000\frac{Qp}{Jw\ Su}$$

with:

– N modules: number of modules;

– Qp: permeate flow rate ($m^3 \cdot h^{-1}$);

– Jw: average hydraulic flow of sizing ($l \cdot m^{-2} \cdot h^{-1}$);

– Su: unit area of the modules (m^2).

The number of pressure tubes to be implemented is then:

$$N\ tubes = N\ modules\frac{1}{n}$$

with:

– N tubes: number of pressure tubes;

– N modules: number of modules;

– n: number of modules per tube.

The number of modules per pressure tube depends on the conversion rate chosen. The modules are distributed in the different stages.

19.6. Implementation

19.6.1. *Membranes*

The description of the membranes is largely detailed in section 18. In contrast to microporous ultrafiltration (UF) or MF membranes, RO and nanofiltration membranes are known to be "dense". They have a very different and tighter porous structure. The membrane is made up of several layers of polymers: it is called a composite membrane (or sheet).

The dimensions of these modules are very similar from one supplier to another. The dimensions of the SWC3+ module are given in Table 19.10, as an indication.

Module reference	Length A inches (mm)	Diameter B inches (mm)	Internal diameter C permeate tube inches (mm)	Weight (kg)
SWC3+ (hydranautics)	40 in (1,016)	7.95 in (201.9)	1.125 in (28.6)	36 lbs (16.4)

Table 19.10. *Dimensions of the SWC3+ membrane*

There are also standard 4 and 2.5 inch diameter modules that are used on small capacity installations or pilot units.

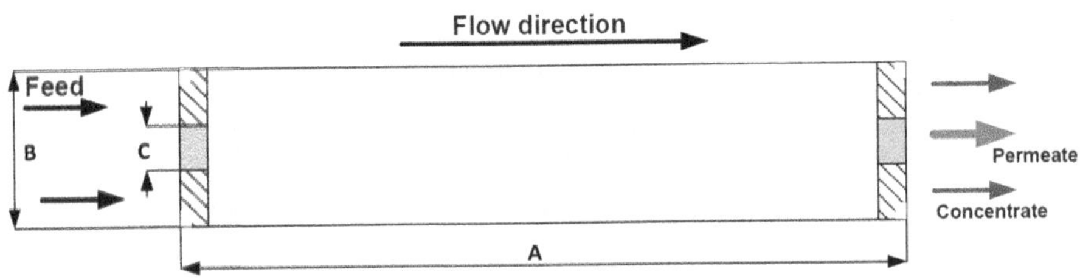

Figure 19.27. *Standard hydranautics module. For a color version of this figure, see www.iste.co.uk/gaid/watertreatment4.zip*

19.6.2. *Pressure tube*

The spiral modules are connected to each other in pressure tubes, with a maximum of one to eight modules per tube. The conversion rate of a pressure tube depends on the number of modules placed in series.

The hydraulic principle of the pressure tube consists of feeding the module placed in position (N + 1) with the concentrate of the module placed in position N, the permeate of each module being collected in the central permeate collecting tube.

The permeate circuits of each module are hydraulically connected to each other by means of interconnectors. The modules located at the ends of the pressure tube are connected to the pressure tube via an end connector. There are two main families of pressure tubes:

– "front port" pressure tubes: all connections (Figure 19.28) are at the front (feed, permeate and concentrate);

– side port pressure tubes: the supply and concentrate connections are lateral, and the permeate connection(s) is (are) at the front (Figure 19.29).

Due to the very similar dimensions of the modules from one supplier to another, the same pressure tube can receive modules from different suppliers.

The pressure tubes assembled in parallel establish a stage that have a conversion rate between 40 and 47% for seawater.

Two parameters will limit the choice of the conversion rate of the installation (and therefore the number of stages): the salinity of the water to be treated, on the one hand, and the risk of salt precipitation, on the other.

Figure 19.28. *Pressure tube with front connection. For a color version of this figure, see www.iste.co.uk/gaid/watertreatment4.zip*

Figure 19.29. *Pressure tube with lateral connection. For a color version of this figure, see www.iste.co.uk/gaid/watertreatment4.zip*

19.6.3. *Pass*

A pass is an osmosis unit. In seawater, the salinity of the permeate from a first pass may be higher than required. In this case, the permeate undergoes a second pass through a second osmosis unit. A system with two osmosis units in series is also known as a double pass.

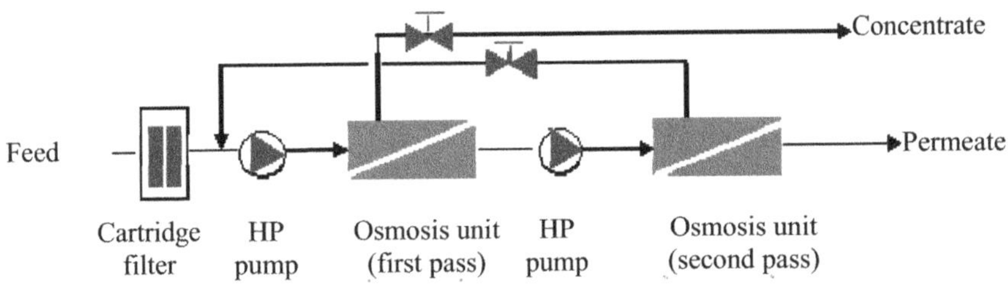

Figure 19.30. *Diagram of a membrane plant with a double pass*

In some cases, the quality of the permeate may require three or four passes.

19.7. Pre-treatment

19.7.1. *Pre-treatment selection*

The characterization of the seawater, which indicates the main parameters that are removed during the pre-treatment such as Fe, Mn, natural organic matter, SDI, bacteria and algae, is necessary for several reasons. It makes it possible to:

– acquire good knowledge of seawater quality and its seasonal variation;

– use predictive tools to assess the impact of the variation of the different parameters on the pre-treatment;

– acquire additional indicators, compared to conventional parameters, in order to quantify the risks of membrane fouling.

Parameters to be removed or reduced during pre-treatment	**Recommended limits for RO membrane feed**
Iron	<50 μg·L^{-1}
Manganese	<50 μg·L^{-1}
Aluminum	<50 μg·L^{-1}
Turbidity	<0.2 NTU
Suspended matter	<1 mg·L^{-1}
SDI	<4 (95% of the time)
Algae	<100 u·L^{-1}
Chlorophyll	<1 μg·L^{-1}
Dissolved organic carbon	<2 mg·L^{-1}
Hydrocarbons	<0.1 mg·L^{-1}

Table 19.11. *Recommended limits for RO membrane feed*

The absence of residual oxidant (except monochloramine) and chlorinated solvents must be ensured.

Before water is fed to the membranes, it is important that undesirable substances or elements are removed or reduced to acceptable levels. Without proper pre-treatment, desalination facilities are destined to have short life spans, shortened operating periods and require high maintenance.

Once the physicochemical and bacteriological composition of the water is known, the type of pre-treatment is selected to produce water of acceptable quality for the membranes.

One of the most important factors in the successful operation of an RO desalination plant is the pre-treatment system's ability to reliably and stably produce water that is relatively free of particles of various kinds.

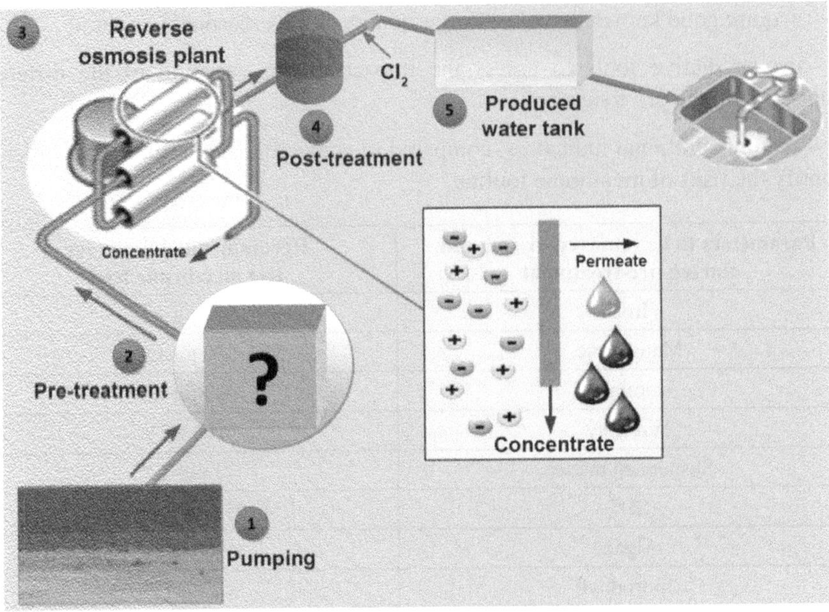

Figure 19.31. *Which pre-treatment should be chosen to optimize the operation of the RO plant? For a color version of this figure, see www.iste.co.uk/gaid/watertreatment4.zip*

This first stage is therefore essential, because it directly impacts the operation of the membranes through operating costs (chemical reagents for cleaning, higher feed pressure and reduced membrane life). In addition, a decrease in the quality of the water produced, as well as its quantity, can be impacted.

Various technologies can be implemented alone or combined with other processes in series, depending on the quality of the seawater.

The purpose of pre-treatment is to:

– remove turbidity and SS;

– remove algae and other similar species;

– reduce DOC;

– remove hydrocarbons and industrial chemicals;

– remove elements such as iron and manganese;

– condition pre-filtered water to optimize membrane performance.

	Turbidity NTU	TOC $mg \cdot L^{-1}$	Algae $u \cdot L^{-1}$	SDI	Hydrocarbons $mg \cdot L^{-1}$	Bacteriology	Pre-treatment
Seawater							
Coastal wells	<0.2	<1	Absence	SDI_{15} < 2	Absence	Low	Not necessary Fe and Mn to be checked
Taken directly	1–10	2-5	Possible presence with seasonal blooms	10→33	0.1–2	Medium to abundant	Necessary One to two stages
Affected by estuaries	10–100	>>5	Possible presence with seasonal blooms	33	0.1–2	Abundant	Necessary One to three stages
Brackish water							
Groundwater	2–10	<1	Absence	SDI_{15} < 2	Absence	Low	Necessary One to two stages Presence of Fe, Mn, Al, H_2S T > 45°C
Surface water	5–30	2–6	Possible	15–30	–	Medium to abundant	Pre-treatment of conventional surface water

Table 19.12. *General characteristics of seawater and brackish water*

19.7.2. *Pre-treatment systems*

To achieve the quality objectives upstream of the membranes, a large number of treatment processes that use various physical and chemical mechanisms can be used. Each unitary stage plays an important role, and the role of the design engineer is to select and design the most suitable process, while considering the characteristics of the water and its variations over the past years. Even the smallest event must be considered.

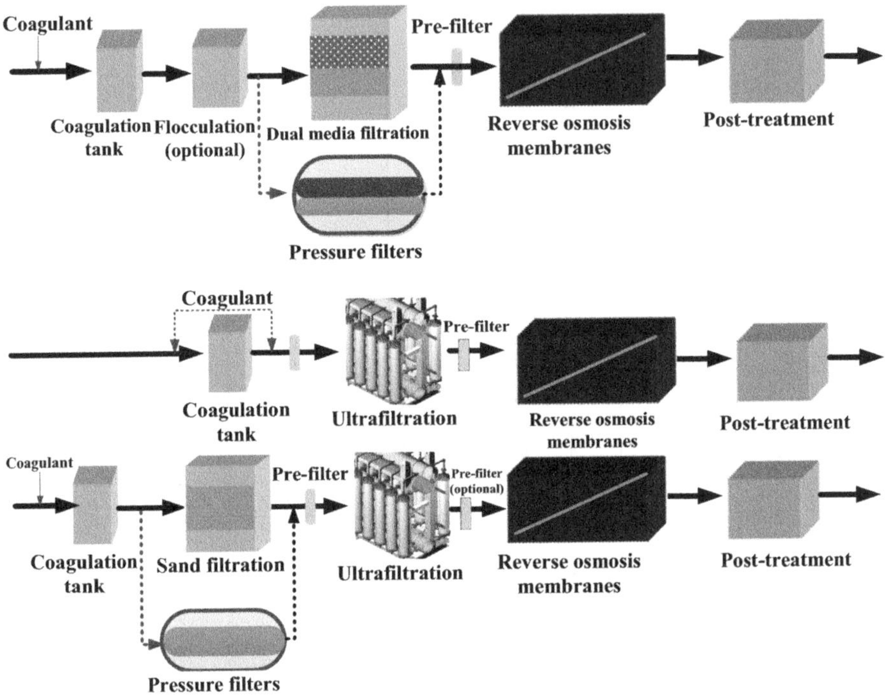

Figure 19.32. *Treatment of low turbidity water with filtration and/or direct membranes. For a color version of this figure, see www.iste.co.uk/gaid/watertreatment4.zip*

19.7.2.1. *Low turbidity surface water (<10 NTU), SDI < 25, TOC < 2 mg·L^{-1}, no algae, no hydrocarbons*

For water quality in which only the removal of microparticles (turbidity and SDI) is sought, media filtration or membrane technology (submerged or pressure internal-external or external-internal filtration) can be proposed. Coagulation and an optional flocculation (if the temperature drops below 15°C) must be provided with media filtration. This can be a gravity or pressure type of filtration. A micro-coagulation upstream of the membranes (addition of a very low dosage of coagulant) can be provided in order to improve the filterability of the water. The dosage of coagulant is low (about 0.1–0.3 ppm metal).

19.7.2.2. *Moderately turbid water (<15 NTU), SDI > 30, TOC < 3 mg·L^{-1}, few algae, no hydrocarbons*

Several solutions are possible:

– a two-stage filtration in series, comprising a dual-media filtration followed by a mono-media filtration. The coagulant is injected upstream of the first stage and a micro-coagulation is injected on the second stage filter;

– high-speed filtration (15–20 m·h^{-1}) on a dual-media filter that has a total height of 3 m. It is preceded by a mechanical filtration at 50 μm;

– sand filtration with addition of coagulant upstream and UF downstream of the sand filter.

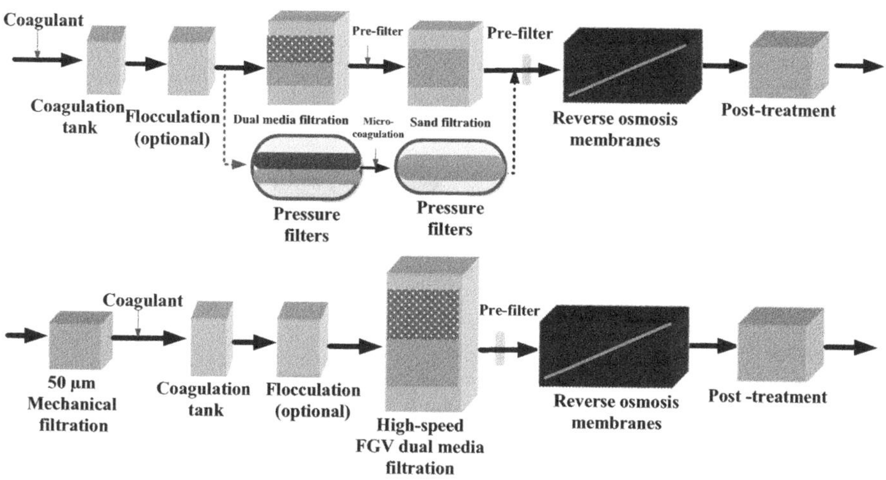

Figure 19.33. *Double stage filtration in series. For a color version of this figure, see www.iste.co.uk/gaid/watertreatment4.zip*

As the water is not very turbid, it is possible to use both submerged membranes (external–internal filtration) and pressurized membranes with internal–external filtration.

19.7.2.3. *Moderately turbid water (<30 NTU), highly loaded with algae (>5–200 million·L^{-1}), TOC > 3 mg·L^{-1}, SDI > 33.3, possible hydrocarbons (1 mg·L^{-1})*

In this case, flotation (Spidflow®) followed by dual-media filtration is recommended. An alternative is the Spidflow® Filter (a combination of Spidflow® and dual-media filtration in a single structure).

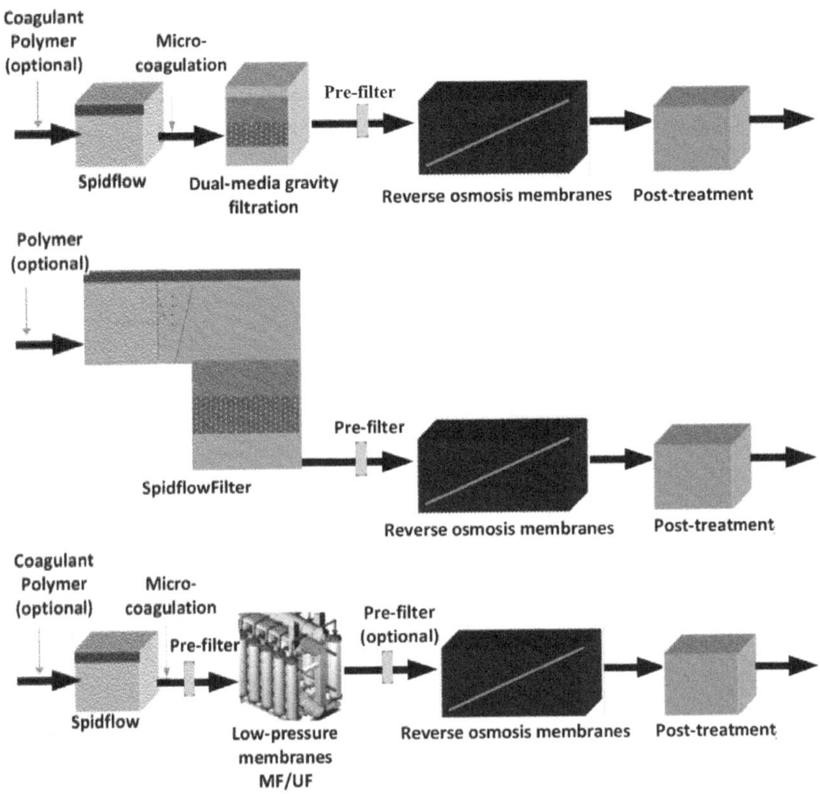

Figure 19.34. *Different possibilities for algae and TOC-loaded waters. For a color version of this figure, see www.iste.co.uk/gaid/watertreatment4.zip*

Lastly, flotation followed by UF (Spidflow® UF) is possible if the specifications require an SDI at the inlet of the RO membrane pre-filters that is permanently below 3.2.

19.7.2.4. *Highly turbid water (>50 NTU with peaks at 100 NTU), SDI > 33.3, TOC > 3 mg·L^{-1}, algae (5–20 million·L^{-1})*

The processes implemented for this type of water quality involve coagulation–flocculation–settling because of the high turbidity. A dual-media filtration is installed after the settler (Actiflo or Multiflo). With a Multiflo, a UF unit can replace the dual-media filtration.

In summary, pre-treatment must be tailored to address the potential for membrane clogging based on the characteristics of the seawater (Table 19.13).

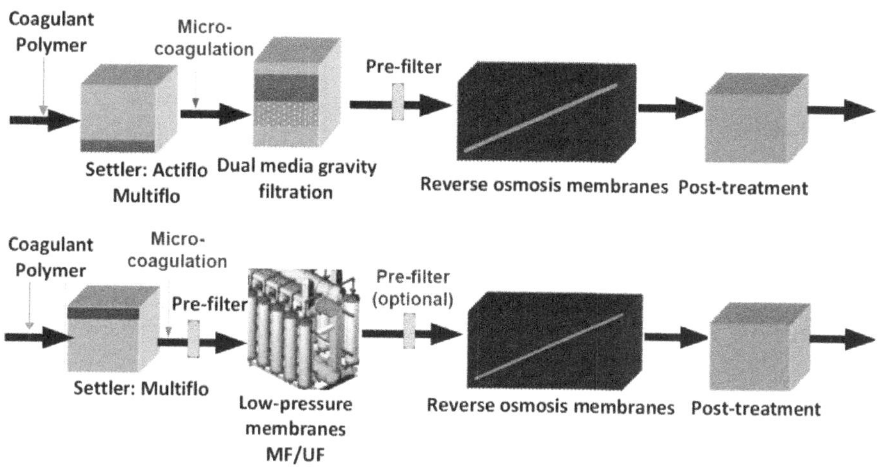

Figure 19.35. *Filtering system including settling upstream of a dual-media filtration or a MF/UF. For a color version of this figure, see www.iste.co.uk/gaid/watertreatment4.zip*

Clogging potential	Causes	Pre-treatments
Biologicals	Micro-organisms Bacteria Viruses Protozoa	Shock pre-chlorination Coagulation–flocculation–filtration Microfiltration–ultrafiltration (with coagulant)
Particles	Sand, clays (high turbidities and SS)	Conventional filtration Coagulation–flocculation–settling–filtration
Colloids	Organic and mineral complexes Colloidal particles Microalgae Phytoplankton and others	Coagulation–flocculation–flotation–filtration Coagulation–flocculation–flotation–ultrafiltration Coagulation–flocculation–filtration–ultrafiltration
Organic	Natural organic substances Biopolymers	Coagulation–flocculation–filtration Coagulation–flocculation–filtration (one or two stages)
Mineral	Calcium, magnesium, barium, sulfates, strontium, carbonates	Acidification Antiscalant

Table 19.13. *Different pre-treatments adapted according to the clogging potential*

19.8. Pre-chlorination

19.8.1. *Pre-chlorination and development of micro- and macro-organisms*

The addition of chemical oxidants, such as chlorine or ozone, can create a pre-disinfection of the pre-filtered water.

Pre-chlorination is very often practiced on desalination sites. Unfortunately, it involves many disadvantages: the reaction of chlorine with organic substances produces short-chain substances, part of which can be removed biologically during filtration, and the other part is deposited on the membranes and is fixed there. These substances then become a source of nutrients for the bacteria and promote the creation of biofilm.

The first stage of fouling formation is uncontrolled growth of microbial organisms on surfaces, with preliminary slime formation, which results in a biofilm produced by living cells and their metabolic by-products. Biofilm formation occurs in three stages:

– a preliminary covering of the surface by bacteria and other organisms present in the water;

– a transition stage with several layers of cells that integrate into their own polymer material;

– a final mature biofilm development with high population density.

These short-chain organic substances are hardly removed by coagulation–flocculation (see Volume 2, section 7.4), and therefore end up in the filtered water if no biological growth is present on the filter media. On the other hand, natural substances (HS) are well removed by coagulation–flocculation.

The pre-chlorination of organic substances generates halogenated compounds, such as trihalomethanes (THM), which are also a source of nutrients for bacteria.

In all cases, the use of chlorine must be carefully monitored to keep the chlorine concentration below 0.1 $mg \cdot L^{-1}$ of free residual chlorine, at the risk of irreversible damage to the membranes. The addition of a bisulfite reagent injected upstream of the pre-filters allows the residual chlorine to be reduced to zero. The redox potential is measured upstream and downstream of the pre-filters, so as to ensure that the chlorine is removed.

Despite its pre-disinfection power, there are still enough bacteria in the filtered water to grow on the membrane surfaces, especially if organic carbon is present in the water.

The development of marine organisms in seawater pumping equipment and structures is common, especially since this type of bio-clogging is dynamic and evolves according to seasonal variations in water quality parameters, such as nutrient loading, wastewater inflow, algal blooms and so on.

A majority of these marine organisms are cirripides (or *Cirripedia*), which are a class of marine animal belonging to the crustacean sub-branch. *Cirripedia* are crustaceans, which include lepadomorphs, balanomorphs (such as barnacles) and rhizocephalan parasites (such as the sacculine, *Sacculina carcini*, a parasite of the *Carcinus maenas* green crab). Their larvae attach themselves by the antennae to any support and begin a complete metamorphosis, which produces the very particular adult morphologies of this group. The animals thus fixed will never move again. When they are not parasitic, the adults can be pedunculated or operculated and sessile. Cirripedes are equipped with whips adorned with bristles (the "cirres"), which are there to collect organic particles that are suspended in water.

Cirripedes are all attached to rocks, to various floating objects, to living organisms (turtles, marine mammals), or embedded in mollusk shells or in the skeleton of a coral reef. The balanus is the most important species of crustacean that develops on media.

The term *biofouling* refers to the final deposit, resulting from the mixture of biofilms (microbial and their extracellular polymeric substances [EPS], SS, corrosion products and macro-organisms that adhere to and grow on the surface). The fouling layer reaches maximum development with the adhesion of marine animals (Figures 19.36 and 19.37) such as crustaceans (barnacles), mollusks (mussels and clams), anellids (serpulids) and coelenterates (hydroids).

Mussels are also considered to be the most characteristic species of macro-fouling. They are bivalve species, having two shells and forming large colonies on a large number of materials. They are the main species responsible for the clogging of industrial pipes. It is very difficult to destroy and detach mussel shells from pipe walls because of their strong adhesion.

In order to minimize the problems related to micro- and macro-fouling in desalination plants, a punctual injection of oxidant is carried out at the pumping level. This shock treatment is effective in limiting the development of these marine macro-organisms. Chlorine concentrations can vary between 5 and 10 mg·L^{-1}, injected for 1–2 h per day.

a) b)

Figure 19.36. *Mussel development (a) in seawater pumping equipment and structures and (b) in pipelines. For a color version of this figure, see www.iste.co.uk/gaid/watertreatment4.zip*

Figure 19.37. *Photographs of biofilm and barnacles in pipes. For a color version of this figure, see www.iste.co.uk/gaid/watertreatment4.zip*

Figure 19.38 shows electron microscope images of microbial cells that have grown on organic matrices.

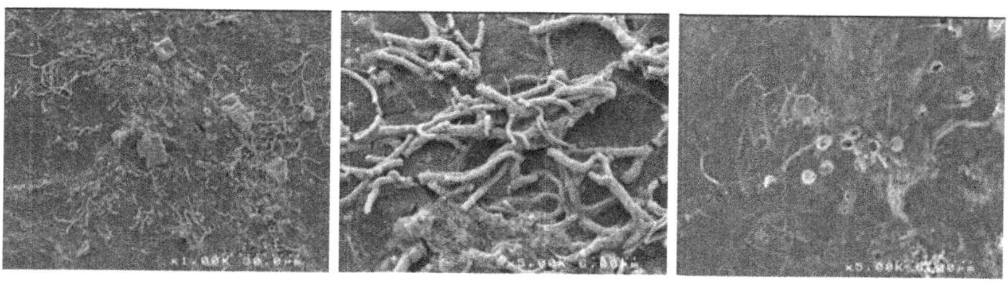

Surface × 1,000: bacteria and mineral crystals	Surface × 5,000: bacteria	Surface × 5,000: microbial cells embedded in an organic matrix

Figure 19.38. *Electron microscope observations of various media*

19.8.2. *Implementation of chlorination*

Chlorine is generally used in the form of chlorine gas, or sodium or calcium hypochlorite. These reagents hydrolyze rapidly in water:

$$Cl_2 + H_2O \rightarrow HOCl + HCl$$

$$NaOCl + H_2O \rightarrow HOCl + NaOH$$

$$Ca(OCl)_2 + 2H_2O \rightarrow 2HOCl + Ca(OH)_2$$

Hypochlorous acid dissociates into H^+ ions and hypochlorite ions:

$$HOCl \leftrightarrow H^+ + OCl^-$$

Cl_2, NaOCl, $Ca(OCl)_2$, HOCl and OCl^- are expressed as free residual chlorine (mg $Cl_2 \cdot L^{-1}$). As will be discussed later, chloramines are formed by the reaction between chlorine and ammonia:

$$HOCl + NH_3 \leftrightarrow \underset{\text{monochloramine}}{NH_2Cl} + H_2O$$

There are three types of chloramines (mono, di and tri) and they are referred to as combined available or residual chlorine.

Today, the main question is, continuous chlorination or intermittent chlorination?

Depending on the seasonal and/or daily parameters (temperature, organism population, sunlight), chlorine is dosed continuously or intermittently. It is currently accepted that, in order to maintain the best control of the biological fouling process, as well as to reduce revival phenomena, chlorine dosages should be carried out in the form of shocks (higher dosages for shorter periods of time at fixed time intervals).

Shock dosages can be extremely effective and provide a high rate of organism inactivation. In the continuous or intermittent method, chlorine dosage is essentially determined on the basis of the "chlorine feed water demand", taking into account the transport of the disinfectant to the structures and the contact time in the structures.

19.8.2.1. *Continuous chlorination*

Continuous chlorine dosing has been accepted as an effective method in the long history of RO membrane system operation, while requiring strict control.

Chlorine is added continuously to the water inlet with a minimum reaction time of 20–30 min, before it reaches the pre-treatment unit. A free chlorine residual concentration of 0.5–1.0 mg$\cdot$L^{-1} must be maintained throughout the pre-treatment line.

However, it is well known that particular attention must be paid to minimize oxidative deterioration of polyamide membranes. And these have often been deteriorated by a failure of the sodium bisulfite (SBS) dosage means or monitoring of the redox potential.

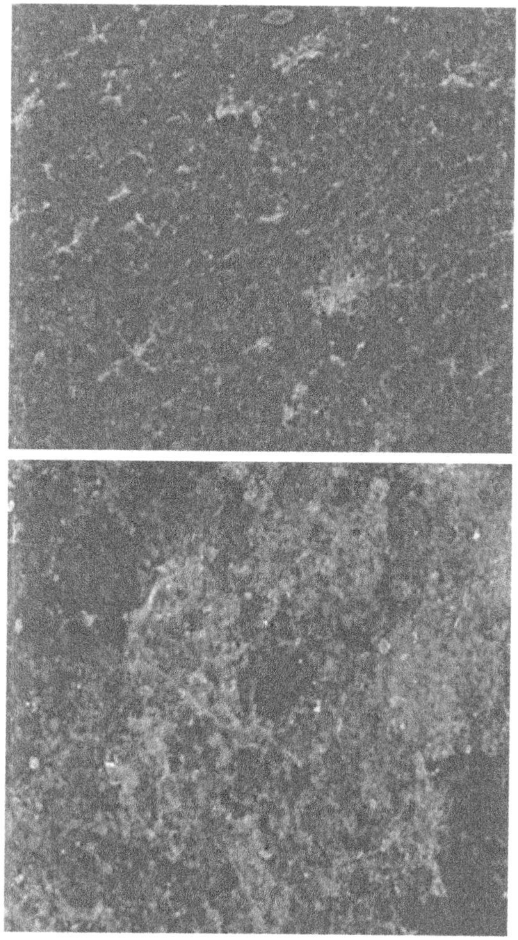

Figure 19.39. *Organic deposition on RO membranes responsible for bacterial growth, despite continuous chlorination*

Problems of biological fouling downstream of the dechlorination point are, however, quite common. It is believed that chlorine reacts with the organic matter in the water and breaks it down into more biodegradable fragments. Given that there is no chlorine present on the membranes, micro-organisms thrive in the presence of these nutrients. Therefore, it is accepted that bio-fouling refers to the unwanted accumulation of a biotic deposit on a membrane surface. It is composed of an organic layer of micro-organisms integrated in a polymeric matrix. As a result, the chlorination-dechlorination protocol has become less popular.

19.8.2.2. *Intermittent or shock chlorination*

As an alternative to continuous chlorination, chlorine is increasingly applied intermittently or as shock chlorination.

Thus, chlorine is added a few hours per day to the water inlet, at concentrations of about 5–10 $mg \cdot L^{-1}$. If there is any residual chlorine at the end of the reaction time, the feed water is rejected. Before the system is put back into operation, all feed water containing chlorine is thoroughly rinsed and checked to be chlorine free (oxidation reduction potential).

The best approach to minimizing the adverse effects of chlorination is to reduce chlorine dosages within the plant and reduce residual chlorine rejection levels, preferably to zero, at the point of rejection. This can be achieved by establishing an optimized dosage regime based on the situation on a case-by-case basis depending on the feed water quality. In order to "neutralize" any residual chlorine levels prior to rejection, various chemicals can be used. Sodium bisulfite is most commonly used in SWRO plants to dechlorinate water upstream of RO membranes.

Several alternative pre-treatment methods are proposed, but chlorine remains the most widely used oxidant.

Thus, optimization of the chlorine injection method is necessary in order to avoid membrane malfunctions and to increase their life span.

For this reason, the intermittent or shock chlorine injection method is increasingly recommended instead of continuous chlorination.

19.8.2.3. *Dechlorination*

Sodium metabisulfite is commonly used to neutralize free chlorine. Other reagents such as sulfur dioxide or SBS can also be used. SBS is produced from metabisulfite:

$$NaS_2O_3 + H_2O \rightarrow 2NaHSO_3$$

Bisulfite reduces hypochlorous acid according to the following reaction:

$$2NaHSO_3 + 2HOCl \rightarrow H_2SO_4 + 2HCl + NaSO_4$$

In theory, 1.34 mg of sodium metabisulfite removes 1.0 mg of free chlorine. In practice, however, 3.0 mg of sodium metabisulfite is normally used to remove 1.0 mg of chlorine.

With the addition of chlorine, various organic and inorganic materials such as iron, hydrogen sulfide and so on are oxidized, and all pathogenic and other micro-organisms are killed or inactivated. More or less effective control of bio-fouling is achieved at concentrations between 0.5 and 1 $mg \cdot L^{-1}$ when used in a continuous procedure, and between 5 and 10 $mg \cdot L^{-1}$ when used intermittently for a few hours per day. However, dechlorination upstream of the membranes is necessary to protect the membranes from the chemical oxidation of chlorine.

19.8.2.4. *Chlorine dioxide (ClO_2)*

Particular interest in chlorine dioxide as an alternative to chlorine has been noted in some desalination plants in the Persian Gulf or in Florida. Chlorine dioxide is chosen as a pre-treatment disinfectant because of its effectiveness as an oxidant and biocide, as well as to minimize the formation of disinfection by-products (DBP) such as THM and haloacetic acids.

Chlorine dioxide is a yellow-green gas, highly soluble in water, which is prepared on site by reacting sodium chlorite with chlorine or hydrochloric acid:

$$2NaClO_2 + Cl_2 \rightarrow 2NaCl + 2ClO_2$$

$$5NaClO_2 + 4HCl \rightarrow 4ClO_2 + 5NaCl + 2H_2O$$

ClO_2 is a strong oxidizer that decomposes into chlorides and chlorites:

$$ClO_2 + 4H^+ + 5e^- \rightarrow Cl^- + 2H_2O(Eo = 1.51\ V)$$

$$ClO_2 + e^- \rightarrow ClO_2^-\ (Eo = 0.95\ V)$$

$$ClO_2 + 4H^+ + 4e^- \rightarrow Cl^- + 2H_2O(Eo = 0.78\ V)$$

The main characteristics of chlorine dioxide are as follows:

– it does not react with bromides to form hypobromites, which contribute to the formation of organobromine compounds;

– it does not react with ammonia;

– it is highly reactive to a number of compounds, including secondary and tertiary amines, organosulfur compounds and activated aromatic cycles;

– it is not a chlorinating agent and does not promote addition and substitution reactions, nor does it lead to the formation of THM-type of halogenated organic compounds. In the 6–8.5 pH range, chlorine dioxide remains in solution as a dissolved gas.

The limitations of chlorine dioxide observed on site are the production of chlorites, which can represent 30% of the ClO_2 concentrations. Since chlorites are not removed during pre-treatment, their impact on RO membranes through possible oxidation is possible but not clearly proven. The second limitation is often the operating cost. ClO_2 is more expensive than sodium hypochlorite.

In plants such as Palm Jumeirah (Dubai), ClO_2 shock dosing is performed for 10 min every 12 h, with a dosage of 3 ppm. No algae or mussel growth has been observed in the seawater inlet, and the process is currently working well.

19.8.2.5. *Conclusions on pre-chlorination*

In order to control the development of shellfish in the seawater inlet pipes, continuous chlorination treatments are commonly practiced. This type of treatment is also a common practice on the backflow of brackish water wells, so as to "protect" the bacteriological quality of the water during its transport in the pipes.

Given the high sensitivity of the membranes to oxidants, it is imperative to carry out a joint dechlorination treatment before passage through the membranes.

Continuous chlorination-dechlorination treatment is less and less recommended for water supplied to SWRO plants. The proposed explanation is that chlorine reacts with organic matter, algae and bacteria to form biodegradable organic carbon (BDOC). This BDOC is the source of very fast bacterial redevelopment (bacterial overgrowth phenomenon) on the membranes straight after the dechlorination point.

In order to avoid this risk, it is recommended to carry out shock-chlorination treatments with sewage of chlorinated water downstream of the pre-treatment stage and upstream of the membranes.

When the discharge of pre-treated and chlorinated water is not possible, be it for environmental reasons or simply because of the size of the plant, a joint dechlorination treatment with sodium bisulfite ($NaHSO_3$) will be provided. The injection point of the sodium bisulfite will be located downstream of the cartridge filter.

Implementation of shock chlorination:

– indicative chlorine dosage (Cl_2, 100%): 10 $g \cdot m^{-3}$ (default dosage, since the dosage is a function of the chlorine demand of the water between the point of injection and the point of discharge of the chlorinated water. As an example, the chlorine demand on Abu Dhabi seawater was 7 mg of 100% chlorine/L of water to be treated);

– injection point for seawater: at the water inlet of the supply line;

– injection point for brackish water: at the well discharge;

– frequency: every 2 days;

– average duration: 2 h.

19.8.3. *pH adjustment*

The pH adjustment stage during pre-treatment allows the feed water to be regulated to an optimal pH. Once the coagulants are added, the pH is often changed significantly. In most cases, the pH is reduced to a neutral or slightly acidic level. Adjustment chemicals to lower the pH include carbon dioxide, sulfuric acid and hydrochloric acid.

However, CO_2 is not recommended for lowering the pH after lime addition, due to scaling potential.

19.8.4. *Direct filtration*

The type and choice of pre-treatment depends on the characteristics of the water to be treated. With the many technologies available for pre-treatment, design engineers can ensure acceptable SDI (or MFI) values for membrane systems.

The most common pre-treatment used in seawater desalination is multimedia filtration.

It is an essentially mechanical operation with the objective of trapping particles larger than 10 μm. Direct filtration can be used in different configurations associated with the qualities of seawater:

– mono-media, dual-media or three-media filtration;

– single-stage or two-stage filtration in series;

– gravity or pressure filter.

19.8.4.1. *Mono-media or dual-media direct filtration*

19.8.4.1.1. Mono-media gravity filters

Direct mono-media filtration is proposed when the water quality has characteristics like those shown in Table 19.14, such as turbidity < 2 NTU and SDI < 10.

Direct dual media filtration is offered for slightly more degraded water.

Seawater parameters	Mono-media filtration	Dual media filtration
Turbidity, NTU	1–2	<10
Suspended solids, mg·L^{-1}	<3	<15
SDI	10	30
Algae, u·L^{-1}	–	2,000
Chlorophyll, µg·L^{-1}	–	<5
DOC, mg·L^{-1}	<1.0	<2.0

Table 19.14. *Recommended limits for direct filtration*

Mono-media filtration consists of a filter media, most often silica sand (Figure 19.40), with an effective size of 0.8 mm and a uniformity coefficient ≤ 1.3. The sand height varies between 1.0 and 1.5 m. Coagulation (and often flocculation) is performed with the injection of an Fe- or Al-based coagulant. The filtration rate is between 6 and 9 m·h^{-1}.

This type of filter is mainly proposed when the SDI is less than 25 and the SS are in low concentrations. For coastal wells, where natural filtration has already occurred, the need for a filter at the pump outlet should be considered on a case-by-case basis. Coagulant is not necessary for coastal wells, even if the equipment is provided in the case of a punctual degradation of the raw water. This filtration acts as a "second stage" and is only reasonably suggested if sand particles can be drawn in during pumping, or if iron and/or manganese are present.

The filter media must meet various specifications such as grain size, grain surface condition, density, void index, acid loss and so on.

The shape of the grains used in filtration media mainly depends on the origin of the material. Grains collected from river beds are generally rounded and smooth. Grain resulting from the crushing of larger pieces is jagged and angular. The performance of the crushed particles compared to that of the rounded grains is slightly better, but this is not a determining factor in the choice of material.

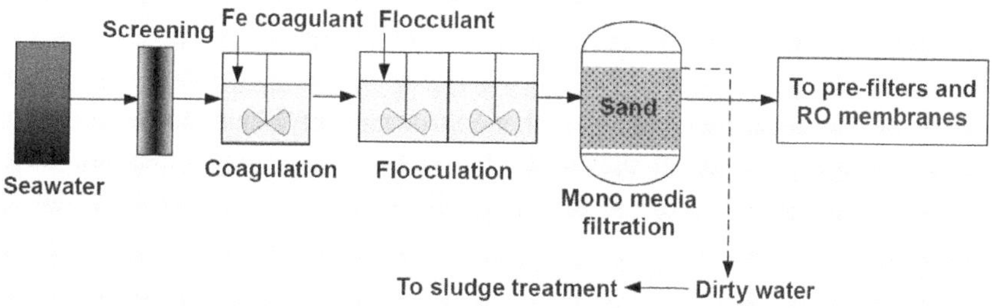

Figure 19.40. *Principle of coagulation–flocculation on a mono-media filter. For a color version of this figure, see www.iste.co.uk/gaid/watertreatment4.zip*

19.8.4.1.2. Diatom filters

Diatomaceous earth filters are rarely used in direct filtration because of the high pressure drop and short operating times. They are more often used in tertiary wastewater treatment.

19.8.4.1.3. Dual media filters

Dual media filtration involves the combination of two media with different specific densities. This difference allows the media to be classified after a wash in which they have been mixed. The top layer is composed of the lighter media, has a larger effective size and a higher void index, giving it a greater particle holding capacity. This very robust layer is capable of operating at high solids loads (SS, turbidity, algae), especially during certain seasonal periods. This layer can receive higher coagulant dosages. This media is most often anthracite or pumice.

Figure 19.41. *Diatom filters. For a color version of this figure, see www.iste.co.uk/gaid/watertreatment4.zip*

The lower layer is made up of sand with a smaller diameter (TE: 0.65 mm), which acts as a final barrier to the fine particles responsible for clogging.

The use of two supports allows higher SS (or turbidity) loads to be filtered and reduces pressure loss during operation. This means that cycle times are longer, implying a progressive maturation that refines the quality of the water produced.

In-line coagulation (often combined with flocculation) is used depending on seawater quality, temperature, algae content, etc., and dosages are directed to the inlet turbidity or SDI.

The conditions for in-line coagulation using metal salts are pH, dosage, velocity gradient (G and Gt) and temperature. The pH affects the surface charge of the colloids and determines the predominant coagulant species. As a result, floc size and structure (porosity and density) may differ with pH. Flocs formed at low pH and coagulant dosage are denser and less porous than those formed at higher pH and dosage. The physical properties of flocs are sensitive to flocculation conditions such as G and Gt. An example of this is the increasing shear stress, which reduces the average floc size.

The performance of the dual media filter is based on the combination of flocs trapped on the upper layer, and the refinement of the particle cut-off involved in the lower layer (Figure 19.42).

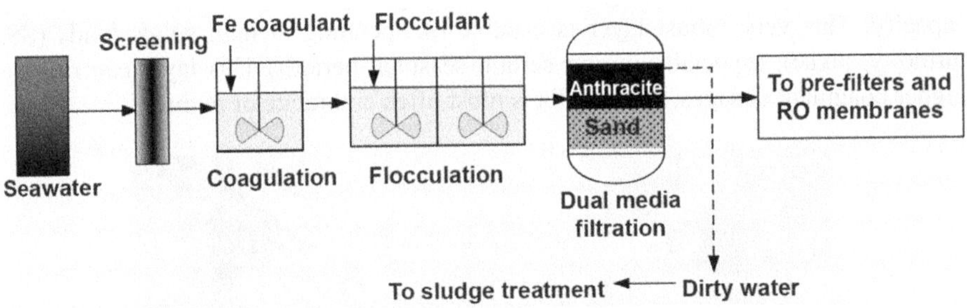

Figure 19.42. *Coagulation–flocculation and dual media filtration. For a color version of this figure, see www.iste.co.uk/gaid/watertreatment4.zip*

Anthracite or pumice can be replaced by filtralite. This material is made from clay at about 1,200°C, followed by crushing and sieving. The material has a high pore structure compared to other substrates, with varying densities depending on the application. The advantage of this material lies in the slightly longer cycle times.

However, its cost is higher than that of the other two materials. The SDI at the outlet of an anthracite or pumice-sandstone bilayer is identical to that at the outlet of filtralite.

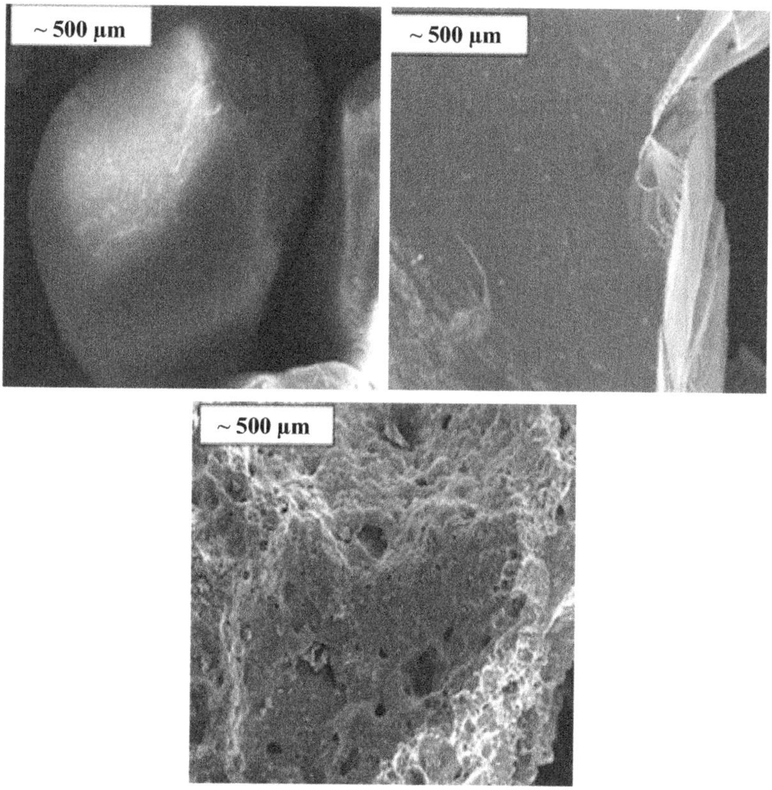

Figure 19.43. *Electron microscope views of sand (0.8 mm), anthracite (1.2–2.5 mm) and NC filtralite (1.5–2.5 mm)*

The total height of the two media varies from 1.4 to 1.6 m, and it is based on equivalent proportion in the filter. The filtration rate is between 7.5 and 10 $m \cdot h^{-1}$. A new (Veolia) development of a dual media filter for seawater pre-treatment is based on a total height of 3–4 m of the two media, and a high filtration speed (15–20 $m \cdot h^{-1}$). A linear relationship has been demonstrated between the total height of the media, the filtration rate and the ultimate performance of the dual media filter. The filters are equipped with a backwash system.

Seawater parameters	Mono-media filtration	Dual media filtration	
	Anthracite		
Height, m	–	0.7–0.8	1.5–2
Effective size, mm	–	1.0–1.5	1.4–2.5
Uniformity coefficient	–	≤1.4	≤1.4
	Sand		
Height, m	1.0–1.5	0.7–0.8	1.5–2
Effective size, mm	0.8	0.6	0.8
Uniformity coefficient	≤1.3	≤1.3	≤1.3
Velocity, $m \cdot h^{-1}$	6–9	7.5–10	15–20

Table 19.15. *Characteristics of mono- and dual-media filters*

Unlike coastal well water, surface water is, in all cases, chemically conditioned with coagulation and flocculation of suspended matter (colloids, particles, algae), followed by removal by filtration. Coagulation is a process in which a coagulant (Al or Fe salt) is added to the water to destabilize the suspended particles. Coagulation is done by means of a static mixer or a coagulation tank with a suitable mixer. An acid (mainly H_2SO_4) is added when necessary at the coagulation stage, so as to maintain optimal pH conditions. The coagulated water then flows into the flocculation tanks. These are recommended when the water temperature is cold at certain times of the year and are associated with the addition of a flocculant (polyelectrolyte). The flocculated water is then fed to the downstream filters.

Algae cells are difficult to remove by direct filtration. Filter clogging (pressure loss) varies with the concentration of algae. The higher the algae concentration, the shorter the cycle times. For this reason, the acceptable algae concentration for a dual media filter is 2,000,000 $u \cdot L^{-1}$. In addition, algal removal depends on the species present.

Filtration problems caused by algae are grouped into three categories:

1) deep diffusion of algal cells into the filter media, with leakage into the filtered water;

2) interference with coagulation–flocculation caused by extracellular algal material;

3) too short cycle times (increased pressure loss due to filter clogging) with increased need for backwashing.

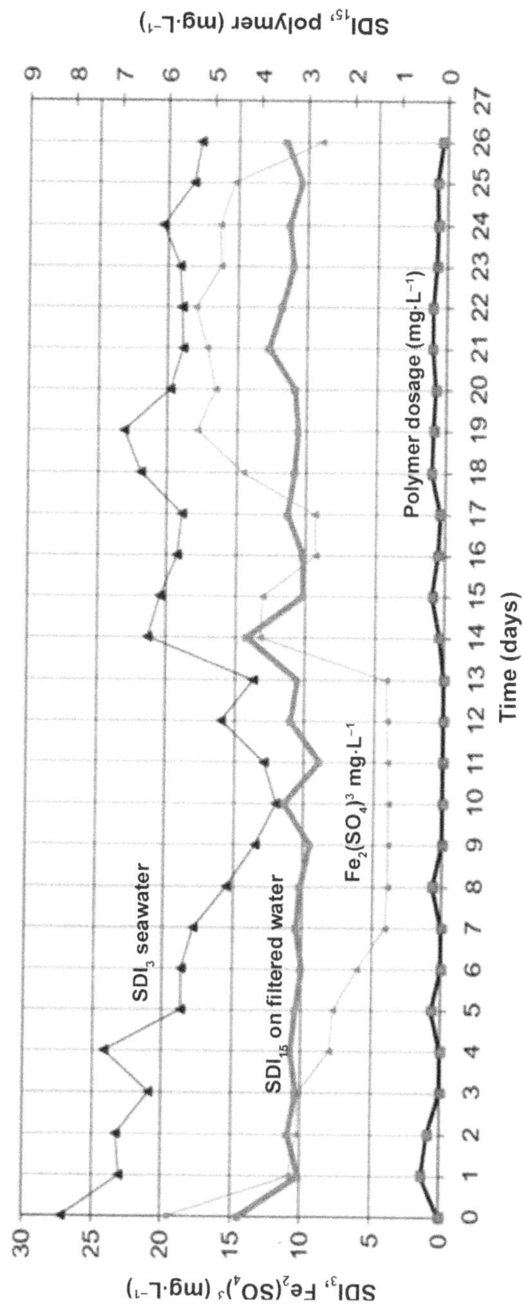

Figure 19.44. *SD15 of filtered water according to reagent dosages (Gold Coast-Australia). For a color version of this figure, see www.iste.co.uk/gaid/watertreatment4.zip*

Figure 19.45. *Mono-media pressure filters (Oman Sur desalination plant). For a color version of this figure, see www.iste.co.uk/gaid/watertreatment4.zip*

19.8.4.1.4. Three-media filters

When three layers of media are used in a filter, the distribution consists of anthracite (or pumice) in the top layer, sand in the middle layer and garnet in the bottom layer. This stratification is justified by the respective densities of the materials (0.5–0.7 $g \cdot cm^{-3}$, 1.5 $g \cdot cm^{-3}$, and 4 $g \cdot cm^{-3}$). This solution is valid for waters with very low turbidity and does not require backwashing with air and water. This is the case because this phase contributes to the mixing of the media and does not make it possible to acquire the initial classification. The reason for this is that the garnet particles are very fine and remain mixed with the sand. However, this solution is valid when a water wash alone is required.

Figure 19.46. *Dual-media gravity filters (desalination plant in Australia). For a color version of this figure, see www.iste.co.uk/gaid/watertreatment4.zip*

19.8.4.2. *Two-stage filtration*

Pre-treatment may require two-stage filtration (Figure 19.47). The first-stage filter is a dual-media filter with a total height of 1.2–1.6 m. The specifics of the two media are summarized in Table 19.16. The top layer consists of the lighter media, which has a larger effective size and a higher void index allowing for greater particle holding capacity. This very robust layer is capable of operating at high solids loads (SS, turbidity, algae), especially during certain seasonal periods. This layer accepts higher coagulant dosages. This media is most often anthracite or pumice. The lower layer is made up of sand with a smaller diameter (ES: 0.65 mm), which serves as a final barrier to the fine particles responsible for clogging.

The first-stage filters are designed to remove the majority of particles present in the feed water.

The second-stage filters are mono-media filters with a sand height of 1 m and an effective size of 0.4–0.5 mm, which maximizes the possibility of attaching the finest particles. The second stage is a refining filter, generally operating between 12 and 15 $m \cdot h^{-1}$.

In all cases, the seawater is chemically conditioned by coagulation (and flocculation, if necessary) at the first stage. The first-stage filters operate as dual media filters, as described above. The second-stage filters are refining filters and operate at high speeds (Table 19.16). Experience shows that the second-stage filter can remove 0.3–0.4 SDI.

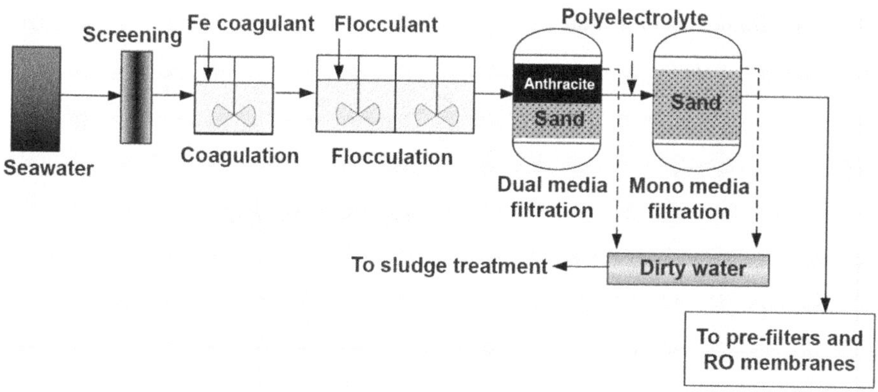

Figure 19.47. *Two-stage filtration. For a color version of this figure, see www.iste.co.uk/gaid/watertreatment4.zip*

Parameters	First-stage filtration	Second-stage filtration
	Anthracite	
Height, m	0.5–0.7	–
Effective size, mm	1.0–1.2	–
Uniformity coefficient	≤1.4	–
	Sand	
Height, m	0.5–0.7	1.0
Effective size, mm	0.6	0.4–0.5
Uniformity coefficient	≤1.3	≤1.3
Velocity, m·h^{-1}	10–13	14–18

Table 19.16. *Characteristics of two-stage filtration*

Figure 19.48 shows the performance of two-step filtration for SDI_{15} removal (Millipore membrane). SDI_{15} after two-step filtration is better than that recorded after dual media filtration. A difference between 0.4 and 0.5 is observed between the two configurations. The SDI_3 of the raw water was between 26 and 33% minimum. Typically, the two-stage filtration implementation is designed such that the filtered water from the first stage flows by gravity to the second filtration stage. The filtered water from this second stage is then sent to the pre-filtration unit, and then to the membrane workshop.

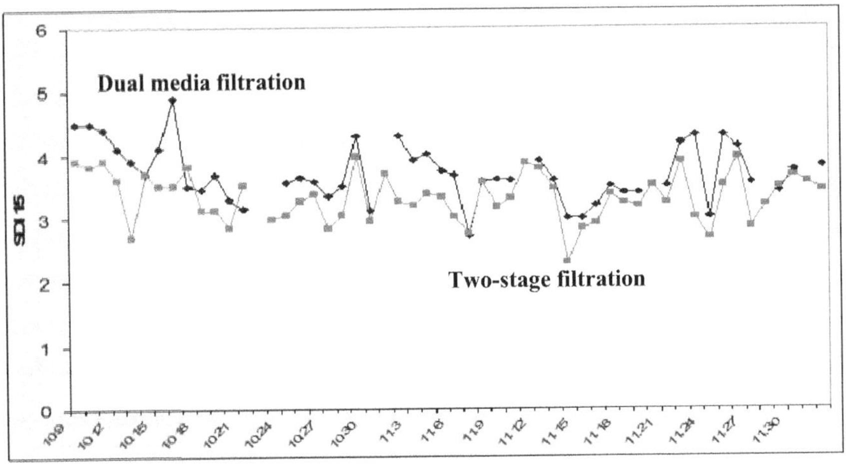

Figure 19.48. *Comparison of SDI reduction between a dual-media filter and a two-stage filtration. For a color version of this figure, see www.iste.co.uk/gaid/watertreatment4.zip*

An intermediate water tank is installed between the filters and the pre-filters. Vertical turbine pumps are installed in the tank to transfer the water to the pre-filters, and then to the high-pressure (HP) pumps.

Both stage filters are equipped with backwash equipment. An in-line SDI analyzer is installed at the outlet of the filtered water collector. In two-stage configurations, SDI meters are installed at the outlet of the first-stage filter and at the outlet of the second-stage filter. If the SDI is higher than the required value, the operators are made aware by an alarm, so as to manually isolate the filter train with a high SDI. In addition, an in-line turbidimeter is installed in the first- and second-stage filtered water collectors.

19.8.4.3. *Gravity filters or pressure filters*

Gravity filters are either covered or installed in a building in order to minimize algae growth. A protective coating suitable for seawater applications is used in the filter cells to protect the concrete structure of these filters.

Pressure filters are proposed for economic reasons, or for ease of implementation, or to avoid intermediate pumping and send water that is already under pressure to the pre-filters (Figure 19.45).

Pressure filters are also widely used in small plants because they are cost competitive, space saving, easier and faster to install and operate than granular media gravity filters. In the case of two-stage filtration, the use of this type of filter avoids intermediary pumping between the two stages.

19.8.5. *Chemical conditions of implementation*

The principles of coagulation and flocculation are detailed in Volume 2, Chapter 4. The basic principle is a destabilization of the colloidal (negatively charged) particles, with the addition of a coagulant. The optimal pH range for adsorption coagulation with iron salts is between 6 and 8, while the optimal pH range with aluminum salts is narrower and is about 6.5.

For seawater pre-treatment, ferric chloride ($FeCl_3$) is most often used as a coagulant. Alternatively, aluminum-based coagulants can be applied and the Al residual in the filtered water must be controlled, so as to not damage the membranes. In addition, since the pH of seawater is often around 8, the water should be acidified to operate in the optimal pH range for the coagulant used.

Ferric chloride is well soluble in water, with a solubility product (Ksp) of 27.9. Thus, 162 g of $FeCl_3$ dissolved in 1 L of water, gives 55.8 $mg \cdot L^{-1}$ of Fe^{3+} and 106.5 $mg \cdot L^{-1}$ of Cl^-.

Fe^{3+} and OH^- ions precipitate rapidly because the solubility product of iron hydroxide is low ($Ksp\ Fe(OH)_3 = 10^{-38}$). Only 7.8×10^{-10} $mol \cdot L^{-1}$ Fe^{3+} and 2.34×10^{-9} $mol \cdot L^{-1}$ OH^- can be residual in the interstitial water. When the pH of the seawater is known, the concentration of iron ions (Fe^{3+}) can be calculated using the solubility product of $Fe(OH)_3$.

$$Fe^{3+} + 3OH^- \rightarrow Fe(OH)_3; K_{sp} = 10^{-38}\ H_2O \leftrightarrow H^+ + OH^-;\ K_w = 10^{-14}$$

Or: $Kw = 10^{-14} = [H+] \cdot [OH^-]$ and the OH^- concentration is therefore equal to $10^{-14}/ [H+]$.

The combination of these equations leads to:

$$[Fe^{3+}] \times [OH^-]^3 = 10^{-38}$$

$$[Fe^{3+}] = (10^{-38} \times [H+]^3)/(10^{-14})^3 = 10^4 \times [H+]^3$$

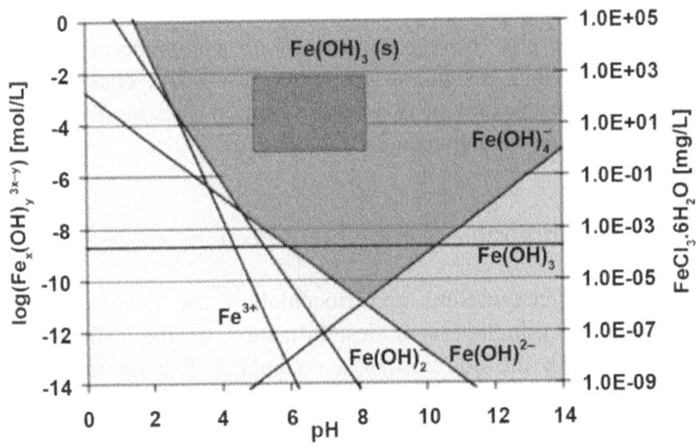

Figure 19.49. *Hydrolysis products of $FeCl_3$. For a color version of this figure, see www.iste.co.uk/gaid/watertreatment4.zip*

And since $[H+] = 10^{-pH}$:

$$[Fe^{3+}] = 10^4 \times [10^{-pH}]^3$$

In addition to iron hydroxides, the following hydrolysis products are present: $Fe(OH)_2^+$, $Fe(OH)^{2+}$, $Fe(OH)_4^-$.

Table 19.17 shows the Ks values for the different hydrolysis products.

Hydrolysis reactions	Ks
$Fe(OH)_3 \leftrightarrow Fe^{3+} + 3OH^-$	10^{-38}
$Fe^{3+} + 2H_2O \leftrightarrow Fe(OH)^{2+} + 3H^+$	6.8×10^{-3}
$Fe(OH)^{2+} + H_2O \leftrightarrow Fe(OH)_2^+ + H^+$	2.6×10^{-5}
$2Fe^{3+} + 2H_2O \leftrightarrow Fe_2(OH)_2^{4+} + 2H^+$	1.4×10^{-3}
$Fe(OH)_3 + OH^- \leftrightarrow Fe(OH)_4^-$	10^{-5}

Table 19.17. *Hydrolysis products and their solubility product*

As a result of $FeCl_3$ dosage, the OH^- ions are neutralized and the pH decreases. The magnitude of the pH decrease depends on the buffering capacity of the water. One of the alternative coagulants is aluminum sulfate, which is well soluble in water, with the formation of aluminum hydroxide $Al(OH)_3$ and the release of sulfates into the water.

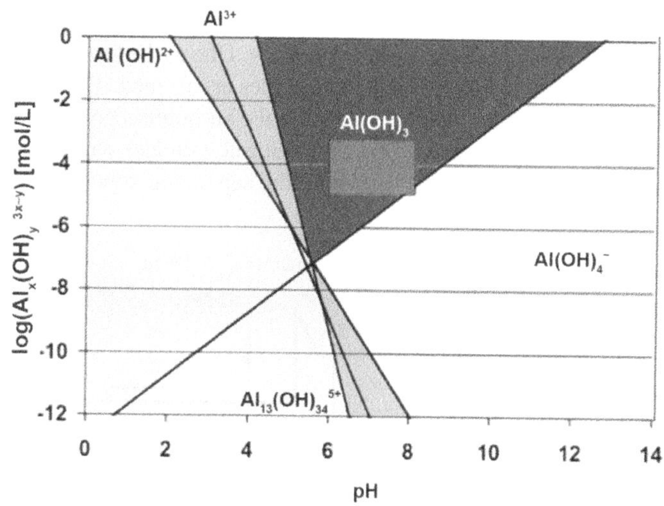

Figure 19.50. *Hydrolysis products of aluminum sulfate. For a color version of this figure, see www.iste.co.uk/gaid/watertreatment4.zip*

$Al(OH)_3$ has a low solubility product and precipitates quickly (Figure 19.50).

$$Ks = [Al^{3+}]\,[HO^-]^3 = 10^{-33} = [Al^{3+}]\,(10^{-14})^3/[H+]^3$$

The dissolution of aluminum sulfate leads to the formation of different hydrolysis products, depending on the pH.

19.8.6. *Flotation*

As opposed to settling, flotation is a solid–liquid separation technique that is applied to particles whose density is lower or has been made lower than the liquid in which they are found. As a pre-treatment, the dissolved air flotation (DAF) process has proven to be very effective in the presence of algae. The operating mechanisms and sizing parameters are well described in the "flotation" chapter.

Applied in seawater, the DAF process (Figure 19.51) is an effective solution for the separation of SS (turbidity, algae, etc.) and SDI from seawater, after the addition of a coagulation reagent. This process utilizes the adhesion property of micro-bubbles to SS, thus increasing the particles' tendency to float. The water is coagulated in line or in a coagulation tank, then passes into the flocculation tank where it has a short residence time (<10 min), but long enough to form flocs. The flocculated water passes to the solid–liquid separation zone, which is divided into two zones. The first is the contact zone, in which air bubbles are injected into the flocculated water by means of specific nozzles. A flow of water with supersaturated air is distributed through the nozzles. This water flow is called the saturation flow. Due to the loss in pressure of the supersaturated water at the nozzles, small air bubbles are formed. The speed of the air bubbles is higher than the speed of the water, and the air bubbles collide with the flocs in the flocculated water. They then form an aggregate (sludge) with the air bubbles, which rises to the surface. In the second zone (the separation zone), this aggregate is then separated from the interstitial water.

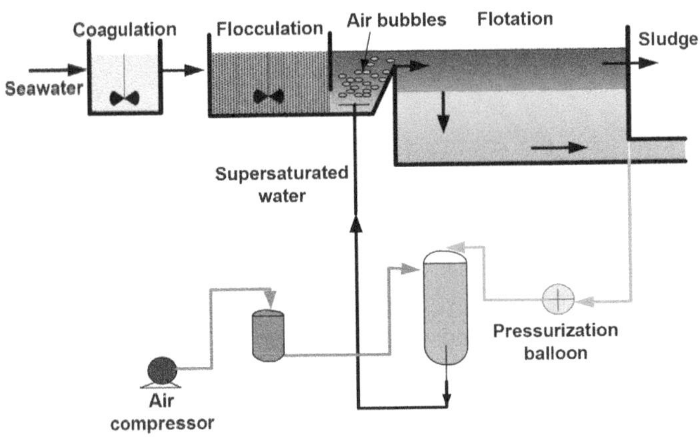

Figure 19.51. *Operating diagram of a conventional dissolved air flotation system. For a color version of this figure, see www.iste.co.uk/gaid/watertreatment4.zip*

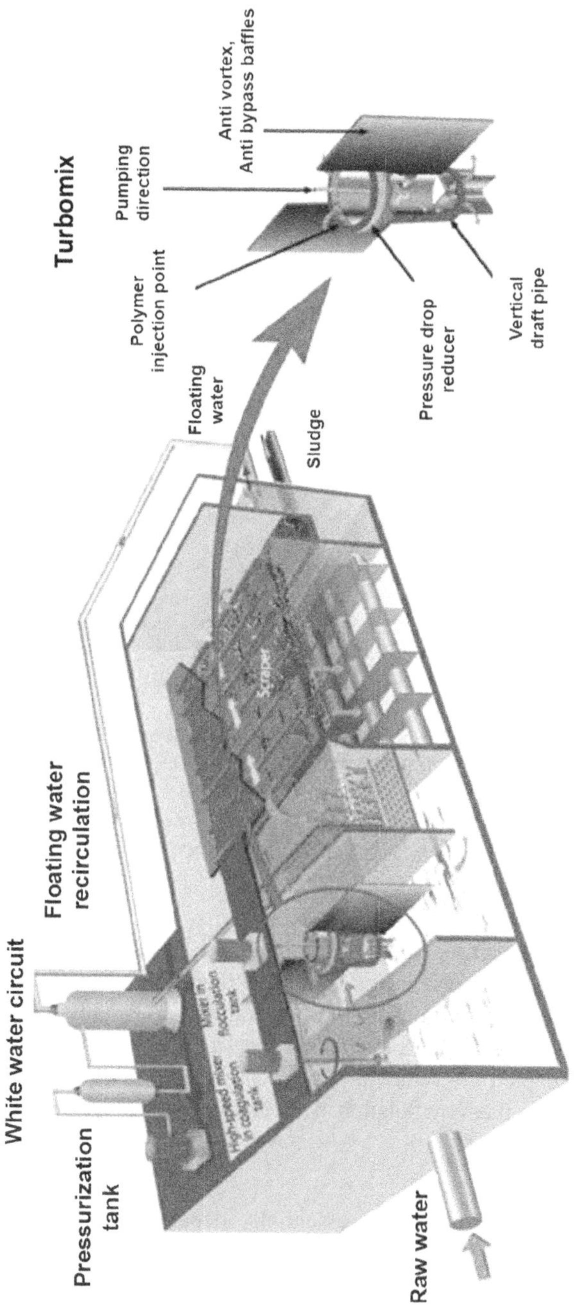

Figure 19.52. *Spidflow® process (Veolia). For a color version of this figure, see www.iste.co.uk/gaid/watertreatment4.zip*

The Spidflow® process (a Veolia process) comprises a coagulation stage, a flocculation stage and a clarification stage by rapid flotation (Figure 19.52). The fine air bubbles, formed by pressurizing the air in the water (at pressures of 5–6 bars) during the production of white water, are injected through specific nozzles (Figure 19.53) via a dedicated distribution system. This ensures the separation of SS, microalgae and hydrocarbons that are trapped in hydroxide flakes, formed by the addition of coagulant.

Figure 19.53. *Specific nozzles for white water distribution in the Spidflow®. For a color version of this figure, see www.iste.co.uk/gaid/watertreatment4.zip*

19.8.6.1. *Spidflow® performance applied as pre-treatment in seawater*

19.8.6.1.1. Hydrocarbons

The accepted limit for RO membranes is 0.1 $mg \cdot L^{-1}$. In many cases, desalination plants are built in or near industrial areas, and draw their raw water from sometimes very busy navigation areas. Hydrocarbon concentrations are highly variable at the pumping depth (0.01–3 $mg \cdot L^{-1}$), depending on the site and geographical area. The Spidflow® is well adapted to removing hydrocarbons by trapping them during coagulation and removing them by flotation. The removal percentage is around 90%.

19.8.6.1.2. SDI removal

The removal efficiency of the SDI essentially depends on the dosage of the chemical coagulant, which allows the microparticles within the metal hydroxides to be retained. In case of a very high SDI, a flocculant is added in order to form larger flocs and retain more microparticles.

19.8.6.1.3. Removal of DOC

DOC is partially removed using enhanced flocculation (Figure 19.54), under pH conditions < 6.5 and in the presence of an iron-based metal coagulant ($FeCl_3$). Iron salts are generally considered to be more effective in removing organics compared to aluminum salts. Figure 19.55 shows a chromatogram of the raw water and the Spidflow® outlet water. High molecular weight substances (polysaccharides, HS, building blocks) are well removed and confirm the effectiveness of coagulation-flocculation, with respect to NOM.

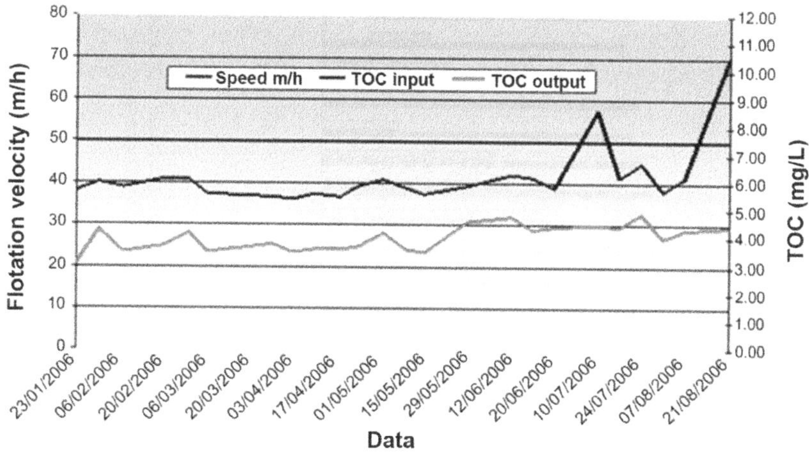

Figure 19.54. *DOC removal in a Spidflow® process (FeCl3 and pH < 6.5). For a color version of this figure, see www.iste.co.uk/gaid/watertreatment4.zip*

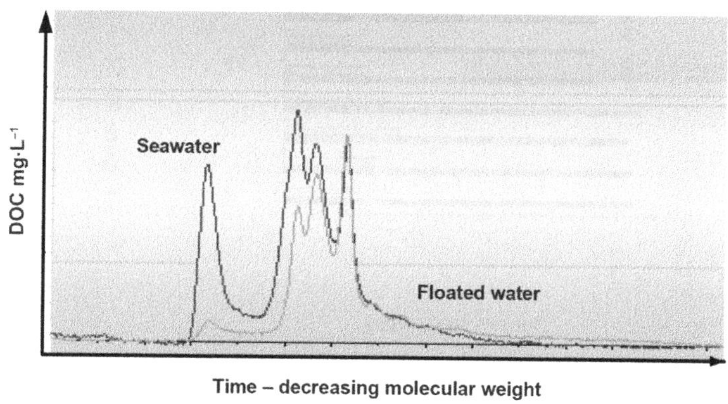

Figure 19.55. *LC-OCD of raw water (seawater) and floating water. For a color version of this figure, see www.iste.co.uk/gaid/watertreatment4.zip*

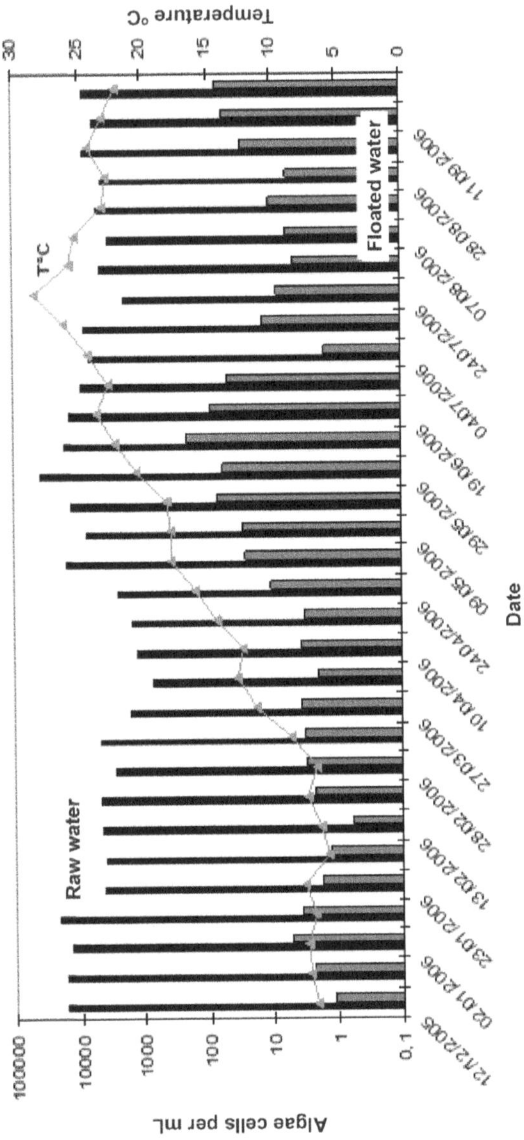

Figure 19.56. *Algae removal results using a Spidflow® process. For a color version of this figure, see www.iste.co.uk/gaid/watertreatment4.zip*

19.8.6.1.4. Algae

One of the main applications of DAF flotation is algae removal. Algae are difficult to remove by a conventional treatment such as sedimentation, because they are naturally less dense than water and yields are lower than those obtained with flotation, even with high dosages of coagulant and flocculant.

Species of algae are generally characterized by a negative surface charge in natural aquatic environments, and their destabilization by charge neutralization can be achieved by chemical coagulation and flocculation.

Poor removal of algae (diatoms, green algae, dinoflagellates, blue-green algae) leads to early filter clogging and short cycle times. Removal of 99–99.9% can be achieved with the Spidflow® process (Figure 19.56). The removal of algae by flotation is more efficient than that achieved by conventional sedimentation (60–90%), the efficiency of which also depends on the concentration of algae in the raw water.

When implementation conditions are well respected (sizing, reagents, contact time), the quality of the water after flotation and filtration is better than after other processes. This is because parameters such as turbidity, DOC and algae are well removed, whereas an MF or UF membrane does not remove DOC or algae in high concentrations.

The probability of catching microalgae and other small particles is higher with a flotation device because the particles are small and light, and can easily float.

The desired outlet SDIs are thus ensured if all operating conditions are met, taking seawater variations into account.

19.8.7. *Settling*

The Actiflo process is a settling process that combines the advantages of ballasted flocculation and lamellar clarification. The settling chapter gives a detailed description of the Actiflo process. The use of microsand gives the Actiflo process the advantage of being able to treat a wide range of raw water by operating at mirror speeds of over 60 $m \cdot h^{-1}$. The density and shape of the microsand particles increase the flocculation efficiency and accelerate the settling. Coagulants applied in seawater can be iron or aluminum based. An anionic polymer is injected into the flocculation tank. Figure 19.57 shows SDI results for water treated with Actiflo and sand filtration. The SDI is less than 3 and the turbidity varies between 0.02 and 0.03 NTU.

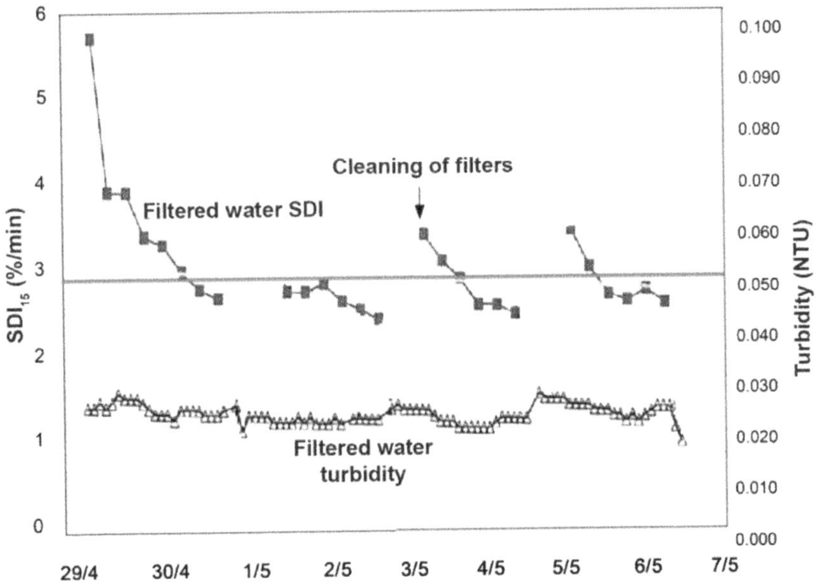

Figure 19.57. *SDI and turbidity at the outlet of Actiflo seawater filters*

19.8.8. *Membranes (UF and MF)*

The majority of desalination plants use conventional pre-treatment processes (in other words, dual-media filtration, preceded by coagulation and sometimes sedimentation or flotation with dissolved air for the more difficult marine waters). These conventional processes are effective in reducing the potential for clogging of seawater, as well as obtaining an SDI of less than 5 (100% of the time) and less than 4 (90% of the time). In addition, one of the advantages of conventional pre-treatment processes is the use of a variable coagulant dosage that is tailored to the raw water conditions, allowing a reduction in the high organic matter that controls biofilm on RO membranes.

In drinking water plants fed by surface water or groundwater, low-pressure membrane processes such as MF or UF are used to produce high-quality water. In the last decade, pre-treatment by UF or MF has garnered wide attention as a potential pre-treatment for seawater desalination by RO.

Table 19.19 summarizes the different specifications of membranes used for seawater pre-treatment. However, this technology also has its own limitations. In the case of treatment of degraded seawater (presence of DOC, microalgae and exopolysaccharides [EPS]), MF and UF processes are susceptible to clogging, for which chemical-free backwashing is less effective. Only chemical backwashing is effective and helps to regain the correct permeability.

The decrease in the flow observed under these operating conditions is explained through different kinetic models that aim to describe the fouling mechanism. These models are based on either full blockage, intermediate blockage, standard blockage or cake filtration mechanisms. The complete blockage model occurs when particles arrive at the membrane and block the pores, without superposition of particles. An intermediate blocking model occurs when particles deposit on other particles present on the membrane surface, blocking the pores by superposition. A standard blocking model occurs when particles arrive on the membrane and settle on the inner pore wall. The cake filtration model is justified when particles settle on top of others particles already present on the membrane surface that are already blocking some pores (Table 19.18).

Types of models	Formulas	Model mechanisms
Complete blockage	$\ln Jw = -kb\,t + \ln J0$	No superposition of particles
Intermediate blockage	$Jwo/Jw = (1 + kit)$	Superposition of particles
Standard blockage	$Jwo/Jw = (1 + ks\,t)^2$	Particles deposited on the inner walls of pores, reducing the pore volume
Cake filtration	$Jwo/Jw = (1 + kc\,t)^{1/2}$	Particles accumulating on other particles that are already present, blocking the pores

Table 19.18. *Description of the proposed models*

We have:

– Jw: permeate flux at time t;

– Jwo: permeate flux at time to;

– kb: kinetic constant related to complete pore blockage;

– ki: kinetic constant related to intermediate pore blockage;

– ks: kinetic constant related to standard pore blockage;

– kc: kinetic constant related to cake filtration.

During membrane filtration, suspended matter is retained on the surface of the membrane. They accumulate and cause the membrane to clog. In this case, the filtration process cannot be maintained for very long because the pressure on the membrane must be constantly increased to maintain the flow rate. Backwashing removes a large part of the accumulated solids, but it is not sufficient. Chemical cleaning is then used because the energy of backwashing alone is not sufficient to dislodge the retained matter. These materials, often of organic and biological origin, tend to block at a slow rate but diffuse deep into the pore structure. In-line coagulation helps to stabilize the UF process between two chemical cleanings.

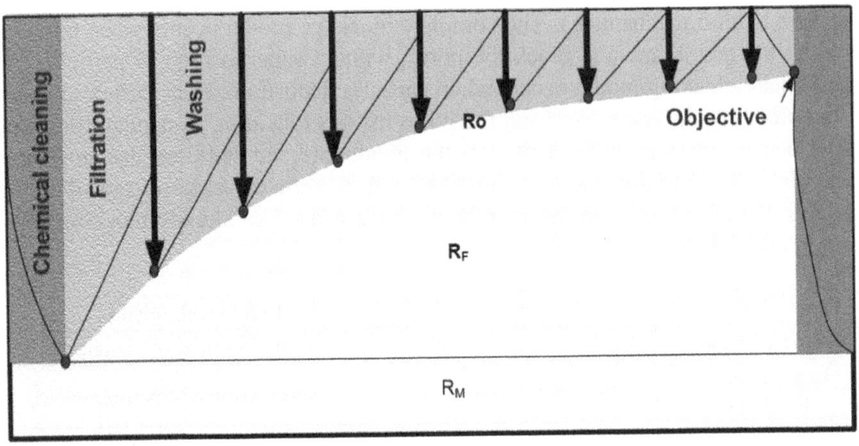

Figure 19.58. *Series of filtrations and backwashes between two chemical cleanings. For a color version of this figure, see www.iste.co.uk/gaid/watertreatment4.zip*

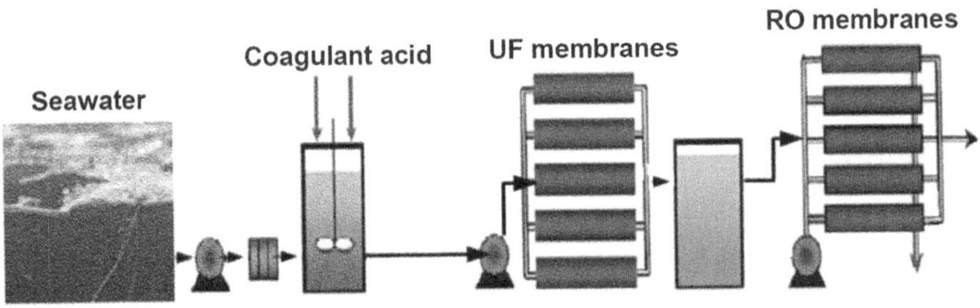

Figure 19.59. *Treatment plant with ultrafiltration (UF) followed by reverse osmosis. For a color version of this figure, see www.iste.co.uk/gaid/watertreatment4.zip*

The desalination process comprising a UF membrane in the first stage and an RO membrane in the second stage (Figure 19.59) is very interesting in the case of water with a low load of algae, DOC and hydrocarbons. This configuration consisting of UF pre-treatment, followed by an RO membrane applied to desalination, is often referred to as the "double membrane process".

Membrane characteristics	ZeeWeed 1000 ultrafiltration	Norit ultrafiltration	DOW ultrafiltration	Hydranautics ultrafiltration	Pall microfiltration
Active surface (m^2)	55.8	40	51	46	50
Filtration direction	External–internal	Internal–external	External–internal	Internal–external	External–internal
MWCO Cut-off point	100,000	150,000	150,000	150,000	–
Nominal pore diameter (µm)	0.02	0.05	0.05	0.05	0.10
Absolute pore diameter (µm)	0.1	0.075	0.075	0.075	–
Material	PVDF	PES	PVDF	PES	PVDF
Type	Hydrophilic	Hydrophilic	Hydrophilic	Hydrophilic	Hydrophobic
Membrane charge	Slightly negative	Neutral	Neutral	Neutral	Slightly negative
pH range	5–10	2–12	2–12	2–12	1–10
Maximum TMP (bar)	0.9	2.5	2.1	2.5	3
Chlorine or oxidant resistance	Resistant to NaOCl, ClO_2, $KMnO_4$	200 ppm continuously	> 5,000,000 ppm·h	200 ppm	Chlorine 10,000 ppm
Required pre-filtration (µm)	100	100	100	100	200

Table 19.19. *Main characteristics of some membranes used in pre-treatment*

19.8.8.1. *UF and MF membrane performance*

19.8.8.1.1 Biofilm control on RO membranes

Organic and biological clogging of RO membranes is mainly due to the accumulation of organic substances initially present in the seawater that promote bacterial growth, leading to biofilm formation. A membrane biofilm is known to have good backwash resistance and even, in some cases, tolerance to biocides. The physical structure of a biofilm is compact and similar to an adhesive gel on the membrane. The importance of extracellular substances such as EPS or transparent exopolymers (TEP) explains the cohesion of microbial cells, as well as adhesion on membrane surfaces. Unlike most EPS, TEPs exist as individual particles rather than as cellular linings. They are characterized as transparent, sticky, gel-like substances and are mainly composed of acidic polysaccharides. They are hydrophilic and are known to exist in different shapes and sizes (~0.4–200 µm). Most TEPs are derived from polysaccharides released by phytoplankton and bacteria, which then coagulate to form TEPs. They are also lysed by macroalgae and some higher marine organisms. The majority of TEPs are sized between 1 and 3 nm in diameter and one to several hundred nanometers in length. They are flexible enough to pass through 8 kDa pore size membranes. Thus, TEPs can easily pass through the pores of an MF or UF membrane.

19.8.8.1.2 Algae and membranes

Seaweed species are larger than the nominal pore size of MF and UF membranes, and algae removal occurs by the steric exclusion mechanism.

For example, dinoflagellates are between 10 and 50 µm in size and easily pass through the pre-filters installed on the raw water. However, they are rejected by MF and UF membranes. Nevertheless, in high concentrations (in the range of 10^4 cells/mL), it quickly forms a thick cake layer that then impedes the passage of water. If the cells are damaged, either by natural mortality cycles or by shearing in the pumping system, they release organic matter that passes through the pre-treatment system to the RO membranes. The organic matter can foul the RO membrane directly and/or serve as a substrate for bacterial species in bio-fouling.

This being said, the retained live algae also release extracellular substances that block membrane porosity, resulting in a progressive decrease in flux and/or an increase in TMP.

Therefore, clogging of these membranes often occurs as a result of accumulation and/or adsorption of these substances, as this material layer should not be considered as a non-adhesive cake layer.

In the absence of pre-treatment upstream of the membranes, that is to say, in direct filtration on MF/UF membranes, algae and organic matter are in direct contact with the surface and the logic of pore blockage is easily explained. The released polysaccharides can bind to other organic species and increase the filtration resistance. For this reason, a direct UF system is not suitable for the treatment of algae-rich water.

In addition, the accumulation of extracellular substances (mucilaginous or gelatinous materials) is likely to serve as "cement" for the particles rejected by the membranes. Additionally, backwashing with water and chlorine have proven to be ineffective in several cases.

The different situations in which the membranes are clogged are:

– the presence of Fe and Mn, which causes mineral clogging that can be removed with citric acid or oxalic acid;

– EPS and TEP, generated by the presence of algae and bacteria, form a hydrated gel that can be partially removed with concentrated oxidants (e.g. NaOCl/NaOH) in a basic medium and at a temperature of 35°C;

– NOM adsorbed on the surface is cleaned with caustic soda (temperature 35°C) by hydrolysis and dissolution mechanisms;

– particles and colloids are simply removed during hydraulic backwashing.

The addition of coagulation upstream of the membranes can improve the permeability of the MF/UF by flocculating the deposited substances and forming a loose cake. Low molecular weight organic matter can be trapped in this layer and clogging is reduced. When treating algae-rich water, coagulation–flocculation can only adsorb and flocculate the algae cells, but cannot render them inactive. The algae can still release extracellular material and affect the membrane filtration. Coagulation pre-treatment alone cannot satisfy the requirements of the UF application in the treatment of algae-rich water.

The two water quality parameters showing the highest level of variation and sensitivity between conventional pre-treatment processes and MF/UF processes are bacteria counts and phytoplankton counts. The main observations are as follows:

– pre-treatment processes with direct dual media filtration provide limited removal of pico phytoplankton and bacteria (1 log), while membrane-based pre-treatment (MF/UF) provides high removal of both types of micro-organisms;

– while chlorophyll removal by granular filtration and membranes (MF and UF) are above 90%, pico phytoplankton and bacteria counts allow a better understanding of the performance of both systems, and can be a tool for the selection and optimization of the pre-treatment according to the quality of the seawater;

– SDI_{15} min values are lower after MF or UF pre-treatment, compared to dual media filtration. SDI_{15} min values lower than 3 can be achieved by coagulation–flocculation–dual media filtration pre-treatment when the raw seawater quality is quite good (seawater SDI_{15} min values lower than 25, no algae and TOC lower than 1.2 $mg \cdot L^{-1}$), and with optimal operating conditions (type and dosage of coagulant, type and dosage of polymer, optimal filtration speed).

With Cabot-Norit membranes (X-Flow SXL225), CEB sequences use acid and/or caustic soda/chlorine for the removal of reversible clogging, such as incipient biological growth or precipitated minerals, when it is not simply removed by hydraulic backwashing. The CIP procedure uses a 1,000 ppm citric acid solution for 1 h. A second citric acid solution is sent in a similar manner.

The seawater from the Palm Jumeirah plant is characterized by a very high seawater SDI, measured in $SDI_{2.5}$ min. Moreover, as is common in the Gulf marine waters, the contribution to SDI is given by very small particle sizes, and this explains the relatively low turbidity values observed upstream of the Norit UF membranes.

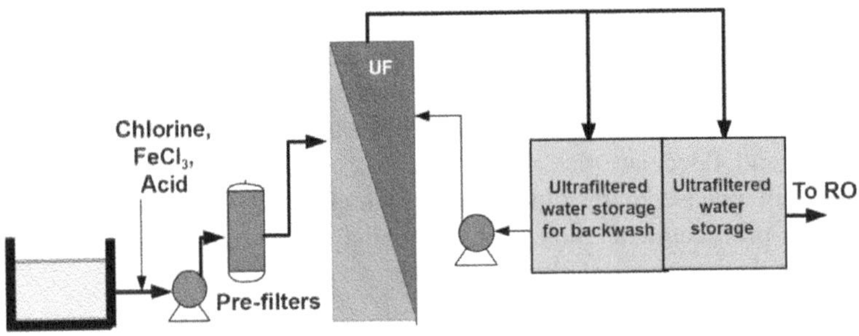

Figure 19.60. *Pre-treatment including Cabot-Norit UF membranes in Palm Jumeirah (Dubai). For a color version of this figure, see www.iste.co.uk/gaid/watertreatment4.zip*

On the other hand, excellent permeate quality was obtained during plant operation with excellent SDI permeate, regardless of the very high quality of the feed seawater (about 30 NTU). Measured SDI permeate values below 3 were recorded. Due to the presence of microparticles, a dosage of ferric chloride was given (0.25–0.50 mg $Fe \cdot L^{-1}$) and this led to deposits forming on the membrane surface, as well as an increase in operating pressure. Given that the CEBs could not completely restore the TMP, a cleaning-in-place was implemented with a solution containing a mixture of oxalic acid (1% $H_2C_2O_4$) and ascorbic acid (0.25% $C_6H_8O_6$), heated to 36°C. The results (Figure 19.59) showed good membrane permeability.

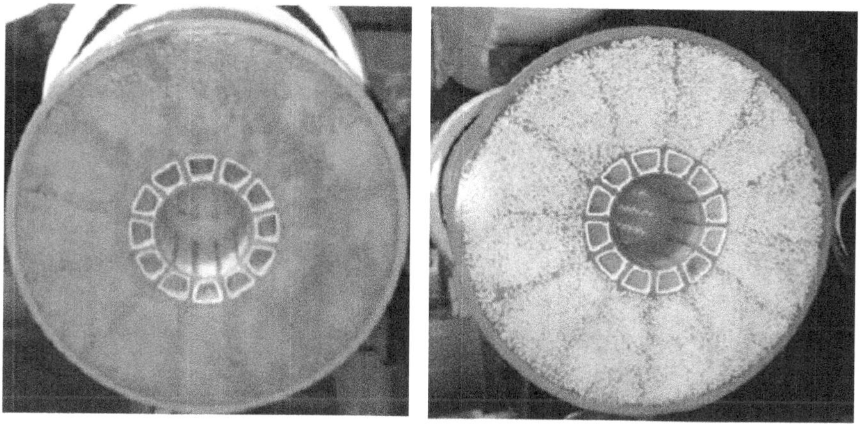

a) b)

Figure 19.61. *UF membrane (a) before and (b) after cleaning-in-place. For a color version of this figure, see www.iste.co.uk/gaid/watertreatment4.zip*

The UF pre-treatment on highly contaminated water proved to be particularly difficult. The autopsy of the modules showed irreversible clogging (brown to red color), which is explained by the high concentration of HS and the presence of iron. The use of a 2% oxalic acid solution heated to 35°C, which was circulated for 2 h and soaked for 3 h, restored the permeability.

In the same order of tests and autopsies on UF membranes operating directly on seawater, an autopsy of a Seaguard UF membrane (Cabot-Norit) revealed a dark red coloration caused by iron, with traces of silica and aluminum present. The most likely cause of this performance degradation is clogging with excess ferric chloride. Specific acid cleanings have restored the permeability.

Figure 19.62. *Autopsy before chemical cleaning. For a color version of this figure, see www.iste.co.uk/gaid/watertreatment4.zip*

Figure 19.63. *Cleaned module (oxalic acid). For a color version of this figure, see www.iste.co.uk/gaid/watertreatment4.zip*

Figure 19.64. *UF membrane with red coloration due to coagulant injection ($FeCl_3$). For a color version of this figure, see www.iste.co.uk/gaid/watertreatment4.zip*

19.8.9. *Conclusions on pre-treatment with UF membranes*

If there is any doubt about the long-term quality of the feed water, the following should be considered when designing a desalination plant with a UF pre-treatment:

– a UF membrane can provide effective treatment, but the RO membrane unit must be designed with a large enough safety factor;

– turbidity alone is not a good indicator of clogging potential;

– the UF unit should be designed using safe assumptions about flow and conversion rate. Space should be provided for additional membrane trains in case the predicted flow and conversion rate cannot be met;

– a CIP system must be installed along with conventional backwashing and CEB systems for the UF phase.

However, an irreversible clogging of the membranes must be considered and could be a limitation in the application of this technology, depending on the quality of the seawater.

19.9. Energy consumption

The improvement of the energy balance of a desalination plant has a direct impact on the cost per m^3 of water produced, but also has a positive impact on global warming because of the limited greenhouse gas emissions from power stations. Thus, for a desalination plant producing 100,000 $m^3 \cdot d^{-1}$, a gain in consumption of 0.1 $kWh \cdot m^{-3}$ makes it possible to save 3,468,000 $kWh \cdot year^{-1}$ with a utilization rate of 95%.

19.9.1. *Energy consumption without recovery*

For a flow rate Q of a liquid of density d at pressure P, the pumping power is equal to:

$$P \text{ pumping} = \frac{Qp\ Pr\ d}{36.7\ \eta}$$

with:

– P: pumping power (kW);

– Qp: permeate flux rate ($m^3 \cdot h^{-1}$);

– Pr: discharge pressure (bar);

– η: efficiency of the electro-pump unit (%);

– d: density of the liquid at temperature T (°C).

Taking the conversion rate Y into account, the energy spent per m^3 of permeate is equal to:

$$W\ (kWh \cdot m^{-3}) = \frac{Ppumping}{Qp} = \frac{\text{Pr}\, d}{36.7\ \eta\ Y}$$

The energy consumption is therefore proportional to the operating pressure, inversely proportional to the conversion rate and the efficiency of the electro-pump unit. For an applied pressure of 65 bars, a permeate flux rate of 5,000 $m^3 \cdot h^{-1}$, a density of 1,029, a unit efficiency η of 95% and a conversion rate of 42%, the energy consumption is 4.56 $kWh \cdot m^{-3}$.

19.9.2. *Energy consumption with recovery*

Energy recovery systems are implemented to recover the energy still present in the concentrate, at the outlet of the RO membranes. Energy recovery units can be hydraulic or the positive displacement type. It can be a Pelton turbine that is directly or indirectly coupled to the feed pump, or positive displacement systems such as ERI-PX, or Calder Dweer, or other similar systems. In these latter cases, the pressurized concentrate transfers its pressure to a portion of the feed water by direct or indirect contact.

The knowledge of the energy balance of the process makes it possible to choose the equipment and to design the installations by considering the energy parameter. This design is based on the hypotheses of salinity, temperatures and hydraulic conditions, taking into account the variations. Control loops are used to stabilize the system. The regulation of the energy recovery is part of the design parameters.

Considering the pressure drop ΔP in the modules and the efficiency of the recovery system, ηrec, the recovered energy is equal to:

$$W\,recov \cdot (kWh \cdot m^{-3}) = \frac{(\mathrm{Pr} - \Delta\mathrm{P})\eta rec}{36.7\,Y}\,\frac{(1 - Y)}{Y}$$

The energy actually consumed is therefore equal to the difference between the pumping energy and the energy recovered by the recovery system:

$$W\,recov \cdot (kWh \cdot m^{-3}) = \frac{\mathrm{Pr}\,d}{36.7\,\eta réc\,Y} - \frac{(\mathrm{Pr} - \Delta\mathrm{P})\eta rec}{36.7}\,\frac{(1 - Y)}{Y}$$

19.9.3. *Hydraulic exchanger systems*

19.9.3.1. *Pelton turbine*

The principle of the Pelton turbine is to transform hydraulic energy into mechanical energy, then back into hydraulic energy.

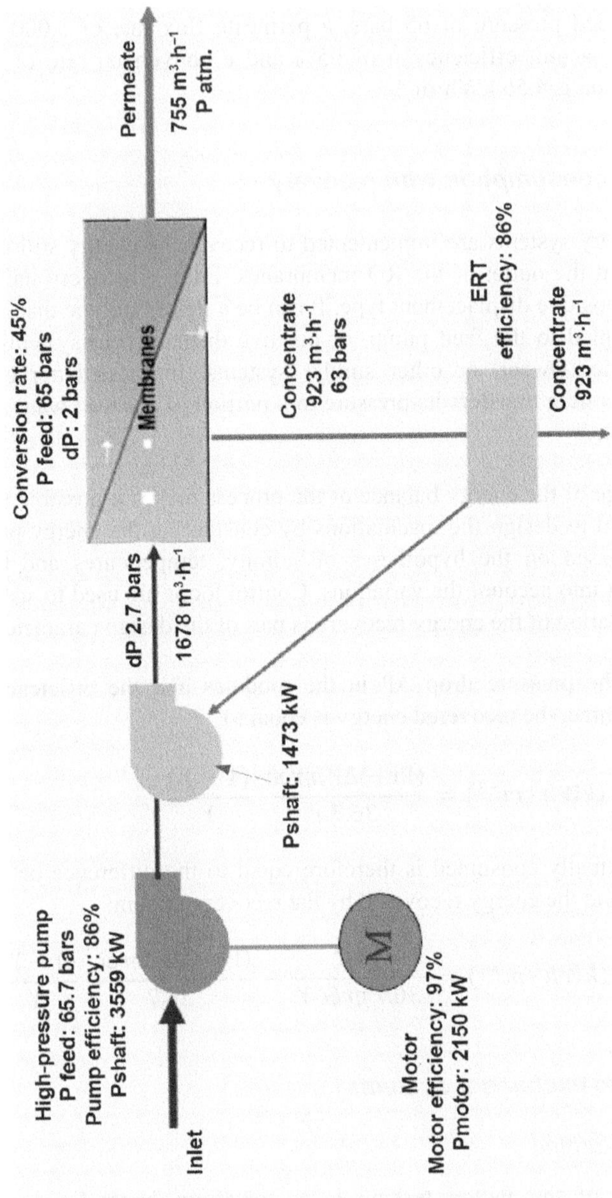

Figure 19.65. *Energy balance on a membrane unit with Pelton turbine. For a color version of this figure, see www.iste.co.uk/gaid/watertreatment4.zip*

The Pelton turbine is a partial injection, free stream turbine. The water's expansion to atmospheric pressure is entirely carried out in the machine's distributor. Energy is available at the wheel inlet, only in the form of kinetic energy. This turbine does not have a diffuser and the water flows freely out of the wheel. The Pelton turbine distributor is formed by one or several injectors. The injector consists of a nozzle in which the passage section is adjustable by the displacement of an injection needle or pin. The Pelton turbine wheel is a disc carrying double spoon-shaped buckets with a median edge and an indentation. The axis of the jet coming out of an injector is tangent to the circumference of the wheel, and the wheel diameter is, by definition, the nominal diameter of the machine. The water hits the half-buckets symmetrically with respect to the edge that separates them. It leaves the injector at about twice the average tangential velocity of the buckets: the shape of the indentation and the number of buckets are determined, so that the entire flow works in the buckets without loss. Efficiencies are around 80–85%.

Figure 19.66. *Pelton turbidity (Morocco-Laayoune site). For a color version of this figure, see www.iste.co.uk/gaid/watertreatment4.zip*

19.9.3.2. *Systems with pressure exchangers*

Pressure exchangers directly transfer the residual energy from the concentrate to the feed water via a booster pump, which compensates for the pressure losses in the modules, in the connecting pipes and in the energy exchangers.

19.9.3.2.1. Dweer (Dual Work Exchanger Energy Recovery)

The Dweer (Calder) energy recovery system is based on linear pressure exchangers. Schematically, these systems can be considered as two piston pumps operating alternately, using the concentrate discharge (which is around 60–70 bar in the case of seawater) to "assist" the piston movement. The principle of the Dweer pressure exchanger is based on a cylindrical exchanger, in which the pressurized concentrate drives the seawater into the membrane plant and, alternatively, the seawater drives the concentrate into the discharge.

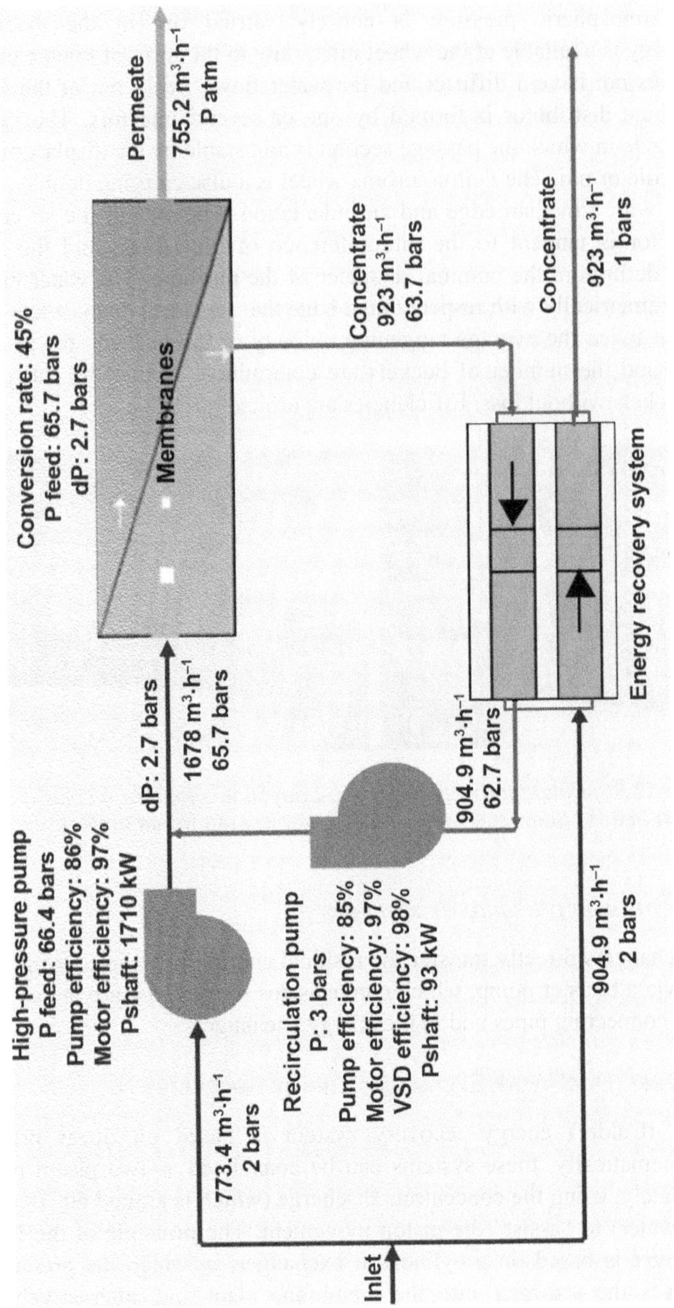

Figure 19.67. *Energy balance on a membrane unit with the Dweer system (Calder). For a color version of this figure, see www.iste.co.uk/gaid/watertreatment4.zip*

In a Dweer, the HP concentrate (60–70 bar) from the membranes is directed to an exchanger filled with seawater, which pressurizes this water to the same pressure as the concentrate. A recirculation pump is used to equalize the pressure of the seawater leaving the exchanger with that of the HP pump. The seawater then joins the flow from the HP pump to form the feed flow to the membrane. A set of valves and piping ensures the alternate distribution of the fluids, and a piston in the exchanger separates the fluids.

As the end of the half-cycle approaches, the LinX valve diverts the HP concentrate to the second exchanger. The low-pressure seawater then fills the exchanger, discharging the concentrate that was there. A new cycle then begins again. Two cylinders thus work in tandem to ensure continuity of all flows.

The unit flows treated with the Dweer are high (250 $m^3 \cdot h^{-1}$) and involve less equipment than other systems that deliver five times less flow per unit. Moreover, the Dweer 1100 can handle an additional flow without adjusting its operating parameters. On the other hand, one of the disadvantages of the Dweer is a salinity increase in the membranes, which results in an increase in pressure in the event of a machine shutdown. For example, with 3 Dweer, shutting down one of them would lead to a 16% pressure increase on the membranes.

Figure 19.68. *Calder-Dweer energy recovery system. For a color version of this figure, see www.iste.co.uk/gaid/watertreatment4.zip*

Mixing between the HP concentrate inlet flow and the seawater (HP) feed to the membranes is very low, due to a strong physical barrier (piston) that separates these two flows. Any failure can be detected by sensors positioned on the LinX valve, as well as on the vessels, which allow continuous monitoring of the exchanged flows. As a result, overflow levels are controlled but not zero, according to feedback from the field.

Projection software from membrane manufacturers shows that mixing levels are less than 1.5%. Therefore, there is little impact on membrane feed pressure to achieve a given permeate flow rate. However, feedback from the field shows that in some cases, a 1.2% increase in salinity on the membranes or about a 1.2% increase in membrane pressure.

With the Dweer 1100, the unit leakage rate is 4.3 $m^3 \cdot h^{-1}$ and corresponds to approximately 1.8% of the inlet flow of the HP concentrate. It represents a direct leakage of water from the system and thus a direct loss of the pressurized feed flow.

The Dweer 1100 indicates hydrodynamic efficiency at high pressure that can reach 97% under optimal conditions. However, in reality, it is more likely to be between 94.5 and 96%.

At the LP differential, the Dweer 1100 only requires 1.6 bar and does not require back pressure to empty the tubes. The noise level associated with the Dweer 1100 is 80 dB, because the mechanism itself is slow-moving. The Dweer are stacked 3 high and occupy significant floor space. A Dweer 1100 is very heavy technology, weighing 5,750 kg.

19.9.3.2.2. ERI-PX (Energy Recovery Inc.)

The ERI energy recovery system is based on rotating pressure exchangers. The ERI-PX system is an isobaric energy recovery device. These pressure exchangers consist of a rotating ceramic cylinder with holes in it, and operate on the principle of a barrel gun. The system has no valves or pistons. The pressure of the concentrate drives the device at a speed of 1,200–1,500 revolutions per minute. It has a HP concentrate inlet port and a low-pressure (LP) concentrate outlet port, as well as HP seawater outlet and LP seawater inlet ports. The ceramic material is chosen for its corrosion resistance and durability.

Recovery efficiency is over 95%. Parallel operation of the PX220 units requires precision flow control and distribution. As additional flow is often required, it implies additional equipment for flow management. However, this additional equipment provides a benefit because, if one of the PX units fails, the salinity of the membrane feed and the membrane pressure only increase slightly. For example, in a plant with 15 units, the PX220 will only experience a 3% increase in salinity if one unit fails.

The mixing between the HP concentrate inlet and the seawater feed (HP) is very delicate, because the PX220 does not have a physical barrier between the concentrate and the feed water (such as a piston). The operation of the PX is based on simple salinity measurements at the outlet of each PX unit.

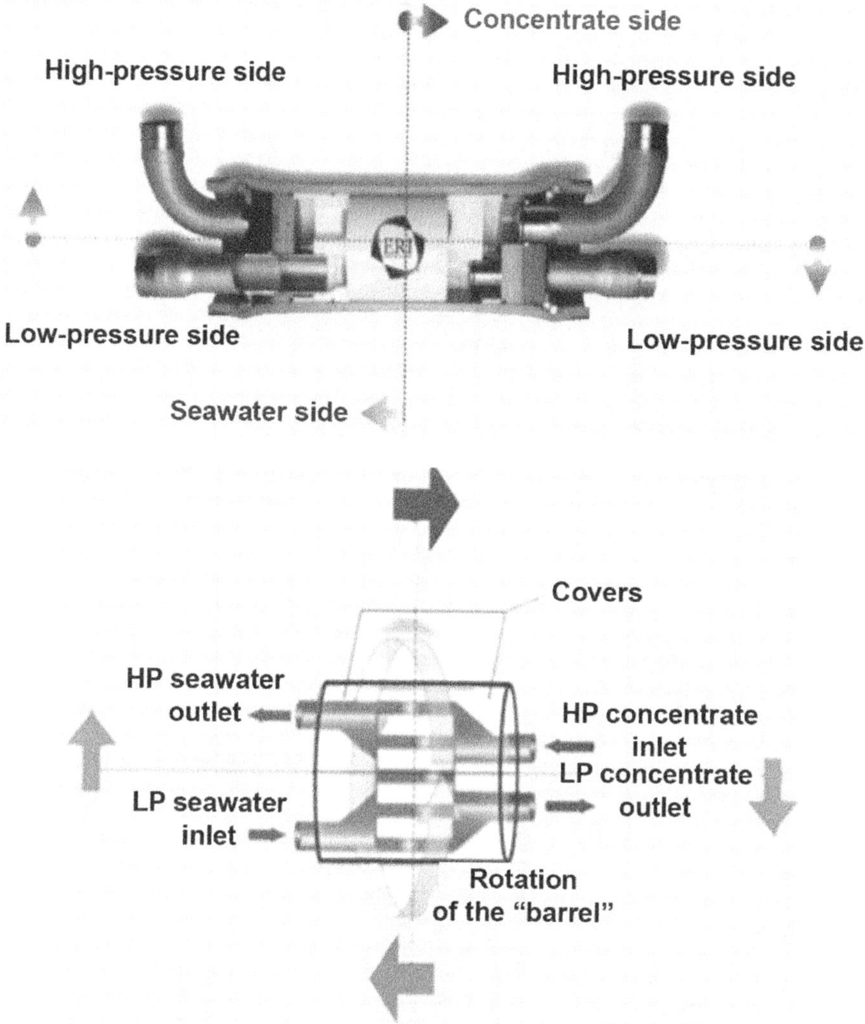

Figure 19.69. *Diagram of the concentrate distribution system. For a color version of this figure, see www.iste.co.uk/gaid/watertreatment4.zip*

Projection software from membrane manufacturers shows that mixing levels are in the range of 5–7%, and can impact the pressure required to feed the membranes and achieve a given permeate flow rate. The additional feed pressure required in a PX220-based plant can be as high as 2 bar.

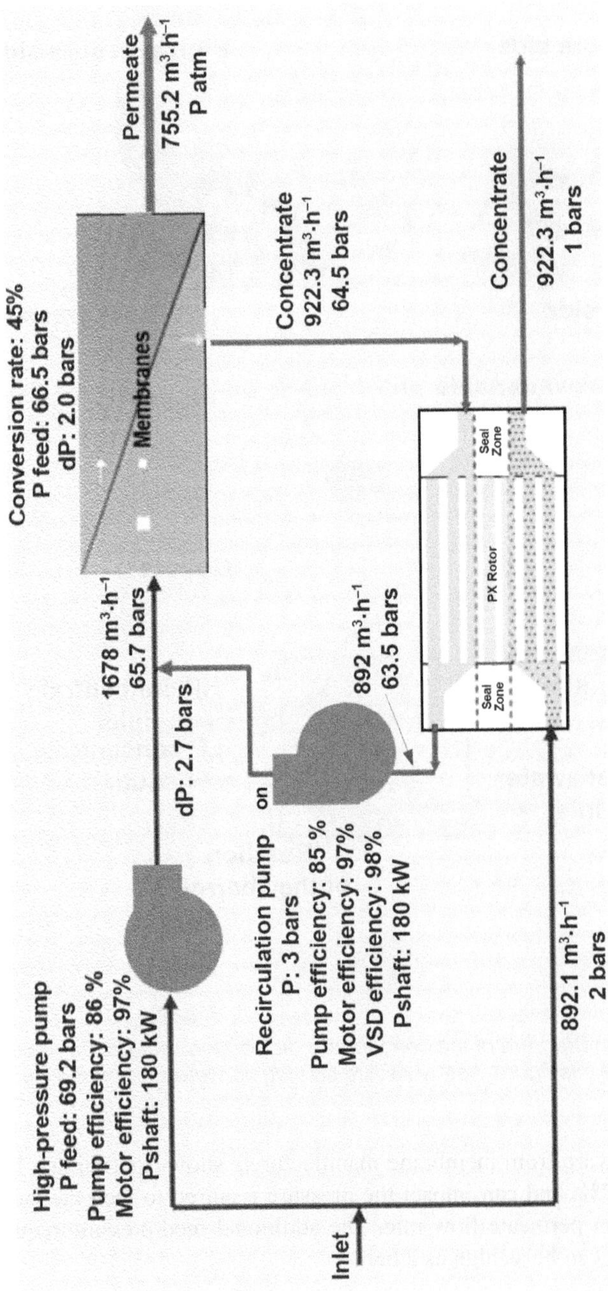

Figure 19.70. *Energy balance on a membrane unit with an ERI system. For a color version of this figure, see www.iste.co.uk/gaid/watertreatment4.zip*

The leakage rate can reach up to 8 $m^3 \cdot h^{-1}$. This would be a direct loss of the pressurized feed flow. This means that the HP pump has to pump more water, and therefore incurs additional costs for the pre-treatment, even if the guarantee given by the supplier is less than 1.8%.

Figure 19.71. *PX energy recovery system (Energy Recovery Inc.). For a color version of this figure, see www.iste.co.uk/gaid/watertreatment4.zip*

The expected differential pressure efficiency of the PX220 is around 95.9%. The noise level associated with the operation of the PX220 units can reach 94 dBA, because the PX rotor rotates at high speed. They therefore require the use of hearing protection.

In terms of footprint, PXs occupy a small area which is, in some cases, an attractive advantage. A PX unit can be lifted manually by two people and is fully repairable on site.

Example of design for a flow rate of 24,000 $m^3 \cdot d^{-1}$

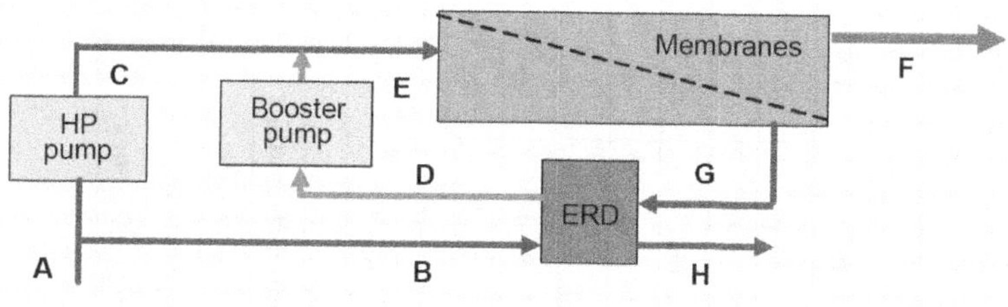

Figure 19.72. *Distribution of flow rates and parameters for each unit. For a color version of this figure, see www.iste.co.uk/gaid/watertreatment4.zip*

ERI-PX		A	B	C	D	E	F	G	H
Flow rates	$m^3.h^{-1}$	1,818	983	835	983	1 818	818	1,000	1,000
	$m^3.d^{-1}$	43,636	23,589	20,046	23,589	43,636	19,636	24,000	24,000
Pressures	**bar**	1.8	1.8	65.0	62.1	65.0	0.0	63.0	1.0
Salinities	$mg \cdot L^{-1}$	40,500	40,500	40,500	42,593	41,615	350	75,378	73,321
Dweer		**A**	**B**	**C**	**D**	**E**	**F**	**G**	**H**
Flow rates	$m^3.h^{-1}$	1,848	1,012	836	982	1,818	818	1,000	1,030
	$m^3.d^{-1}$	44,356	24,288	20,068	23,568	43,636	19,636	24,000	24,720
Pressures	**bar**	2.5	2.5	63.6	60.6	63.6	0.0	61.6	1.0
Salinities	$mg \cdot L^{-1}$	40,500	40,500	40,500	41,000	40,766	350	73,834	72,386

Membrane differentials	**bar**	2.0	**Efficiency booster**	**%**	81
Conversion rate	**%**	45	**DP pre-filters**	**bar**	2.0
Temperatures	**°C**	25	**Additional pump efficiency**	**%**	80
HP pump efficiency	**%**	81			

Characteristics of the energy recovery system (ERD)		Dweer 1100	PX220	Characteristics of the energy recovery system (ERD)		Dweer 1100	PX220
ERD units		5	23	Increased salinity @ membrane	%	0.7	2.8
Flow rate/unit	$m^3 \cdot h^{-1}$	200	43	ERD efficiency	%	94	96
Over-flush	%	3	0	HP pump power	kW	1,759	1,817
Lubricant flow rate/unit	$m^3 \cdot h^{-1}$	18	17	Booster pump power	kW	101	98
Lubricant	%	1.8%	1.7%	Additional pump power	kW	289	239
HP differential	Bar	1.0	0.9	Total power	kW	2,150	2,153
BP differential	Bar	1.5	0.8	Specific energy consumption	$kWh.m^{-3}$	2.63	2.63
Volumetric mix	%	1.5%	6.0%				

Table 19.20. *Energy characteristics of the two energy recovery systems*

Parameters	Pelton turbine	Calder Dweer	ERI
Advantages	References, known and safe technology Simplified automation Low Capex Simplicity and robustness	Fewer units per train, therefore simpler flow distribution The physical barrier (the piston) reduces mixing (mixing of the concentrate with the raw water)	Capex lower than Dweer Flexibility and efficiency Low maintenance: the only moving part is lubricated by the process The new PX220 ($50\ m^3 \cdot h^{-1}$) has now been proven in small and medium-sized plants
Disadvantages	Energy recovery by the turbine < 90 Overall low efficiency (turbine + motor + pump efficiency) < 85 Little flexibility in actual production rate	More expensive to buy, much heavier, takes up more space Probably wears out less quickly than an ERI because of the lower rotation speed So, over the life of the contract, the costs probably balance out	Constraining operation and maintenance The absence of a physical barrier between concentrate and raw water leads to a mixing of about 6% → 1.5 bar additional pressure in the RO feed The life span of the ceramic rotors is not known Estimate: 8 years
Range of flow rates	10–1,200 $m^3 \cdot h^{-1}$ + arrangement of units	250 $m^3 \cdot h^{-1}$ per unit + arrangement of units	from 3.4 to 50 $m^3 \cdot h^{-1}$ per unit + arrangement of units
Overall recovery efficiency	About 85%	Up to 96.5%	95%
Energy consumption (same assumptions)	Around 3.5 $kWh \cdot m^{-3}$	Around 2.7 $kWh \cdot m^{-3}$	Around 2.8 $kWh \cdot m^{-3}$

Table 19.21. *Advantages and disadvantages of the three energy recovery systems*

19.9.3.3. *Comparison of the three main energy recovery systems*

Table 19.21 summarizes the respective advantages and disadvantages for the three energy recovery systems. Figure 19.73 shows an example of an energy balance for an RO plant with a pre-treatment consisting of coagulation–flocculation–dual media filtration, a RO membrane plant including pre-filters and the energy recovery system, post-treatment with limestone filters and disinfection, and the storage tank. Sludge treatment and its removal are also included.

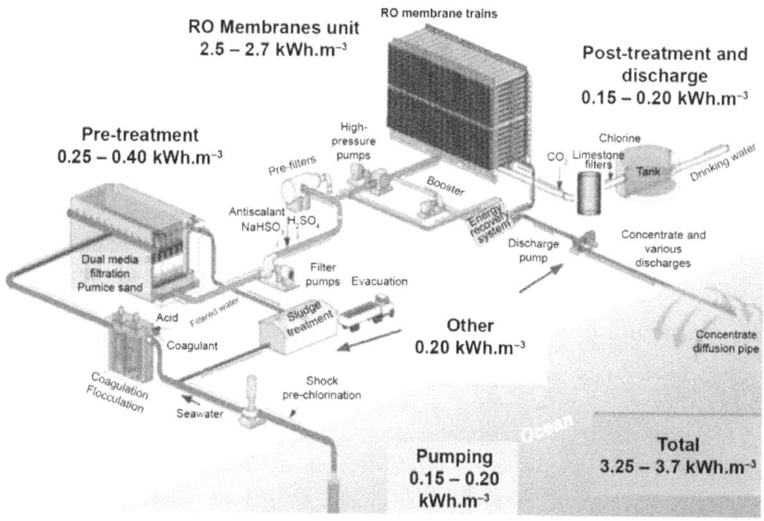

Figure 19.73. *Example of energy balance for a desalination plant. For a color version of this figure, see www.iste.co.uk/gaid/watertreatment4.zip*

It can be seen that the membrane unit accounts for more than 70% of the total energy consumption, while pre-treatment accounts for only 7–10%.

19.10. Operating parameters

19.10.1. *Relationship between conductivity and salt concentration*

A quick approximation (20°C) of the relationship between salt concentration in seawater and conductivity is given by:

$$\text{Total dissolved salts } (\text{mg}\cdot\text{L}^{-1}) = 0.5604\ \sigma^{1.026}$$

where σ is the conductivity in millisiemens per cm.

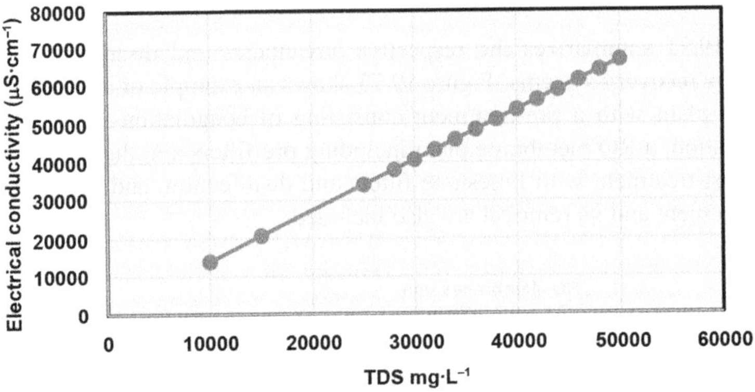

Figure 19.74. *Relationship between total dissolved salt concentration in the feed (and concentrate) according to electrical conductivity. For a color version of this figure, see www.iste.co.uk/gaid/watertreatment4.zip*

19.10.2. *Controlling RO membrane clogging*

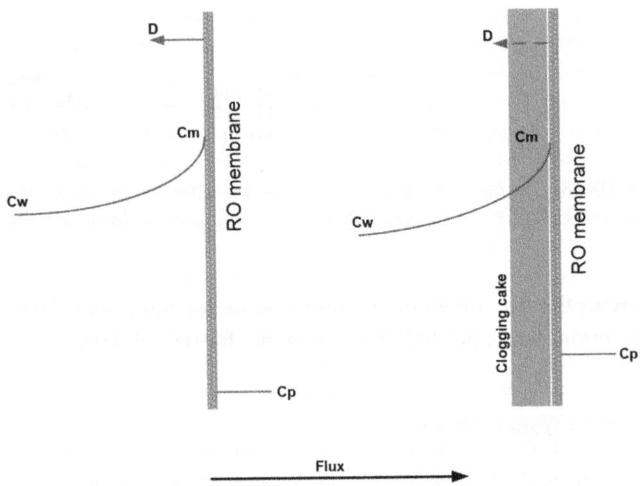

Figure 19.75. *Clogging of a RO membrane. For a color version of this figure, see www.iste.co.uk/gaid/watertreatment4.zip*

Figure 19.75 illustrates the formation of a clogging layer on the membrane, where Cw is the concentration of salts in water and Cm is the concentration at the membrane interface, which is always higher than Cw. D is the salt diffusion coefficient and Cp is the concentration of salts in the permeate. The formation of the cake creates a flow reduction and is a barrier to salt diffusion by reducing the coefficient D.

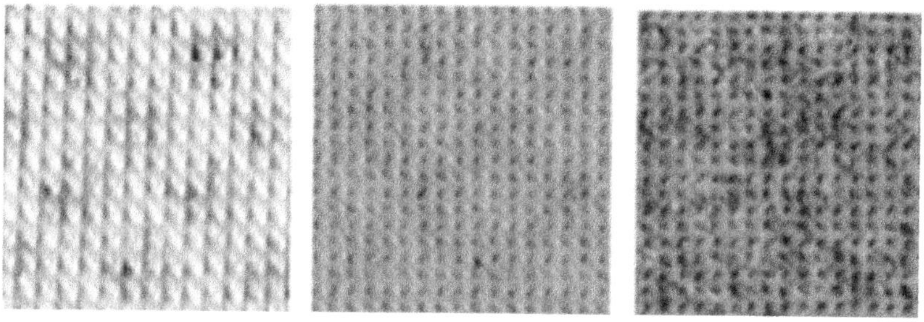

Figure 19.76. *Sections of three RO membranes clogged with microalgae. For a color version of this figure, see www.iste.co.uk/gaid/watertreatment4.zip*

It is important that the pre-treatment is well adapted and well controlled, so as to avoid the cogging of the membranes and their malfunctioning.

19.11. Performance of RO membranes used in desalination

19.11.1. *Boron removal*

The boron issue appeared during the development of desalination plants, whose water treated by RO was used in irrigation. The residual boron concentration in the produced water caused many crops to degrade. Indeed, even if boron is an essential element for plant growth, it is only useful at relatively low concentrations and an increase in these concentrations makes it toxic. For some crops, a concentration between 1 and 2 $mg{\cdot}L^{-1}$ can be toxic. Boron's toxicity to plants occurs by accumulation in the leaves. Symptoms of toxicity are yellowing and/or spotting, or drying of leaf tissue at the tips and edges. On severely affected trees, such as almonds and other tree crops that do not show typical foliar symptoms, a gum on the branches or trunk is often observed. Most crop toxicity symptoms occur when boron concentrations in leaf blades exceed 250–300 $mg{\cdot}kg^{-1}$ (dry weight), but not all sensitive crops accumulate boron in leaf blades. For example, stone fruits (peaches, plums, almonds, etc.) and pome fruits (apples, pears and so on) are easily damaged by boron, but they do not accumulate enough boron in the leaf tissue to make foliar analysis a reliable diagnosis. Its effect depends on the concentration, the sensitivity of the crops and the total volume of water used over time.

Boron toxicity can affect almost any crop but, like salinity, there is a wide range of tolerance between crops.

19.11.1.1. *Presence in water*

In water, most of the earth boron is found in the oceans, with an average concentration of 4.5 $mg \cdot L^{-1}$. Very similar concentrations are found in coastal wells. Estuarine waters are generally rich in boron, due to municipal and industrial wastewater discharges. Concentrations in conventional surface waters are lower (<0.3 $mg \cdot L^{-1}$) and the forms of boron present are not considered harmful to health. The WHO recommends an acceptable daily intake of about 0.4 mg per kg of body weight, which takes into account the physiologically beneficial effects for humans. In recent years, the WHO has revised the limit value for drinking water to 2.4 $mg \cdot L^{-1}$ instead of 0.5 $mg \cdot L^{-1}$, which was one of its recommendations for a long time.

19.11.1.2. *Boron chemistry*

Boron is often present in water as boric acid (H_3BO_3 or structural $B(OH)_3$). It acts as a weak Lewis acid due to the electron deficiency of boron. It tends to combine with the OH^- ion to form the $B(OH)_4^-$ borate ions at high pH. It dissociates according to the following reactions:

$$H_3BO_3 + H_2O \rightarrow H_2BO_3^- + H_3O^+ \ pKa = 9.14$$

$$H_2BO_3^- + H_2O \rightarrow HBO_3^{2-} + H_3O^+ \ pKa = 12.74$$

$$HBO_3^{2-} + H_2O \rightarrow BO_3^{3-} + H_3O^+ \ pKa = 13.80$$

Its concentration is usually expressed as total boron, which includes all species:

$$[B] = [H_3BO_3] + [H_2BO_3^-!] + [HBO_3^{2-}] + [BO_3^{3-}] \text{ in } mgB \cdot L^{-1}$$

In the usual pH operating range, this equation is the most important one because of the presence of both dissociated and undissociated boric acid species in water. At low pH values, the major species is boric acid in molecular form. Due to the absence of ionic charges, hydration of the molecule cannot take place completely. At the pH of seawater (pH 7–8), undissociated boric acid (H_3BO_3) is the dominant form with 99.3% presence at pH 7 and 93% at pH 8 (Figure 19.77). This results in a smaller size of molecule and less rejection by membranes. On the other hand, at higher pHs, the dissociated form is fully hydrated, inducing a larger radius, in turn resulting in better rejection by exclusion and repulsion by negatively charged membranes.

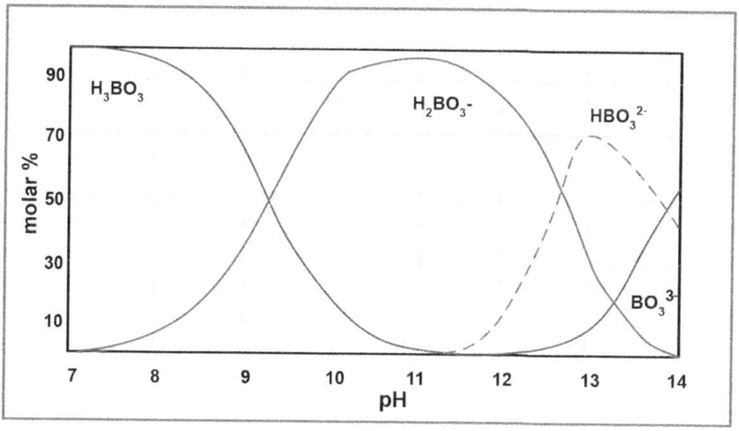

Figure 19.77. *Chemical forms of boron according to pH. For a color version of this figure, see www.iste.co.uk/gaid/watertreatment4.zip*

19.11.1.3. *Boron treatments*

Due to its small size and the configuration in which it occurs naturally in water (non-ionized boric acid), boron removal is quite complex and cannot be achieved by conventional technologies. Conventional water treatment (coagulation, flocculation, settling or flotation and filtration) removes very little boron from any type of coagulant.

Technologies that are not very applicable to municipal desalination, such as magnesium oxide treatment, can reach 90% removal, or activated carbon with a capacity of 1 $mg \cdot g^{-1}$ at pH 8–9, can remove about 60%.

Since it can be removed in its ionized form, the methods used are:

– adsorption and precipitation by mineral compounds;

– adsorption and complexation by organic compounds;

– ionic exchange and HP membranes.

The main methods currently used in desalination plants are ion exchange resins and nanofiltration or RO membranes.

19.11.1.4. *HP membranes*

The problem of boron in RO results from the dissociation of boric acid as a function of pH. At the operating pH of RO (pH: 6–7), boron is in a non-ionized form (H_3BO_3) and is therefore not retained by the membranes.

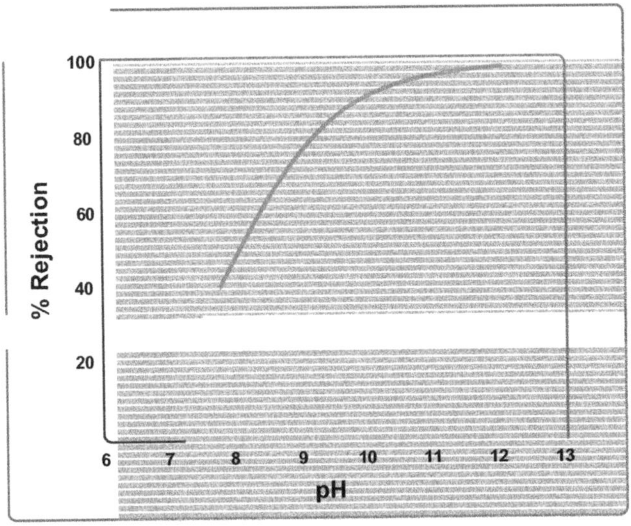

Figure 19.78. *Boron removal on a reverse osmosis membrane according to pH (45% conversion rate). For a color version of this figure, see www.iste.co.uk/gaid/watertreatment4.zip*

The effect of pH (Figure 19.78) shows that it is one of the main parameters controlling rejection of boron removal by the HP membranes (NF and RO). The RO membrane rejects about 46% and 77% of the boron at pH 8 and 9, respectively, because under these conditions, boric acid (a weak acid with pKa = 9.15) is neutral and is not hydrated in water. This species can pass through both RO and NF membranes, due to its small size (60.1 Å) and molar volume (40 $cm^3 \cdot mole^{-1}$). At pH > 10, the percentage of boron removal reaches 96% because it is in the form of $H_2BO_3^-$.

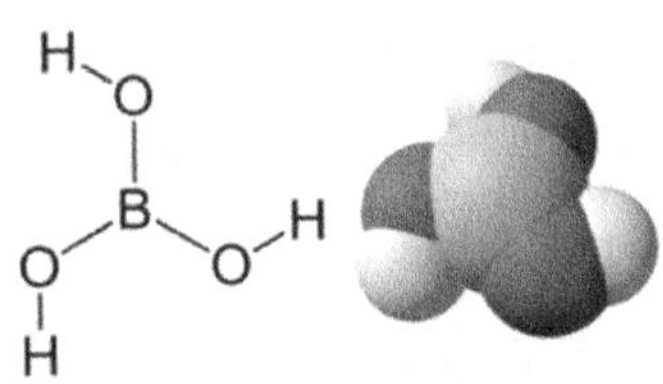

Figure 19.79. *Boric acid. For a color version of this figure, see www.iste.co.uk/gaid/watertreatment4.zip*

Boron release from RO membranes is also affected by permeate flux and temperature. In the case of seawater and brackish water desalination, it is not possible to increase the pH as high to achieve maximum boron removal by the membrane process, due to the problem of salt precipitation leading to membrane fouling.

As a result, the permeate does not comply with the standards when the required qualities are between 0.4 and 1.0 mg·L^{-1} of boron, especially when the temperature of the seawater feeding the membranes is above 25°C, which is very often the case. On the other hand, a value of 2.4 mg·L^{-1} is more easily obtained.

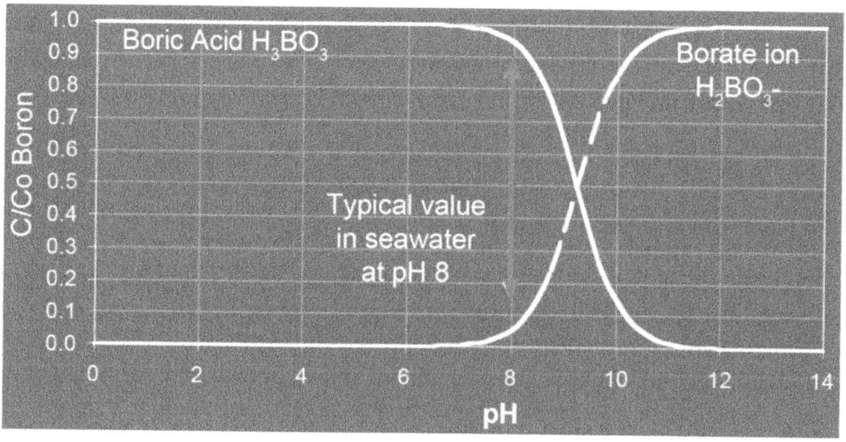

Figure 19.80. *Presence of boron forms in seawater. For a color version of this figure, see www.iste.co.uk/gaid/watertreatment4.zip*

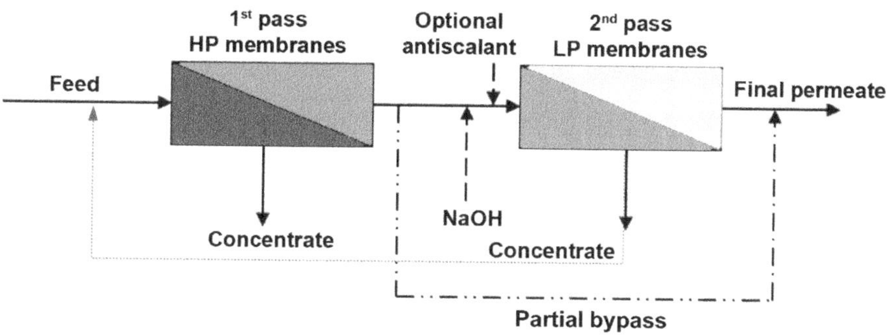

Figure 19.81. *Boron removal with two passes and intermediate pH adjustment. For a color version of this figure, see www.iste.co.uk/gaid/watertreatment4.zip*

In a two-pass configuration, all or a portion of the permeate from the first pass is pH adjusted, then processed by brackish membranes in a second pass unit (Figure 19.81). As with the SWC membrane, the pH increases in the second pass stream, resulting in an increased ionization rate of boric acid and subsequently leading to higher boron rejection by the brackish membranes. Due to risk of scaling at higher pH, the practical pH limit of the second pass is about 10.5, which corresponds to a boron rejection rate of approximately 95% in brackish membranes. In order to obtain low boron values in the produced water, a second pass through the membranes is performed at high pH (pH 10). As the first pass has removed most of the salts, there is no risk of precipitation. The permeate from the first stage is totally or partially fed to a second stage conditioned at a pH between 9 and 10, depending on the boron concentration.

The second pass operates with low-pressure RO membranes at a higher flow (between 25 and 32 Lmh) and a much lower pressure (Figure 19.81). The conversion rate can be 90% or higher.

Depending on boron level requirements and operating conditions, the second pass may either be a full or partial second pass design. As shown in Figure 19.82, the first pass permeate is collected from both ends of the first pass pressure vessels. The low salinity (and low boron concentration) fraction obtained from the feed end is used for blending. The high salinity fraction obtained from the concentrate end is processed with the second pass unit after pH adjustment. Due to high feed pH, special attention should be paid to calcium and magnesium concentrations in the concentrate in order to avoid scaling.

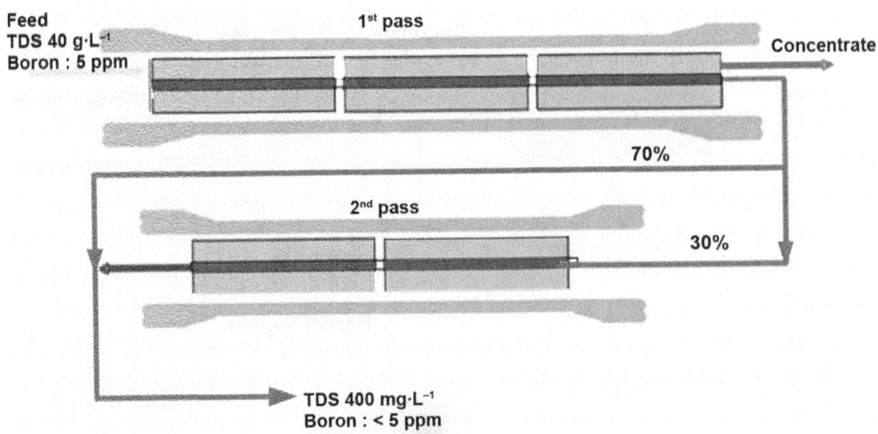

Figure 19.82. *Example of the application of the second pass for boron removal. For a color version of this figure, see www.iste.co.uk/gaid/watertreatment4.zip*

The lack of charge and small size of the boric acid molecule results in poor rejection by an RO membrane. However, at high pH, boron is present as the borate ion, $B(OH)4^-$, which is well rejected by the RO membrane due to its larger radius and negative charge. Figure 19.83 shows estimated nominal boron rejections of the SWC membrane at 25°C and 30°C, as a function of feed pH. Further discussion on how the RO system designer can take full advantage of this change in rejection as a function of pH can be found in the next section.

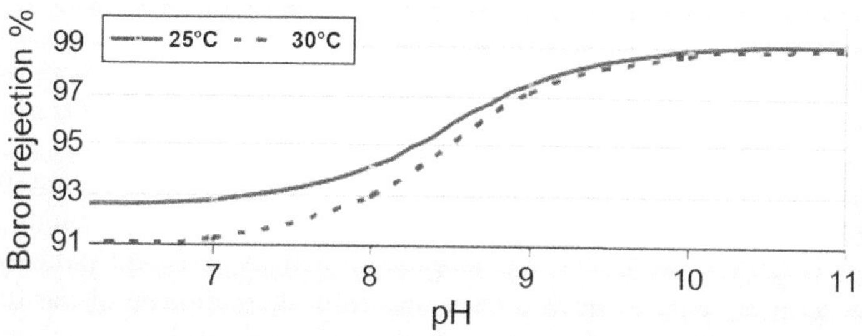

Figure 19.83. *Boron rejection rate according to temperature (Toray membrane). High boron rejection membranes are available on the market (Toray, Filmtec) with over 95% efficiency rates. For a color version of this figure, see www.iste.co.uk/gaid/watertreatment4.zip*

Membranes	SW30-380	SW30HR-380	TM820-370	TM820A-370	SWC3	SWC4
DOW Filmtec	88% (85–90%)	90% (88–92%)	–	–	–	–
Toray	–	–	92% (91–93%)	95% (94–96%)	–	–
Hydranautics	–	–	–	–	89% (88–90%)	92% (91–93%)

Table 19.22. *Boron removal with membranes from three manufacturers (pH 9-10)*

19.11.1.5. *Ion exchange resins*

The use of boron selective resins seems to be at the top of industrial interest. The specific resins consist of macroporous polymeric backbone carriers and a very functional group. The functional groups consist of vis-diols and N-methyl-D-glucamine groups. The vis-diol groups (hydroxyl groups) have a high selectivity toward boron by creating different stable borate complexes. The glucamine group controls the pH of the solution by neutralizing the proton released during the dissociation reaction, and thus keeps it at a stable value. Selective resins such as Diaion-CRB 02 and Dowex-XUS 43594, Amberlite IRA743, Amberlite XE243, Amberlite IRN78, Purolite S108 and S110 meet these criteria.

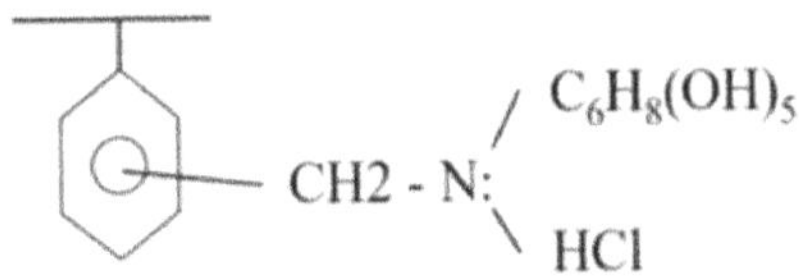

Figure 19.84. *Glucamine functional group of Dowex XUS43594 resin*

Should a low TDS require a second pass RO, a combination of the two treatments can optimize plant economics, while bringing maximum safety and flexibility (Figure 19.85). The alternative solution to a second pass by RO is to totally or partially feed a specific resin unit from the permeate of the first stage of RO membranes. The second pass is much smaller, only reducing salinity while the boron reduction is taken care of by the ion exchange system. The second pass does not have to be operated at a high pH, which saves caustic and antiscalant, allowing the use of a low-energy membrane.

These resins have very good efficiency (>90%), but their regeneration is relatively complex. This solution is not widely adopted in desalination plants.

Figure 19.86 shows that the efficiency of these two resins can reach 98% when injecting resin dosages greater than 2.5 $g \cdot L^{-1}$.

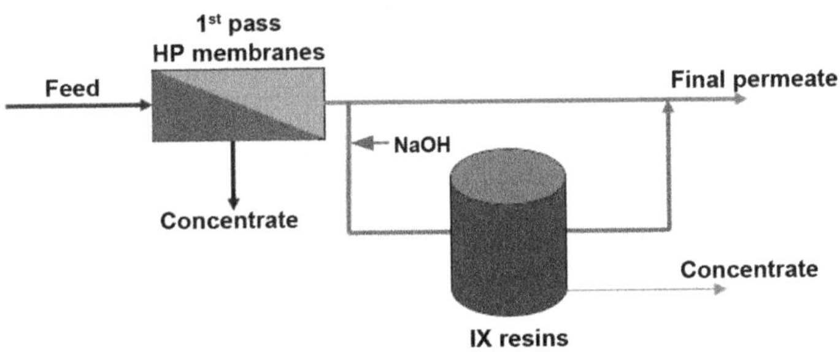

Figure 19.85. *Boron removal diagram with a combination of RO membranes and IX resins. For a color version of this figure, see www.iste.co.uk/gaid/watertreatment4.zip*

Characteristics	Diaion CRB 02 Mitsubishi	Dowex XUS Dow Chemicals
Form	OH	OH
Density, $g{\cdot}L^{-1}$	635	656
Exchange capacity	0.6 eq acid$\cdot L^{-1}$	0.7 eq$\cdot L^{-1}$
Granulometry, µm	355–500	500–600
Uniformity coefficient	1.6	1.1
Max. temperature, °C	100	–
Effective pH	6-10	–
Total boron capacity, $gB{\cdot}L^{-1}$	–	3.4–4.0

Table 19.23. *Characteristics of IX resins for boron removal*

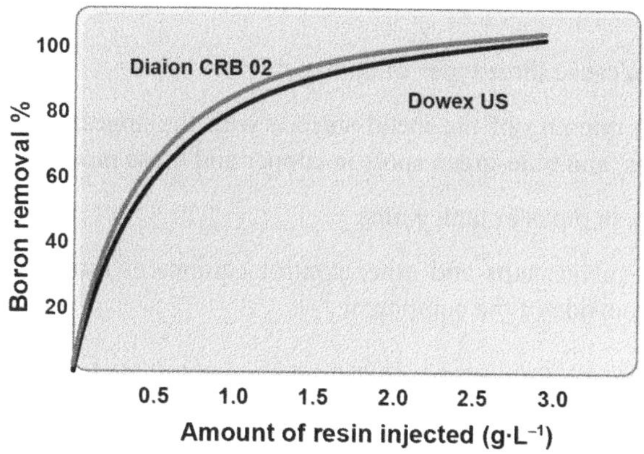

Figure 19.86. *Boron removal efficiency with two specific resins (boron input: 5 mg·L^{-1}, T: 30°C, pH: 8.2–9). For a color version of this figure, see www.iste.co.uk/gaid/watertreatment4.zip*

19.12. Post-treatment

The membrane processes used in desalination produce a permeate which is very poor in minerals and requires a post-treatment, leading to an enrichment in calcium salts in the produced water. The distributed water must meet quality requirements with respect to human health, not generate a change in quality during distribution and not degrade the transport and distribution equipment (no aggressiveness or corrosiveness).

In order to achieve these objectives, water alkalinity, pH and calcium hardness should be considered as the three main parameters of the remineralization process.

Given that desalinated water has no alkalinity or buffering capacity, it has aggressive and corrosive properties. This type of water leads to the dissolution of iron in the pipes, which results in a reddish color at the tap and is the most visual form of degradation. This is because the high corrosive and aggressive potential of the permeate promotes the dissolution of other metals such as lead.

Corrosion is a natural process that occurs when metals react with oxygen, forming metal oxides. The rate of corrosion depends on many factors such as water pH, electrical conductivity, oxygen concentration and temperature, especially since metals dissolve when the water is extremely low in dissolved salts.

The iron dissolution mechanism is based on the presence of a chloride concentration that attacks the metal surface of the pipes, resulting in the release of ferrous compounds into the water. The presence of chlorides or sulfates and low alkalinity and hardness are the predominant features of the corrosion mechanism.

Corrosion can cause three types of damage:

– progressive thinning of the metal surface with the appearance of red spots in steel or iron pipes, and blue-green spots in copper and brass pipes;

– perforations in pipes or tank walls;

– damage to valves, taps and other control equipment through the creation of leaks inside and outside of the equipment.

The consequences of water quality degradation on human health are:

– copper and lead diffusion in water can be toxic;

– copper contamination can cause gastrointestinal problems in the short term and damage the liver and kidneys in the long term;

– lead contamination can cause physical and mental development problems in children. In adults, it can cause high blood pressure and kidney problems.

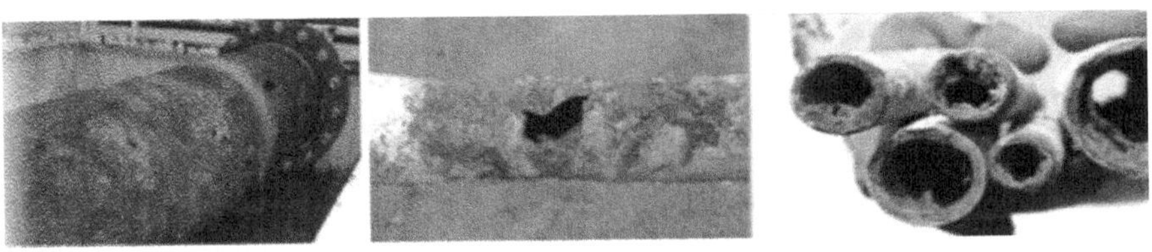

Figure 19.87. *Pipe corrosion. For a color version of this figure, see www.iste.co.uk/gaid/watertreatment4.zip*

19.12.1. *Indicators characterizing the aggressiveness or corrosiveness of the water*

Several models or indicators characterize the aggressiveness or corrosiveness of water.

19.12.1.1. *Langelier index*

This indicates the level of stability of water in relation to its degree of saturation with $CaCO_3$. If the water has a negative Langelier index (LSI) value, it is undersaturated with respect to calcium carbonate and potentially aggressive. Conversely, for waters with a positive LSI, a protective layer of calcium carbonate can form when the water is supersaturated with $CaCO_3$. Water in carbonate-calcium equilibrium has an LSI of zero.

The pH at which water is saturated with $CaCO_3$ is the saturation pH (pHs):

$$LSI = pH - pHs$$

$$pHs = 1.022\ (11.5 - \log (Ca^{2+}) - \log (alkalinity))^*$$

(*quick calculation).

Langelier indices	Tendencies
LSI # 0	Water at equilibrium
LSI < 0	Aggressive water
LSI > 0	Water prone to scaling

Table 19.24. *Tendencies in the Langelier index for water aggressiveness*

19.12.1.2. *Ryznar index*

The Ryznar index (RSI) is derived from an empirical correlation observed between the evolution of the thickness of precipitated $CaCO_3$ in pipes and the qualities of drinking water. It makes it possible to define the corrosive or scaling tendency of a water. This index is based on the fact that Ca^{2+} and HCO_3^- ions are the most important parameters to predict the scaling of a water, and the saturation pH is multiplied by a factor of 2, compared to the LI:

$$RSI = 2\ pHs - pH$$

Ryznar indices	Tendencies
4 < RSI < 5	Significant scaling
5 < RSI < 6	Low scaling
6 < RSI < 7	Balance
7 < RSI < 8	Slightly corrosive
8 < RSI < 10	Highly corrosive
RSI > 10	Very highly corrosive

Table 19.25. *Tendencies in the Ryznar index for water corrosiveness*

Water will be at equilibrium when the RSI, always positive, is between 6 and 7.

19.12.1.3. *Larson index*

The Larson index (LRS) makes it possible to find out the corrosive nature of the water. It is given by the following relation:

$$LRS = \frac{2\,[SO_4^{2-}] + [Cl^-]}{[HCO_3^- + CO_3^{2-}]}$$

Concentrations are expressed in $mmol{\cdot}L^{-1}$.

For a pH d < 8.3, carbonates are very low and this parameter can be neglected.

Larson indices	Tendencies
LRS < 0.5	No tendency to corrode
0.5 < LRS < 0.8	Low tendency to corrode
0.8 < LRS < 1.0	Average tendency to corrode
LRS > 1.0	Clear tendency to corrode

Table 19.26. *Tendencies in the Larson index for water corrosiveness*

19.12.1.4. *Leroy index*

Leroy proposes a second index (complementary index), which makes it possible to differentiate calcium bicarbonate waters from sodium bicarbonate waters:

Leroy index = Alkalinity/Ca hardness

According to Leroy, if the alkalinity is lower than the Ca hardness, it necessarily implies the presence of chlorides or sulfates, whose total concentration will be at least equal to the difference between alkalinity and Ca hardness. If, on the contrary, the alkalinity is greater than the Ca hardness, there will be a significant concentration of alkaline ions. A Leroy index value between 0.7 and 1.3 can be considered acceptable. It is therefore proposed that the assessment of corrosiveness be made by calculating the LRS and Leroy index.

A water is considered mildly corrosive if:

Leroy Index < 0; 7 and 0; 7 < Alkalinity = Ca Hardness < 1; 3

19.12.1.5. *Calcium carbonate precipitation potential*

The calcium carbonate precipitation potential (CCPP) is assimilated to an indicator of the corrosiveness of water, insofar as it allows for estimating the potential of $CaCO_3$ precipitation. It is then possible to adjust the remineralization reagents or to moderate the decarbonation by estimating the quantity of calcium carbonate that will precipitate. The CCPP is not measured directly, but is calculated from models based on the equilibrium associated with calcium carbonate, using parameters such as pH, temperature, calcium and carbonate (or bicarbonate) concentrations. The CCPP is expressed in $mg \cdot L^{-1}$ of $CaCO_3$.

This parameter is often used as an indicator of precipitation in conditions where the water, for example, is heated to 60°C (kettles, water heaters, etc.). As temperature (T, °C) increases and the solubility being lower, the potential for residual $CaCO_3$ in water becomes more significant and precipitation can take place more easily at T = 60°C. This is especially true if the Ca hardness is high and the pH is higher than 7.5.

By raising T, decomposition of HCO_3^- ions into CO_3^{2-}, CO_2 and H_2O can be observed. At 60°C, the decomposition of $Ca(HCO_3)_2$ is 45% after 20 min and more than 70% at 90°C. In contrast, the decomposition of $NaHCO_3$ is lower.

A positive CCPP value around 5 $mg \cdot L^{-1}$ (T < 20°C) is representative of good corrosion protection. As shown in Figure 19.88, a water Ca hardness between 120 and 150 $mg \cdot L^{-1}$ $CaCO_3$ will lead to a CCPP between 15 and 30 $mg \cdot L^{-1}$ at 60°C.

When the Ca hardness is increased to 180 $mg \cdot L^{-1}$ $CaCO_3$, for example, CCPP starts to reach 40 $mg \cdot L^{-1}$ at 60°C, while the water was at equilibrium before the temperature increase.

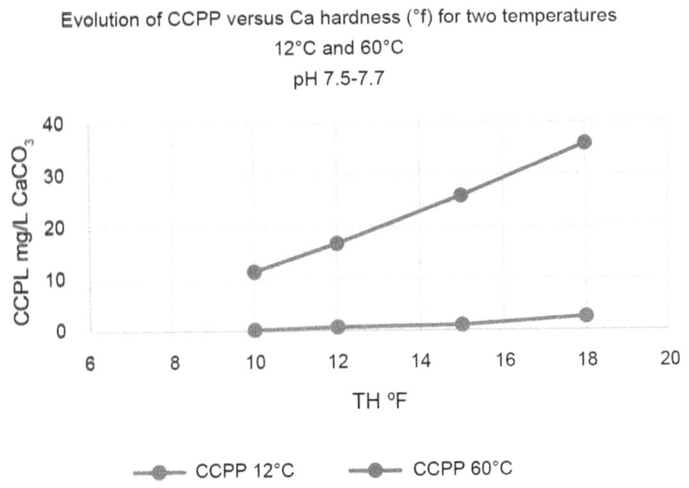

Figure 19.88. *CCPP at 12°C and 60°C as a function of the Ca hardness (1°f = 10 mg·L^{-1} $CaCO_3$). For a color version of this figure, see www.iste.co.uk/gaid/watertreatment4.zip*

Corrosiveness of water	CCPP (mg/L $CaCO_3$)
Non-corrosive	0
Passive	0 to – 5
Moderately corrosive	– 5 to – 10
Corrosive	<–10

Table 19.27. *Water corrosiveness and CCPP*

19.12.2. *Application to desalinated water*

Desalinated water is completely free of calcium and alkalinity salts, and the objective of the post-treatment is to increase these levels to ones that range, depending on the customer, from 60 to 120 mg $CaCO_3 \cdot L^{-1}$.

Moderately alkaline water (60–70 mg $CaCO_3 \cdot L^{-1}$) with a pH between 7.0 and 8.3 is generally not corrosive. Water with a pH below 6.5 is corrosive, especially if the alkalinity is also low, whereas water with a pH above 7.5 can also be corrosive when the alkalinity is low.

Soft water containing sodium salts does not create a protective layer on the pipes and is therefore corrosive, because a simple injection of soda is not enough. Water containing calcium salts is less corrosive, as it creates an internal protective layer on the pipes. Water containing sufficient alkalinity reduces corrosion kinetics.

19.12.2.1. *Water qualities required after post-treatment*

The most frequently requested water qualities after desalination are the following:

– $60 < \text{alkalinity} < 100$ mg $CaCO_3 \cdot L^{-1}$;

– Langelier index > 0;

– Ryznar index < 10;

– 4 < CCPP < 10 mg/L as $CaCO_3$;

– 60 < calcium (Ca) < 120 mg $CaCO_3 \cdot L^{-1}$;

– 7.5 < pH < 8.5;

– $Cl_2 > 1$ mg$\cdot L^{-1}$.

Alkalinity represents the concentration of bicarbonate, carbonate and hydroxide ions in water. It indicates a trend in the buffering capacity of water. In other words, it is the water's ability to resist changes in pH when an acid is added. The higher the alkalinity, the higher the ability of the water to resist a change in pH. However, too much alkalinity and high pH can accelerate the dissolution of copper in the pipes.

The total hardness (calcium and magnesium) represents the overall concentration of divalent ions (calcium and magnesium). Magnesium is rarely added to desalinated water and it is calcium that prevails for remineralization. The lower limit of 60 mg $CaCO_3 \cdot L^{-1}$ is recommended to promote precipitation, but also for nutrient requirements.

The pH is used to express the acid–base intensity of water. It depends on both temperature and ionic strength. In general, low pH values promote the dissolution of lead and copper, while high pH values reduce the dissolution potential of metals. pH ranges between 7.5 and 8.5 are often recommended. Especially since sufficient alkalinity must be present in the water to maintain a stable pH and minimize the corrosive potential of the water.

The LSI and RSI > 0 are considered, while the LRS is not used because remineralized seawater does not contain chlorides and sulfates.

19.12.3. *Treatments*

The simplest post-treatment is to mix desalinated water with spring water or treated water. As this solution is not always available, another approach is to reintroduce some essential ions into the desalinated water in order to obtain a balanced mineral content.

The processes used in post-treatment generally include adjusting the pH of the permeate, using carbon dioxide in combination with lime, using carbon dioxide or acid in combination with limestone (placed in a filter) and adding inhibitors for corrosion control.

The design of the post-treatment is based on the water quality objectives defined in the specifications, and mainly concern calcium hardness and alkalinity.

Application	Post-treatment
Permeate RO membranes	Re-carbonation CO_2-$Ca(OH)_2$ (or lime water) CO_2 or H_2SO_4 and limestone filter Adjustment of pH and alkalinity Addition of corrosion inhibitors

Table 19.28. *Processes used in post-treatment for the remineralization of desalinated water*

The two most common remineralization processes are the CO_2–lime (or lime water) process and the CO_2 (or H_2SO_4) and limestone reaction.

19.12.3.1. *Remineralization on limestone filter*

The process involves bringing desalinated water acidified with CO_2 (or H_2SO_4) into contact with a fixed bed of limestone. Under normal conditions, $CaCO_3$ is only slightly soluble in water. However, its dissolution in water is favored under acidic conditions. The use of CO_2 lowers the pH and converts itself to alkalinity by reaction with the limestone material. The pH of the water rises above 7.5.

The two most commonly used types of limestone filters are upflow and downflow filters. They are based on the same principles and operating mechanisms used for dissolving limescale and remineralizing water.

Veolia uses both processes, depending on the flow rate and type of filter. Downflow filters are most often offered with pressure filters and for low flow rates. Due to the low volumes, these filters require frequent replenishment of calcite to maintain a constant bed height, as well as frequent backwashing to remove impurities.

The main difference between upflow and downflow is the opposite direction of filtration. With upflow, the water enters the bottom layer with a high dissolution rate because the concentration of CO_2 (or acid) is at its highest. This dissolution rate becomes less high as the water moves up the filter. As a result, the $CaCO_3$ grain size should be smaller in the lower layer compared to the upper layer. However, with the rising velocity, the small particles also rise (especially after the washings), leading to a permanent state in which the small particles are present in the upper layer and the larger particles are present in the lower layer.

Upflow filters allow a good distribution of CO_2 or acid at the bottom of the filter, and the chemical reactions appear more homogeneous.

19.12.3.1.1. Materials

Limestone is a chemical compound that must contain more than 95% $CaCO_3$ in order to avoid enriching the water with impurities. Its dissolution is affected by the characteristics of the material's crystals, its internal porosity and its impurity content (inorganic ions such as PO_4, Pb, Cd, Cu, Zn, Fe and Mn and organic matter).

The materials used for remineralization are alkaline-earth type of materials such as limestone ($CaCO_3$), magno ($CaCO_3MgO$) and dolomite (CaOMgO).

These materials are similar to consumable filtration materials with a granular aspect (>1,000 μm), which leads to granulometric control by sieving.

It is important to check the following parameters when selecting and sourcing material:

– the particle size distribution curve represented by the real particle size or effective size d_{10} (opening allowing the passage of 10% [w/w] of the particles);

– the uniformity coefficient CU (d_{60}/d_{10});

– solid density (absolute), dry density (apparent in air), wet density (saturated with water), porosity;

– composition, determined, as the case may be, by the level of purity ($CaCO_3$ content, etc.), solubility in acid or in water and the content of metallic impurities;

– the shape: spherical, round, granular, cylindrical, oval, irregular, angular;

– hardness and friability.

The potential degradation of these materials during the handling or washing phases requires particular attention to the parameters to be respected during their purchase (hardness, friability), as well as to the means to be implemented during the various operations. Loading the filters hydraulically from trucks (maximum capacity of 22 t) requires the use of appropriate hydro-ejectors and a sufficient number of flexible transfer pipes, in order to be able to load the most remote cells. If the cells are covered by a concrete roof, then easy access and a sufficient number of openings (hatches) must be provided to ensure a controlled and homogeneous distribution of the material.

19.12.3.1.2. Mechanisms of remineralization with limestone filters

The H_2CO_3-HCO_3^--CO_3^{2-} matrix

The dissolved inorganic carbon matrix is the main component of weak acid in natural fresh waters, and is also the main one in remineralized waters after desalination. It includes three species, H_2CO_3–HCO_3^-–CO_3^{2-}, whose sum constitutes the dissolved inorganic carbon:

$$CaCO_3 \rightarrow Ca^{2+} + CO_3^{2-};\ Ks = [Ca^{2+}][CO_3^{2-}] = 4.4 \times 10^{-9};\ pKs = -8.36$$

$$CO_2 + H_2O \rightarrow HCO_3^- + H_3O^+;\ K1 = \frac{[HCO_3^-][H_3O^+]}{[CO_2]} = 3.44 \times 10^{-7};\ pk1 = -6.46$$

$$HCO_3^- + H_2O \rightarrow CO_3^{2-} + H_3O^+;\ K2 = \frac{[CO_3^{2-}][H_3O^+]}{[CO_2]} = 3.25 \times 10^{-11};\ pk1 = -10.49$$

$$CaCO_3 + CO_2 + H_2O \rightarrow Ca^{2+} + 2HCO_3^-;\ K2 = \frac{[HCO_3-]^2[Ca^{2+}]}{[CO_2]} = 4.6 \times 10^{-5};$$

$$pk3 = -4.33$$

$$[Ca^{2+}] = \tfrac{1}{2}\,[HCO_3^-]$$

K1 and K2 are the reaction constants at T (°C) and the given ionic strength ($mole{\cdot}m^{-2}{\cdot}s^{-1}$). The concentrations are in $mole{\cdot}L^{-1}$.

Figure 19.89 describes the distribution of species in the carbonate system as a function of pH, according to equilibrium constants. At $pH < 5$, the dominant species is CO_2, whereas at pH 8.2, the dominant species is HCO_3^-.

Dissolution kinetics of $CaCO_3$

The dissolution of limestone provides alkalinity (bicarbonate) and hardness (calcium). The dissolution kinetics of $CaCO_3$ depends on several factors (pH, material characteristics, T, CO_2 or acid concentration, contact time, etc.). Several mechanisms are considered such as diffusion into the media, surface dissolution reaction, diffusion into the boundary layer and upward flow transport.

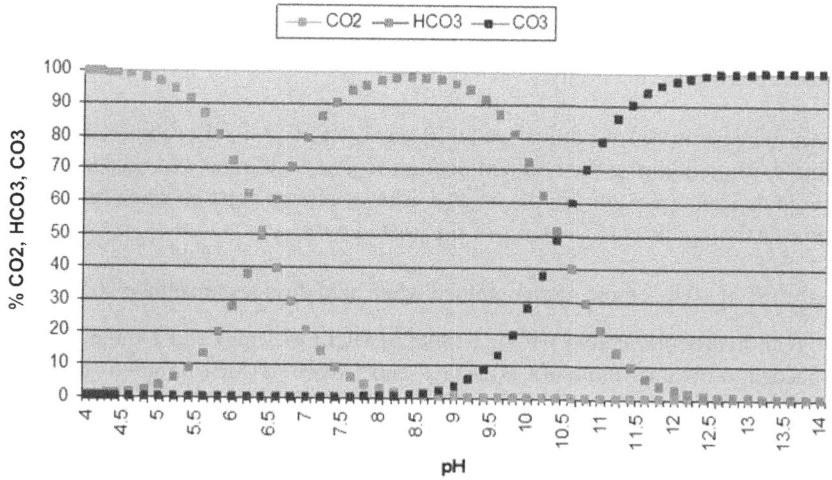

Figure 19.89. *Evolution of CO_2 as a function of pH. For a color version of this figure, see www.iste.co.uk/gaid/watertreatment4.zip*

The reactions are as follows:

– with carbon dioxide (CO_2), it takes 1 mole of CO_2 (44 mg) and 1 mole of $CaCO_3$ (100 mg) to obtain one mole of $Ca(HCO_3)_2$ (162 mg expressed as calcium bicarbonate):

$$CO_2 + CaCO_3 + H_2O \rightarrow Ca(HCO_3)_2$$

Note that 1 mg of CO_2 neutralizes 2.27 mg of limestone, increases hardness and alkalinity by 3.68 mg of $Ca(HCO_3)_2$. There is an increase in 1 $mg{\cdot}L^{-1}$ Ca hardness per $mg{\cdot}L^{-1}$ alkalinity.

An LSI of 0.0 to + 0.1 can be obtained when a sufficient contact time (more than 20 minutes) is considered:

– with sulfuric acid (H_2SO_4), it is necessary to have 1 mole of H_2SO_4 (96 mg) and 2 mole of $CaCO_3$ to obtain 1 mole of $Ca(HCO_3)_2$ and 1 mole of $CaSO_4$. There is therefore a double consumption of limestone material compared to the following reaction:

$$CaCO_3 + H_2SO_4 \rightarrow CO_2 + H_2O + CaSO_4$$

$$CaCO_3 + CO_2 + H_2O \rightarrow Ca(HCO_3)_2$$

$$2CaCO_3 + H_2SO_4 \rightarrow Ca(HCO_3)_2 + CaSO_4$$

Note that 1 mg of H_2SO_4 neutralizes 2.08 mg of limestone, increases hardness and alkalinity by 1.65 mg of $Ca(HCO_3)_2$. There is an increase in 2 $mg \cdot L^{-1}$ Ca hardness per $mg \cdot L^{-1}$ alkalinity.

An increase in SO_4^{2-} equivalent to 10 $mg \cdot L^{-1}$ $CaCO_3$ is also obtained. Calcium sulfate is formed during the remineralization reaction and must be considered, so as to avoid deposits (gypsum) in the reactor. These sulfates must be taken into account when calculating the corrosiveness of the treated water.

An LSI of –0.4 is most often obtained when acid dosages are above 20 $mg \cdot L^{-1}$;

– with hydrochloric acid (HCl), 1 mole of HCl (36.5 mg) and 2 mole of $CaCO_3$ are needed to obtain 1 mole of $Ca(HCO_3)_2$ and 1 mole of $CaCl_2$. There is therefore a double consumption of limestone material compared to the reaction with CO_2:

$$CaCO_3 + 2HCl \rightarrow CO_2 + H_2O + CaCl_2$$

$$CaCO_3 + CO_2 + H_2O \rightarrow Ca(HCO_3)_2$$

$$2CaCO_3 + 2HCl \rightarrow Ca(HCO_3)_2 + CaCl_2$$

Note that 1 mg of HCl neutralizes 2.81 mg of limestone, increases hardness and alkalinity by 2.28 mg of $Ca(HCO_3)_2$. There is an increase in 2 $mg \cdot L^{-1}$ Ca harness per $mg \cdot L^{-1}$ alkalinity.

Calco-carbonic parameters obtained	**Reagents**	**Consequences according to the choice of the reagent**
Final alkalinity $\geq$ 80 mg/L $CaCO_3$	CO_2	Water at calco-carbonic balance or LSI: (0.0) – (+ 0.1)
Final alkalinity < 80 mg/L $CaCO_3$ High calcium hardness	H_2SO_4 HCl	LSI: (–0.03) – (–0.04) Addition of soda to have an LSI $\geq$ 0 Beware of sulfates or chlorides at the outlet: impact on corrosion Higher consumption in limestone materials

Table 19.29. *Water quality criteria according to the acids used*

An increase in Cl^- equivalent to 10 mg·L^{-1} $CaCO_3$ is also obtained. Calcium chloride is formed during the remineralization reaction and must be considered when calculating the corrosivity of the treated water. An LSI of –0.4 is most often obtained when acid dosages are higher than 20 mg·L^{-1}.

As a result, the choice can be made on the following basis, to which the operating cost must be associated.

19.12.3.1.3. Design parameters

In a limestone filter, the mass balance is as follows:

> consumption of $CaCO_3$ material (mole·m^{-3}·min^{-1}) = increase in water ion concentration (mole·m^{-3}·min^{-1}) × water volume (m^3)

Or:

$$\frac{dM}{dt} = V\,\frac{dC}{dt}$$

Furthermore, the mass transfer leads to:

> mass transfer (mole·h^{-1}) = K (exchange rate in mole·m^{-2}·h^{-1}) × surface (m^2)

Or:

$$\frac{dM}{dt} = K\,A = V\,\frac{dC}{dt}$$

The mass transfer, also be expressed by volume of material (V' in m^3):

> mass transfer (mole·h^{-1}) = K' (exchange rate in mole·m^{-3}·h^{-1}) × volume of material (m^3)

Or:

$$\frac{dM}{dt} = K'V' = V\,\frac{dC}{dt}$$

Or even:

$$dC = K\,\frac{A}{V}dt = K'\,\frac{V'}{V}dt$$

with:

– A: surface area (m^2);

– K(A/V) and K' (V'/V): filtration ratios with respect to the surface and to the volume of material.

These parameters depend on the acid concentration at the filter inlet and the temperature.

Veolia has tools and software that indicate the value of the various parameters, and the physical and chemical composition of the remineralized water.

The contact time characterizes the time during which the water passes through the filter material. It is completely related to the volume of material present in the filter and therefore to the concentration of CO_2 in the inlet. A low contact time linked to a high CO_2 concentration leads to a filtered water that is always aggressive (high CO_2 residual), and to a final mineralization that is not met. The contact time varies between 10 and 40 min, depending on the temperature and the $CaCO_3$ objectives in the remineralized water (Figures 19.90 and 19.91). A remineralization target of 50 mg $CaCO_3 \cdot L^{-1}$ requires a shorter contact time than a target of 100 mg $CaCO_3 \cdot L^{-1}$.

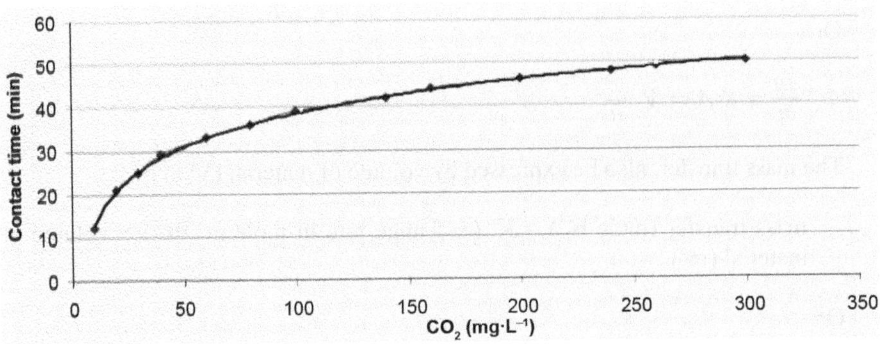

Figure 19.90. *Contact time (min) as a function of CO_2 at the inlet of the limestone filter. For a color version of this figure, see www.iste.co.uk/gaid/watertreatment4.zip*

The contact time is the first design parameter and makes it possible to favor the kinetics that leads to the calco-carbonic equilibrium, and thus to an LSI equal to zero.

The higher the temperature, the faster the CO_2–$CaCO_3$ reaction kinetics and the shorter the contact time, although the dissolution of CO_2 in water increases with increasing temperature.

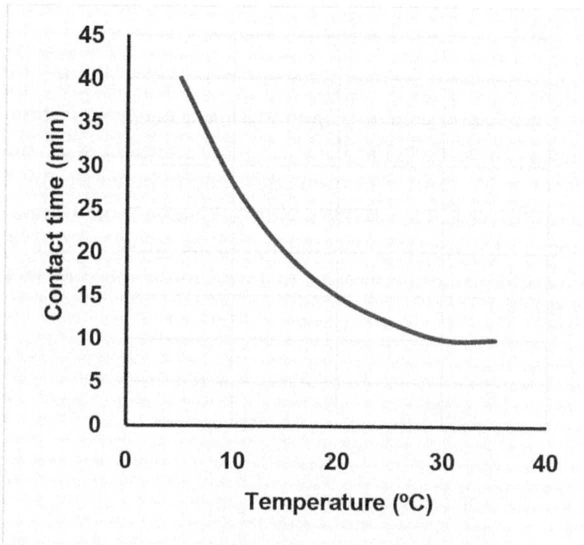

Figure 19.91. *Contact time versus temperature for two quality objectives*

The volume of limestone that has been neutralized, and therefore consumed, depends on the concentration of CO_2, or acid equivalent, present at the filter inlet. The minimum volume of limestone is:

$$\text{Volume (m}^3\text{)} = Q \cdot t \cdot f/60$$

with:

– V: volume of media;

– Q ($m^3 \cdot h^{-1}$): flow rate;

– t (min): contact time;

– f: temperature correction factor.

This volume is calculated to obtain a pH of 8 or to reach the saturation pH, in other words, the calco-carbonic equilibrium.

An additional volume of media is added to the minimum volume to compensate for the media consumed, without affecting performance. This volume is calculated so that the minimum volume is never reduced in the filter. The production of calcium bicarbonates that dissolve in the water implies a constant renewal of limestone in the filters.

There are several $CaCO_3$ grain sizes: 0.71–1.25 mm, 1.0–2.0 mm, 1.6–2.5 mm; 1–3 mm. The 1–2 mm particle size is the most used because it allows fast kinetics with lower pressure drops than the 0.71–1.25 mm particle size. Filtration speeds vary between 4 and 15 $m \cdot h^{-1}$ for a media height between 0.7 and 3 m. As remineralization on filters is carried out within a "reactor", the filtration speed is not a limiting parameter. However, given the hydraulic drive of fine particles in the treated water, we prefer to limit ourselves to a maximum speed of 15 $m \cdot h^{-1}$ in N and 20 $m \cdot h^{-1}$ in N–1 filters.

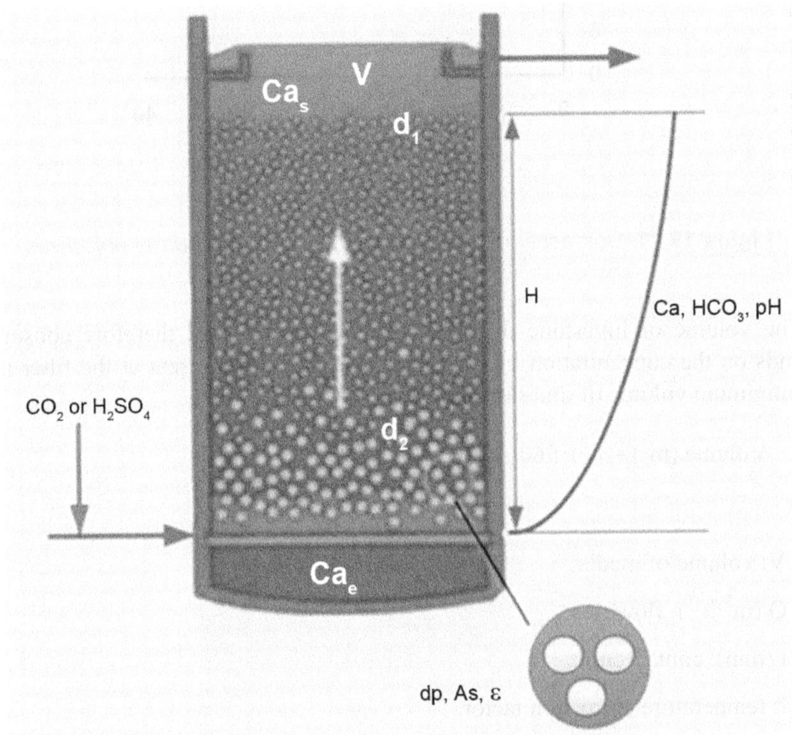

Figure 19.92. *Design parameters of the limestone filter. For a color version of this figure, see www.iste.co.uk/gaid/watertreatment4.zip*

The application of a high filtration rate implies a significant material height, and therefore an increase in the total height of the filter.

When dolomite is used as a media, the first filtration water must be discharged after shutdown periods, due to pH elevations during said periods.

The reagent consumption for the various solutions is shown in Table 19.30.

Remineralization processes	**Chemical reactions**	**Consumption for an increase in 1 mmole (61 mg) HCO_3^-**	**Increase in cations and anions for 1 mmole HCO_3^-**
$CaCO_3 + CO_2$	$CaCO_3 + CO_2 + H_2O \rightarrow Ca(HCO_3)_2$	50.1 mg $CaCO_3$ and 22 mg CO_2	20 mg Ca^{2+}
$Ca(OH)_2 + CO_2$	$Ca(OH)_2 + 2CO_2 + H_2O \rightarrow Ca(HCO_3)_2$	37.1 mg $CaCO_3$ and 44 mg CO_2	20 mg Ca^{2+}
Dolomite + CO_2	$MgOCaCO_3 + 3CO_2 + 2H_2O \rightarrow Ca(HCO_3)_2 + Mg(HCO_3)_2$	35.1 MgO $CaCO_3$ and 33 mg CO_2	6.1 mg Mg^{2+} 10 mg Ca^{2+}
$CaCO_3 + H_2SO_4$	$2CaCO_3 + H_2SO_4 \rightarrow CaSO_4 + Ca(HCO_3)_2$	100.1 mg $CaCO_3$ and 49 mg H_2SO_4	40.1 mg Ca^{2+} 48 mg SO_4^{2-}
$CaCl_2 + Na_2CO_3$	$CaCl_2 + 2NaHCO_3 \rightarrow Ca(HCO_3)_2 + 2NaCl$	55.5 mg $CaCl_2$ and 84 mg $NaHCO_3$	23 mg Na^+ 20 mg Ca^{2+} 35.5 mg Cl^-

Table 19.30. *Consumption of various reagents for remineralization of desalinated water ($CaCO_3$ purity 100%)*

pH adjustment to achieve an LSI greater than 0

An excess of CO_2 is often injected and ends up in the treated water, creating a negative LSI. The acidity caused by the residual CO_2 in the produced water is then neutralized with soda (NaOH) in order to reach the required final pH and respect a positive LI. The dosages are between 1 and 5 $mg \cdot L^{-1}$.

19.12.3.1.4. Limestone filter technology

F-filters are the most commonly used filters, due to their simplicity and low investment costs. They are characterized by an operating water height of generally 1.0 m, the advantage of which is rapid evacuation of the wash water and low water loss. The characteristics are as shown in Table 19.31.

Parameters	Filters F
Objectives	Remineralization
Applications	Fresh or desalinated water
Minimum contact time (min)	12
Filter layer **Height (mm)** **CU**	Neutralizing material 1-3 1,000–4,200 1.4
Support layer **Height (mm)**	Gravel 5–7 mm 100
Water height above the media (mm)	600–1,300

Table 19.31. *Characteristics of limestone filters*

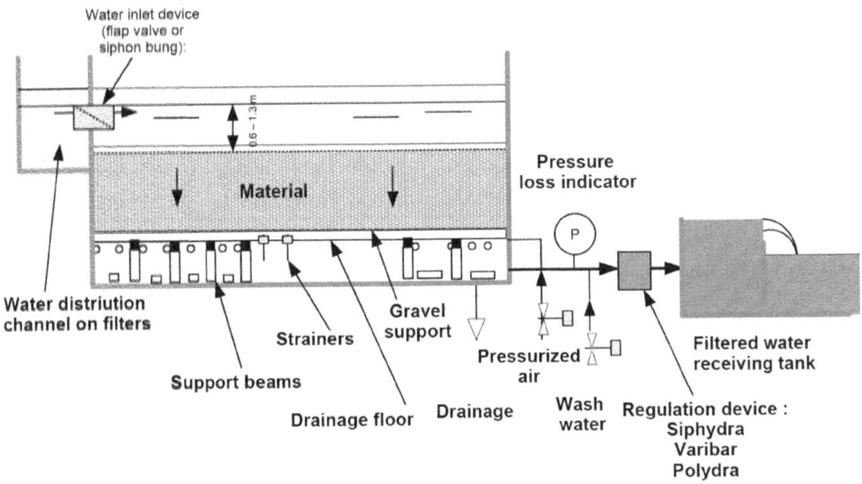

Figure 19.93. *Limescale filter (Veolia). For a color version of this figure, see www.iste.co.uk/gaid/watertreatment4.zip*

In order to reduce the number of filters, desalinated water is often bypassed (40–75% of the total flow) and then mixed with remineralized water. The advantage of bypassing part of the water to be remineralized is that the filtration area and therefore the footprint, as well as the equipment associated with this stage, can be considerably reduced. As the total amount of reagents must be identical with or without bypass, this means that a bypass configuration requires very high reagent dosages and, a fortiori, high limestone consumption. Upflow feeding avoids CO_2 losses, while the same system operating with downflow can generate CO_2 losses by stripping, during injection above the filter. Co-current washes allow for milder conditions, with less frequent air and air/water phases than countercurrent washes.

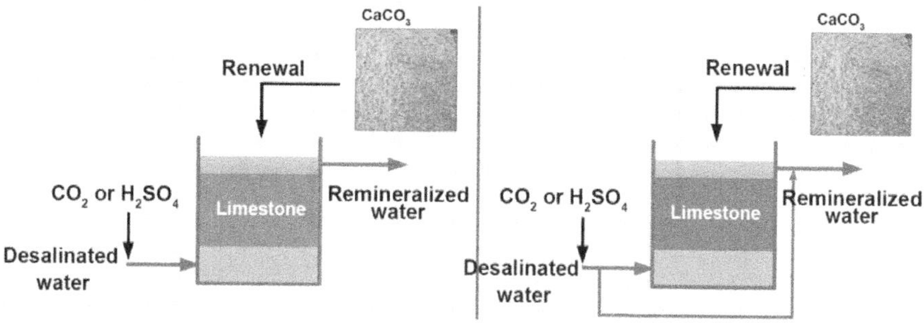

Figure 19.94. *Diagram of a CO2 (or acid) remineralization–limestone filter. For a color version of this figure, see www.iste.co.uk/gaid/watertreatment4.zip*

Particular attention must be paid to the injection and dissolution of reagents (CO_2, acid), to the means of avoiding losses of CO_2, in particular during its mixing, to the protection of the works against aggressiveness and to the good choice of equipment and piping against corrosion.

Example of the advantages of bypassing part of the water to be treated

– Without a bypass:

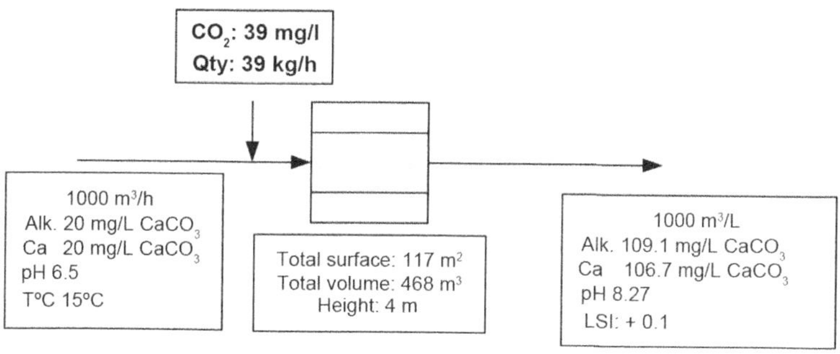

Figure 19.95. *Floor surface and water quality according to the options without a bypass*

– With a bypass:

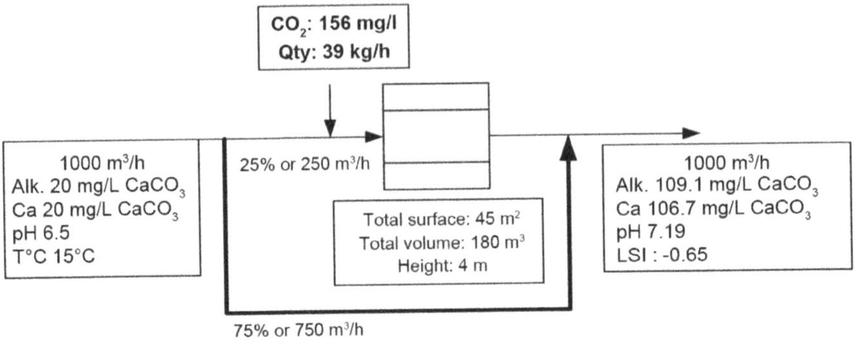

Figure 19.96. *Floor surface and water quality according to the options with a bypass*

The LSI of the final water is often slightly negative and it is important to add soda at the end of the treatment, in order to neutralize the pH and to distribute water at calco-carbonic equilibrium. This has a direct impact on the operating cost because of the amount of soda consumed to reach the equilibrium pH.

On the other hand, there is a significant gain in filtering surface, and this can be a significant advantage in terms of investment.

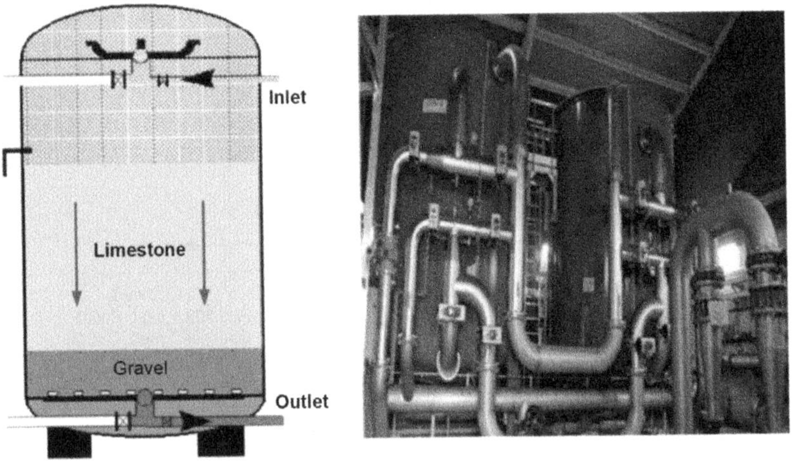

Figure 19.97. *Pressurized limestone filter (Veolia). For a color version of this figure, see www.iste.co.uk/gaid/watertreatment4.zip*

19.12.3.2. *Remineralization with reagents in solution*

19.12.3.2.1. Carbon dioxide–lime

This is one of the remineralization methods that is very often used because of its ease of implementation. It involves a CO_2 injection, followed by the addition of lime or lime water:

$$2CO_2 + Ca(OH)_2 \rightarrow Ca(HCO_3)_2$$

However, as the solubility of lime decreases with increasing temperature, residual turbidities above 2-5 NTU are obtained in the produced water. This is particularly true when lime is introduced directly as quicklime (CaO) or slaked lime ($Ca(OH)_2$). For this reason, this disadvantage is limited by preparing lime water with a saturator. The turbidity of the produced water thus becomes lower than 2 NTU.

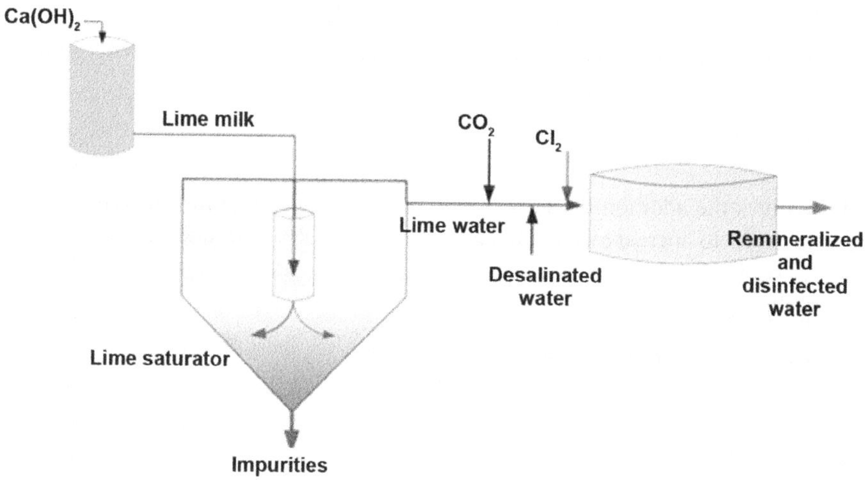

Figure 19.98. *Schematic diagram of the CO_2-lime solution. For a color version of this figure, see www.iste.co.uk/gaid/watertreatment4.zip*

This solution presents both advantages and disadvantages.

Advantages are as follows:

– very good remineralizing power of lime;

– the injection of CO_2 allows good control of the pH and thus the remineralization at the end of the treatment process;

– CO_2 is a weak acid (non-toxic, non-corrosive and harmless) and leaves no polluting residue in the treated water, and therefore environmentally friendly;

– easy to control reaction because of the stability of CO_2.

Disadvantages are as follows:

– solubility of lime at high temperature;

– high residual turbidity with direct addition of quicklime or slaked lime;

– lime must be very pure (>95%) to avoid impurities in the produced water.

Section 20.6 (Volume 5) details the equipment of this CO_2-lime solution, which is identical for the remineralization of fresh or desalinated water.

Alternative solution not widely used: direct dosage of $NaHCO_3$

A minimum level of alkalinity can be obtained by CO_2 action with sodium bicarbonate:

$$NaHCO_3 \rightarrow Na^+ + HCO_3^-$$

It involves the addition of sodium ions in high concentration. Calcium ions are not introduced to increase calcium hardness and CCPP. The addition of calcium chloride ($CaCl_2$) increases the calcium hardness, but increases the chloride concentration:

$$CaCl_2 + 2NaHCO_3 \rightarrow Ca(HCO_3)_2 + 2NaCl$$

19.13. Monitoring and control parameters

The operating parameters (pressure, temperature, longitudinal flow rate, conversion rate), as well as the monitoring and control parameters, are detailed in section 18.4 (Volume 4).

19.13.1. *Standardization of raw data*

Membrane performance is highly dependent on operating parameters such as temperature, water quality, flow rate, and so on.

However, in order to monitor the evolution and the state of clogging of the membranes, it is essential to be able to constantly compare the current performance with the initial performance measured at the time of its implementation. This long-term monitoring is of great importance for the management of the installation.

Types of water	Measured parameters	Means of control and monitoring
Feed water to the membrane unit	SDI Aluminum, Fe, Mn Turbidity Conductivity Residual oxidant pH	Instrumentation to be checked regularly
Water upstream of the pre-filters (after addition of antiscalant) **Water downstream of the pre-filters (high-pressure suction)**	pH Turbidity Conductivity	Measurements carried out by sampling upstream and downstream of the pre-filters with the same apparatus
Water downstream of the pre-filters	Residual oxidant Temperature	Instrumentation to be checked
Permeate from each block	Conductivity Chlorides (optional) Sulfates (optional)	Possibility of manual tube-by-tube measurements Also measures on the general permeate of the line by scanning
Water produced	pH Alkalinity, Ca hardness Conductivity Chlorides Sulfates Boron	Instrumentation to be checked
General concentration	Conductivity pH Iron Suspended matter	Instrumentation to be checked
Cleaning units	pH Turbidity Temperature	To be checked for each cleaning unit
Neutralization	pH Turbidity	pH measurement in each tank

Table 19.32. *Instructions for monitoring water quality at different points of the membrane unit*

To do this, the operating parameters must be standardized, in other words, expressed in relation to reference conditions.

Four criteria are thus considered for performance monitoring:

– the standardized permeability at 25°C: it characterizes the "transversal" clogging of the membrane by very fine particles or by salts;

– the standardized pressure drop factor: it characterizes the "longitudinal" clogging of the membrane, obtained in case of serious failure of the pre-treatment or the pre-filters (algae, filter piercing, large particles, etc.);

– the specific surface flow: makes it possible to know the distribution of the clogging load between the stages;

– the standardized salt passage at 25°C: makes it possible to detect a modification of the membrane's breaking capacity.

The continuous process control points are defined in Table 19.32.

19.13.2. *Bromates*

Bromides are present in seawater at varying concentrations between 50 and 80 $mg \cdot L^{-1}$, and over 98% removal leaves a residual of 1.2–1.5 $mg \cdot L^{-1}$ in the produced water.

Very often, the hypochlorite solution used for disinfection is prepared by electrolysis with sea salt, and bromates are then produced during this oxidization.

In continuous or shock pre-chlorination, the bromates formed by reaction with chlorine are retained by the RO membrane.

Parameters	Chlorination at 2.5 $mg \cdot L^{-1}$			Chlorination at 5 $mg \cdot L^{-1}$		
	RO inlet	RO outlet	Rejection %	RO inlet	RO outlet	Rejection %
TDS ($mg \cdot L^{-1}$)	28,000	170	99.4	29,668	118	99.6
Cl^-($mg \cdot L^{-1}$)	13,900	97.6	99.5	16,482	65.7	99.6
Br^- ($mg \cdot L^{-1}$)	48.5	0.28	99.4	55.5	0.24	99.6
THM ($\mu g \cdot L^{-1}$)	490	2.3	99.5	860	6.4	99.3
TOC ($mg \cdot L^{-1}$)	4.3	0.5	88.4	10.9	0.38	96.5

Table 19.33. *Analysis and rejections after pre-chlorination*

Given that desalinated water only contains a very small proportion of organic carbon, the hypochlorite disinfection stage does not form DBP. However, the bromates present in the hypochlorite solution are found in the water produced. If electrolysis is performed directly with seawater or concentrate, organohalogen compounds and organic bromine compounds can be found in fairly high concentrations.

19.14. Veolia's new processes applied to seawater desalination

19.14.1. *Flotation with the Spidflow® process*

This process was previously detailed in section 6.7 (Volume 2). It is taken up again in this chapter for its application to seawater desalination and its associated performances.

Seawater is characterized by low turbidity, but also by the presence of microparticles of various kinds that are difficult to remove by settling, despite the injections of chemical reagents such as coagulants and flocculants.

These microparticles may include microalgae, phytoplankton, etc., the abundant presence of which in seawater is recognized as a problem that negatively impacts the operation of RO desalination plants. The nature of algal blooms is such that they occur for relatively short periods of time and are, apparently, unpredictable.

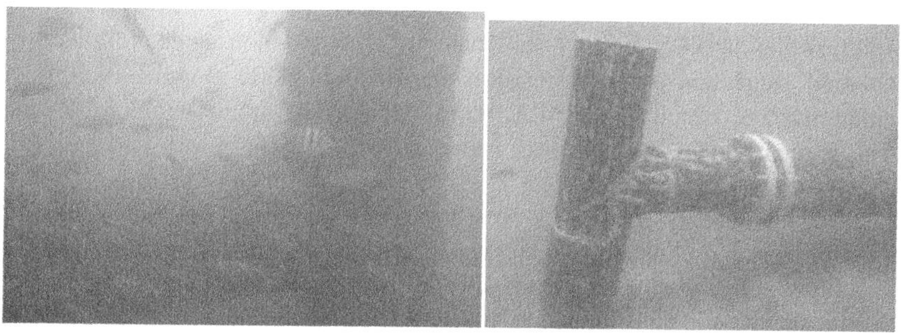

Figure 19.99. *Phytoplankton and other microalgae in seawater. For a color version of this figure, see www.iste.co.uk/gaid/watertreatment4.zip*

Desalination by RO is increasingly favored in water-deficient regions of the world, and this technology requires a feed water quality that is free of particles and clogging substances. Pre-treatment is therefore required to remove these particles and substances, and to allow the membranes to function properly. If this pre-treatment is not adequate in the case of a significant algal bloom, it can be overwhelmed by the concentrations of organic matter and microparticles, leading to the shutdown of the plant. This has been seen many times in the Middle East.

Algal blooms that develop with red or brown algal blooms often consist of *Cochlodinium polykrikoides* (Persian Gulf and Gulf of Oman), and are explained by an increased nutrient influx to coastal areas from municipal and industrial wastewater discharges, as well as ballast water.

During an algal bloom, several parameters change rapidly: turbidity, pH and chlorophyll concentration. It is therefore very difficult to establish a typical seawater analysis for this event. Table 19.34 summarizes the maximum values of several parameters observed during an algal bloom (Persian Gulf).

Parameters	Maximum values (algal bloom)	Typical values
Turbidity (NTU)	20	<2
Phytoplankton (n.L^{-1})	107	<100
Chlorophyll (mg·L^{-1})	700	<2
SS (mg·L^{-1})	30	<5
SDI$_3$	>33.3	15–25

Table 19.34. *Maximum values of seawater parameters during an algal bloom (red tide) in the Persian Gulf*

Spidflow® is a DAF process that includes a coagulation stage, followed by a flocculation stage and a rapid flotation clarification phase (Figure 19.100). The flocculation stage can also use a Turbomix to ensure optimal mixing of water, reagents and particulate constituents.

The fine air bubbles, formed by pressurizing the air in the water (at pressures of 5–6 bar) during the production of white water, are injected into the Spidflow® flotation units via a specific distribution system. The hydraulic sequencing of the different compartments of the Spidflow® process has been designed according to specific computerized fluid dynamics studies. In the contact zone, Spidflow® has a floor for the distribution of flocculated water associated with white water. It also includes anti-spiral flow plates that prevent short-circuiting and float water collectors that uniquely recover the water flow. Applied in the pre-treatment of a desalination plant, Spidflow® guarantees high performance at speeds between 20 and 45 m·h^{-1}.

On a pilot unit installed in a desalination plant in operation, two types of seawater ("normal" and degraded by an algal bloom) were pre-treated, using a stage that included a Spidflow® process, followed by a dual media filtration.

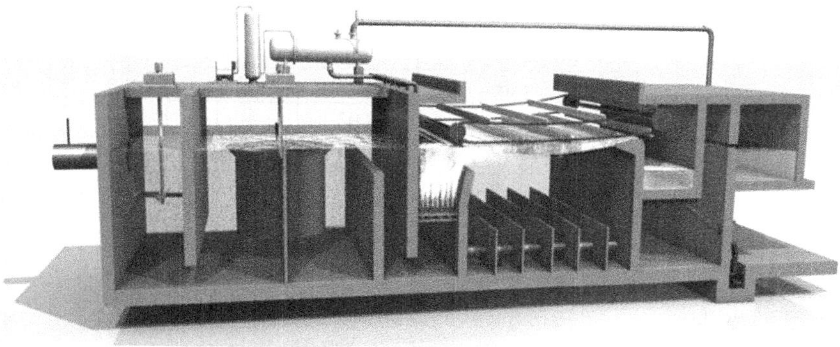

Figure 19.100. *Another schematic view of Spidflow®. For a color version of this figure, see www.iste.co.uk/gaid/watertreatment4.zip*

Normal seawater quality is characterized by an average turbidity below 0.6 NTU, SDI_3 values (%·min^{-1}) around 20, and an absence of phytoplankton. Fe injection at an average dosage of 1 mg Fe·L^{-1} results in filtered water with $SDI_{15} < 3.2$ (%·min^{-1}) and a turbidity < 0.1 NTU. With a mirror speed of 35 m·h^{-1} for Spidflow® and a filtration speed of 10–12 m·h^{-1}, the filtration cycles lasted on average 42 h, and the quality of the pre-treated water always remained very stable.

The seawater then degraded rapidly, due to the development of an algal bloom (red tide), during which the *Cochlodinium polykrikoides* phytoplankton species grew in a very short time. The surface of the seawater turned reddish-brown, emitting an odor and left a layer of mucus on the membranes used for SDI analysis.

The SDI_3 min is extremely high and sometimes impossible to measure, because the SDI membrane clogs very quickly. SS increased spectacularly during this event. Turbidity rose to 10.5 NTU.

Coagulation and flocculation are then optimized, so as to treat the mass and organic load without modifying the mirror speed of the Spidflow® or the filtration speed. The coagulant dosage is increased to between 9 and 11 mg Fe·L^{-1}, and the addition of flocculant is not necessary.

The operation of the Spidflow® and the filters have led to a water quality similar to that obtained during a so-called "normal" period. However, the duration of the filtration cycles is reduced to around 34 h. The filtered water has an average SDI_{15} min value between 3.0 and 4.0 %·min^{-1} and the turbidity remains below 0.1 NTU.

	Average	Maximum
Turbidity (NTU)	2	10.5
SDI $_3$ (%·min^{-1})	26.80	33.33
pH	8.03	8.11
Salinity (g·L^{-1})	37.4	38.1
Conductivity (mS·cm^{-1})	56.0	56.9
SS (mg·L^{-1})	8.4	21.0
Algae (cell·L^{-1})	107	108

Table 19.35. *Degraded water parameters during an algal bloom*

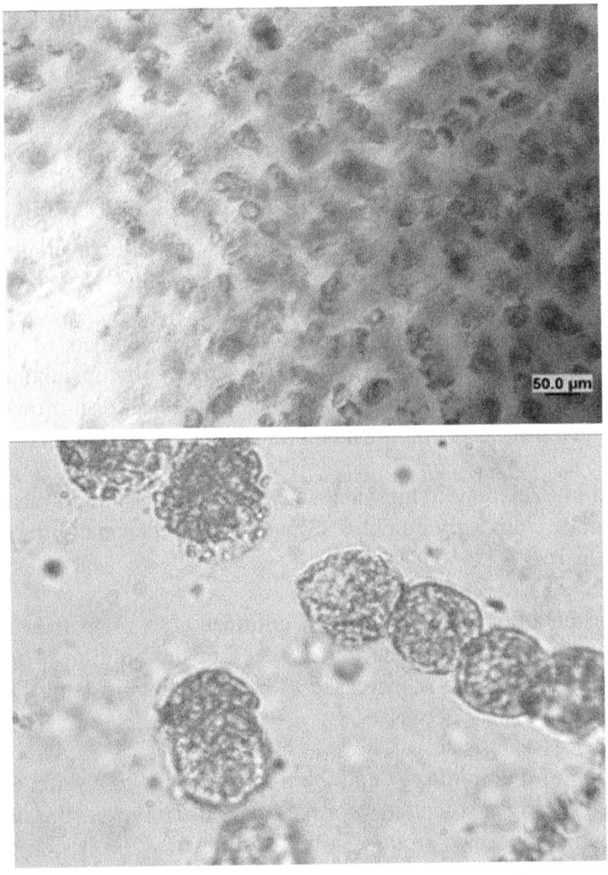

Figure 19.101. *Presence of Cochlodinium polykrikoides during an algal bloom. For a color version of this figure, see www.iste.co.uk/gaid/watertreatment4.zip*

A pH adjustment is not always necessary, because the acidity of the ferric chloride is sufficient to lower the pH to 6.8–7. After applying the proper chemical dosages and the new operation control mode of the pilot unit, the quality of the filtered water meets the required criteria. The evolution of the filtered water quality and the capabilities of the Spidflow® DMF pilot are shown in Figure 19.101. Table 19.36 shows the filtered water quality data during the degraded period.

	Minimum	**Average**	**Maximum**
Turbidity (NTU)	0.031	0.042	0.191 (<2% of the time)
SDI_{15} (%·min^{-1})	1.4	2.9	5 (<2% of the time)

Table 19.36. *SDI and turbidity of filtered water obtained during algal bloom*

The quality of the sludge is different between the normal quality period and the seawater degradation period, as shown in Figures 19.102 and 19.103.

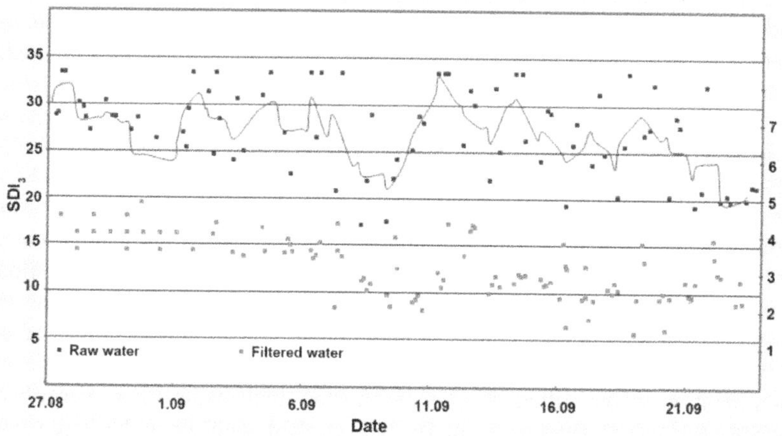

Figure 19.102. *Changes in SDI of raw and filtered water during an algal bloom. For a color version of this figure, see www.iste.co.uk/gaid/watertreatment4.zip*

Spidflow® has proven to be very effective at stopping a strong degradation of the feed water quality by removing more than 98% of microparticles and phytoplankton. The associated filtration allows a refining of the floated water in order to obtain water that conforms with what is expected for the feeding of the RO membranes.

a) b)

Figure 19.103. *Comparison of floating sludge obtained (a) during a normal period and (b) during an algal bloom. For a color version of this figure, see www.iste.co.uk/gaid/watertreatment4.zip*

Spidflow® is proposed as a pre-treatment for desalination plants when the water quality is subject to sudden variations, and when algal bloom events are expected.

19.14.2. *Spidflow® filter process applied to seawater desalination*

The Spidflow® filter is a combination of unitary physicochemical stages combining flotation and gravity filtration. The coagulation–flocculation and flotation stages are similar to those described for the Spidflow® process. The floated water feeds a two-layer gravity filter, located below the flotation zone. The filter materials are anthracite or pumice for the upper layer, and sand for the lower layer.

The operation of this filter is equivalent to the dual media filters described in section 19.8.4.1.3. The filter is regularly backwashed with air and filtered water. Different types of water can be used for backwashing (filtered water or brine).

During the water-only backwash, an additional floc removal flow stage is added to accelerate the removal of flocs present in the flotation zone. The high media height (3 m in total) allows for biological activity within the filter, degrading biodegradable organic molecules that are often responsible for membrane clogging. In addition, this filtration height induces long filtration cycles due to the constant quality of the feed water and thus promotes biological activity.

Alternatively, timers can be used (open time valve – closed time valve). The washing sequence is controlled by three parameters: production time, production volume or increase in pressure loss.

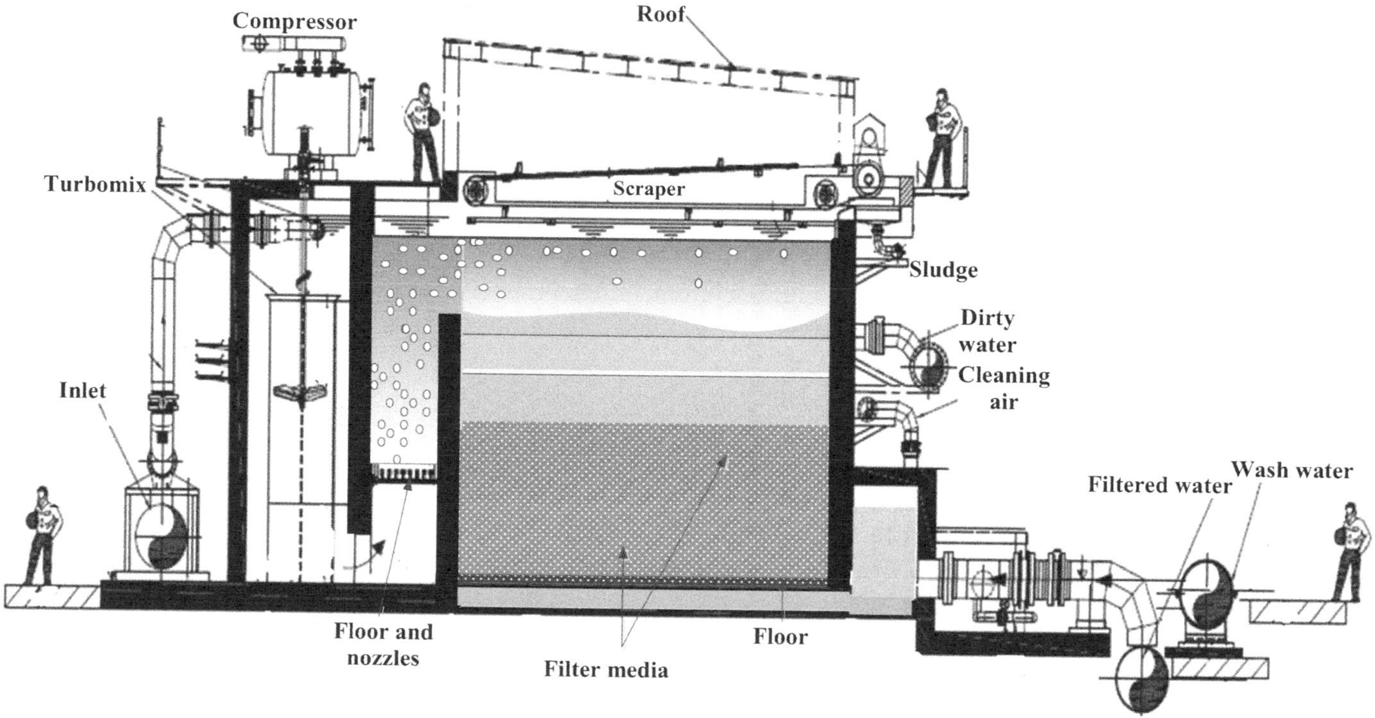

Figure 19.104. *Spidflow® filter. For a color version of this figure, see www.iste.co.uk/gaid/watertreatment4.zip*

This sequence includes a similar procedure to that described for the filters (see section 19.8). Mini-washes are performed periodically during production to avoid embolism in the core of the media due to the trapping of residual air bubbles. This mini-phase lasts between 10 and 20 s.

The coagulation and flocculation functionalities (such as colloid destabilization and agglomeration, microalgae and microparticle flocculation, organic adsorption) offer reliable performance with respect to these constituents. The dosage rates of the chemicals depend on the quality of the seawater.

The flotation features allow the separation of most particles from the water, regardless of the degradation of the raw seawater quality (algal blooms or small mollusk growth, oil pollution, etc.). The retained particles float to the surface, where they accumulate in the thickened sludge before being removed. These functionalities induce less reagents, such as the use of flocculant, which is only useful in the case of very high concentrations (algal bloom), as most of the particles present in raw seawater have a natural tendency to float.

Granular filtration features, such as depth filtration and biodegradation of dissolved organic matter, are reliable against fine particles and biofouling potential.

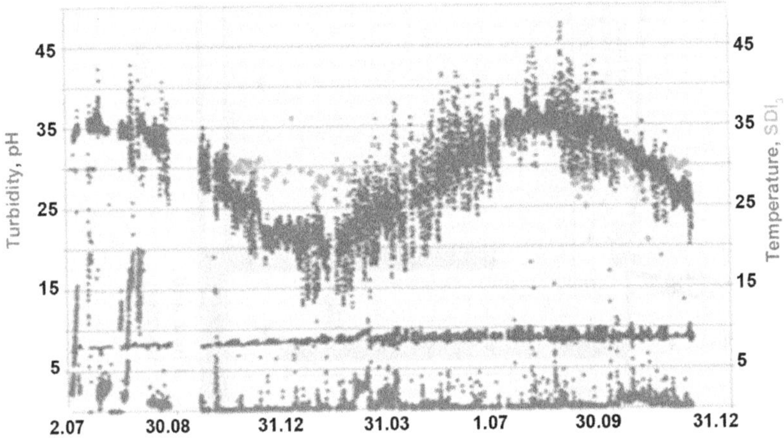

Figure 19.105. *Annual evolution of some parameters in the Arabian Gulf. For a color version of this figure, see www.iste.co.uk/gaid/watertreatment4.zip*

The results obtained on two types of water (Arabian Gulf and Gulf of Oman) show seasonal variations in temperature and SDI3, with periods of the latter parameter being so high that it could not be measured in the Arabian Gulf seawater

(Figure 19.105). Phytoplankton analysis indicated cell concentrations around 10^6–10^7 cells·L^{-1}, with a distribution consisting of about 15% diatoms, 1% dinoflagellates and 84% microflagellates.

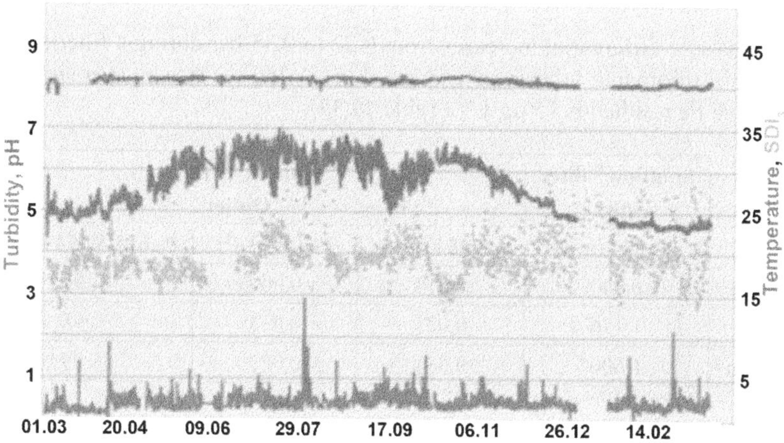

Figure 19.106. *Annual evolution of seawater in the Gulf of Oman. For a color version of this figure, see www.iste.co.uk/gaid/watertreatment4.zip*

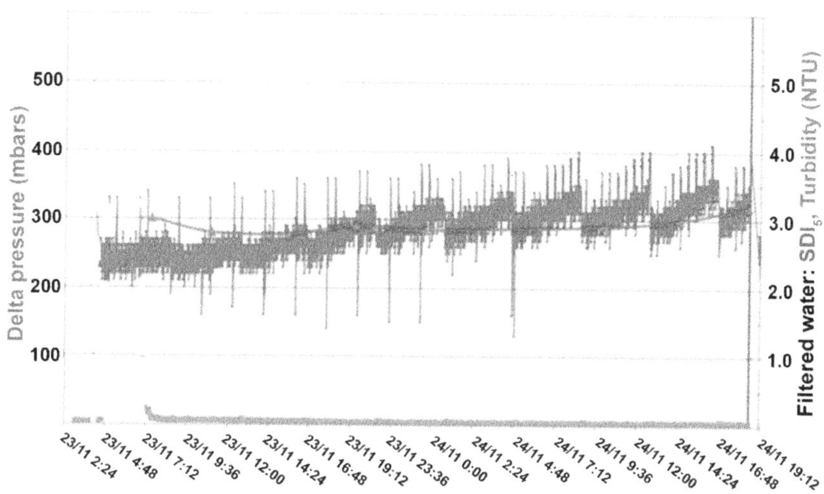

Figure 19.107. *SDI15 and turbidity of filtered water with the Spidflow® filter operating at 30 m·h^{-1}. For a color version of this figure, see www.iste.co.uk/gaid/watertreatment4.zip*

Figure 19.106 shows the evolution of a filtration cycle (an average of 38 h) for the Spidflow® filter operating at 30 m·h^{-1} on seawater from the Gulf of Oman. With a coagulant dosage of 1.07 mg Fe·L^{-1}, the SDI_{15} remains around 2.8–3.2% per minute and the turbidity is less than 0.1 NTU.

The pressure differential increases from 0.22 to 0.35 bar during a filtering cycle. The removal of organic substances expressed in UV254 reaches an average of 98%. The average Fe residual is 15 µg·L^{-1} (Table 19.37).

	Spidflow® filter inlet	Spidflow® filter Outlet		
	UV254 (m^{-1})	UV254 (m^{-1})	Fe (mg·L^{-1})	% of UV254 removal
Minimum	0.740	0.014	0.009	97
Average	1.176	0.021	0.015	98
Maximum	1.600	0.031	0.027	99

Table 19.37. *UV254 and Fe at the inlet and outlet of the Spidflow® filter*

The results obtained over 1 year of operation on seawater from the Gulf of Oman show SDI_{15} values ranging from 1.1 to 3.9, with an average of 2.5.

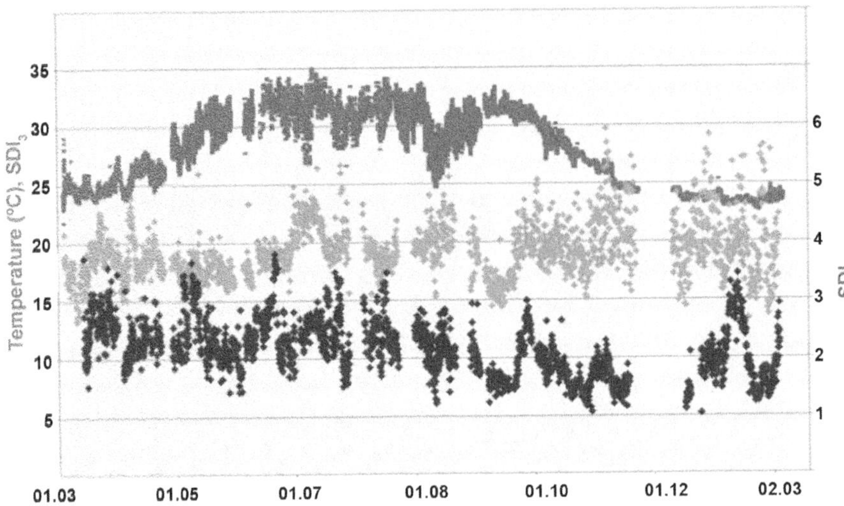

Figure 19.108. *Annual evolution of SDI in seawater (Gulf of Oman) and at the Spidflow® filter outlet. For a color version of this figure, see www.iste.co.uk/gaid/watertreatment4.zip*

During this period, the operation of the RO membranes at the outlet of the Spidflow® filter remained stable. Figure 19.109 shows the results obtained on the first pass during 1 year of operation, with a standardized differential pressure (20°C) of 0.6–0.8 bar and a standardized permeability (20°C) fluctuating between 1.1 and 1.6 $L \cdot m^{-2} \cdot h^{-1} \cdot bar^{-1}$. Only one CIP was performed.

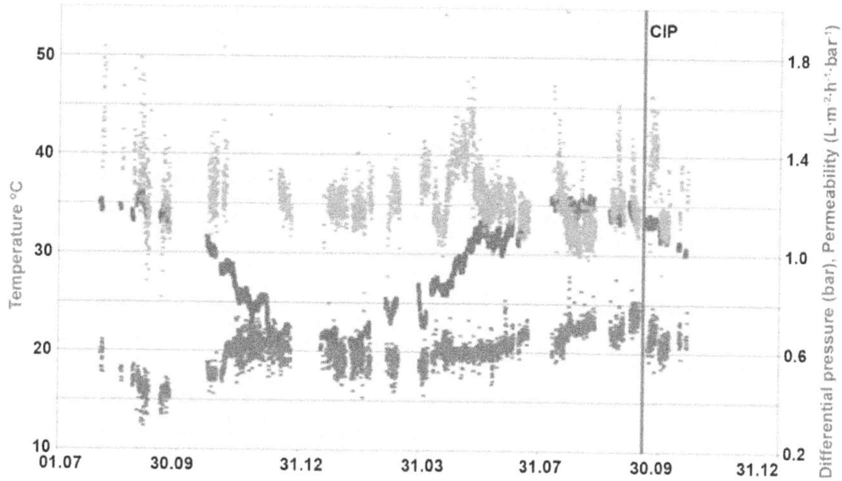

Figure 19.109. *Results obtained on the first pass of RO membranes during 1 year of operation. For a color version of this figure, see www.iste.co.uk/gaid/watertreatment4.zip*

Thus, the results obtained on two seawaters (Arabian Gulf and Gulf of Oman) showed that the Spidflow® filter operating at 30 $m \cdot h^{-1}$ is able to produce filtered water that is compatible with the requirements of an RO membrane, that is to say, SDI_{15} values constantly lower than 4, and turbidity values lower than 0.1 NTU. In addition, the quality of the pre-treated water remained stable despite the seasonal variations in SDI and turbidity.

19.14.3. *BiopROtector*

In order to greatly reduce the biofouling problem in RO membranes, Veolia has developed a process based on the biological removal of carbonated substrates present in the water, before its passage through the RO membranes.

In fact, feed water pre-treatment is essential to protect RO membranes, reduce energy costs and increase salt rejections. It must ensure partial or total removal of particles, organic substances and hydrocarbons. When it is not optimized or when it fails, an organic clogging of the membranes can be observed. This is due to an accumulation and excessive growth of micro-organisms on the surface of the membrane, which is promoted by the organic substances present in the feed water. Critical clogging leads to increased costs for chemical reagents, increased energy consumption due to pressure drops and early replacement of part of the membranes due to irreversible clogging, as well as the need for overly frequent chemical cleaning.

The quality of the feed water is considered to be the main factor contributing to the formation of membrane biofouling, since a continuous supply of nutrients promotes the growth of micro-organisms and leads, in most cases, to an excessive amount on the membrane surface.

The biopROtector affects two major parameters: the concentration of easily biodegradable compounds (organic and ammonia) and the concentration of dissolved oxygen.

The reduction and/or removal of these parameters determine the level of membrane fouling potential. The biopROtector is designed to ensure a significant reduction of bacterial growth in all applications where biofilm can develop: membranes, cartridge filters, heat exchangers or cooling towers. The aim is to reduce or remove all compounds that promote bacterial growth.

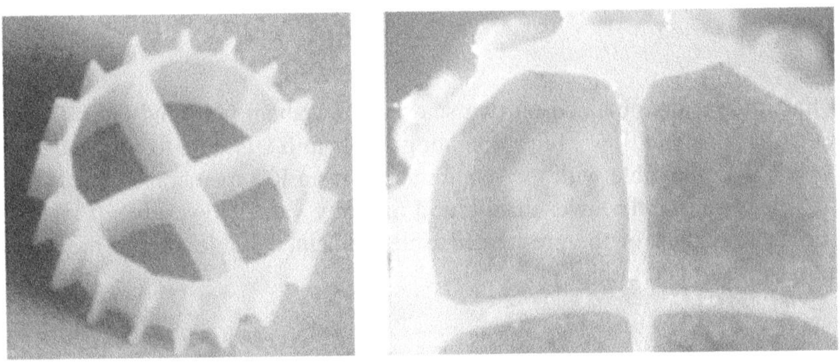

Figure 19.110. *K1 support (Anox Kaldness). For a color version of this figure, see www.iste.co.uk/gaid/watertreatment4.zip*

The biopROtector is a fixed-bed reactor containing plastic supports (such as the K1 Anox Kaldness), whose high surface/volume ratio allows the attachment and development of a relatively high concentration of biomass. Backwashing removes excess biomass, but the supports retain sufficient biomass to ensure the biological treatment process.

The biopROtector operates in a plug flow mode, like all fixed bed reactors, and can be arranged as two reactors in series. The operating parameters are:

– a contact time of 0.5–1.5 min (first reactor at high flow), then 8 min (second reactor at low flow);

– low maintenance;

– no chemical reagents.

Biochemical mechanisms involve significant reductions in dissolved O_2. As most of the particles sized between 5 and 10 µm are removed, a reduction in turbidity (30–50%) can be observed.

Applied in the pre-treatment of seawater desalination, the biopROtector is used in conjunction with a dual media filtration or with a UF membrane.

The pressure differential observed on RO membranes behind a pre-treatment consisting of a dual media coagulation–filtration is greater than that observed behind a pre-treatment consisting of a biopROtector and a UF membrane (Figure 19.111). After 100 days of operation, the ΔP rises to 0.28 bar behind a DMF, while it is only 0.20 bar behind the biopROtector UF combination.

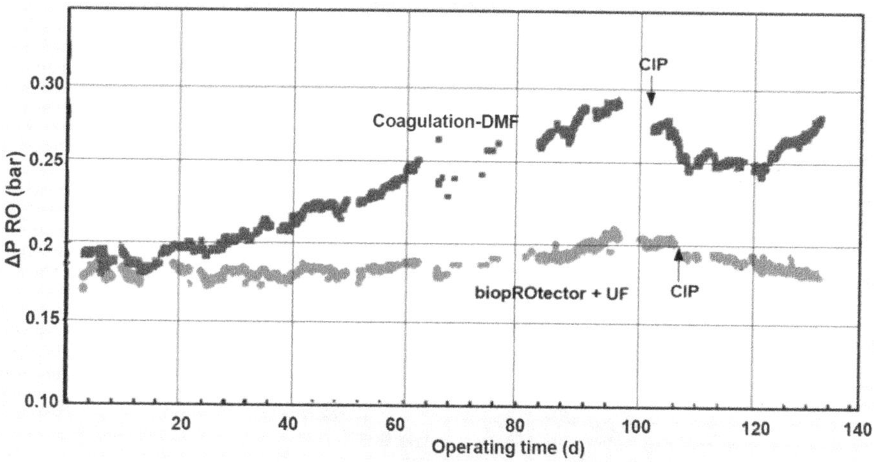

Figure 19.111. *Limiting clogging with biopROtector in pre-treatment on seawater. For a color version of this figure, see www.iste.co.uk/gaid/watertreatment4.zip*

Other results (Figure 19.112) also show less clogging of the cartridge pre-filters.

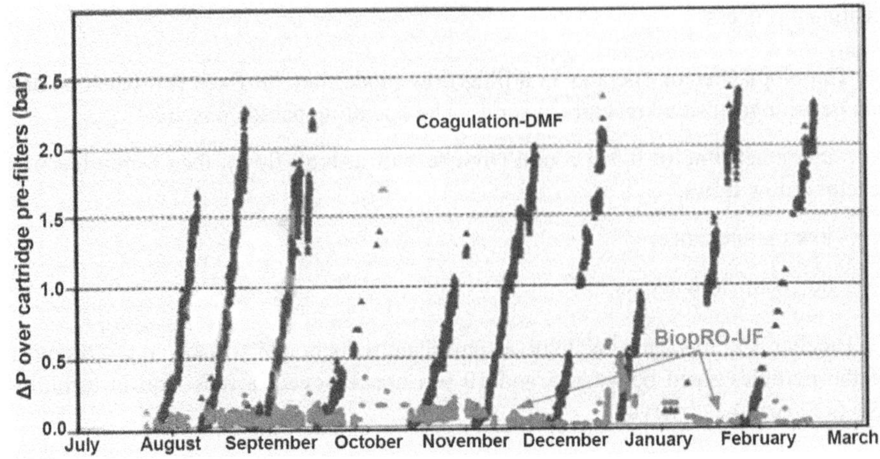

Figure 19.112. *Comparison of DeltaP observed on cartridge pre-filters behind a DMF and biopROtector UF. For a color version of this figure, see www.iste.co.uk/gaid/watertreatment4.zip*

Figure 19.113. *biopROtector reactors. For a color version of this figure, see www.iste.co.uk/gaid/watertreatment4.zip*

19.14.4. *Barrel (SIDEM Veolia)*

The Barrel technology represents the next generation of RO membrane trains. The Barrel is an RO membrane desalination technology that operates on the same principles and mechanisms as a conventional system, producing the same water quality.

Unlike existing systems that require multiple piping inlets, outlets and fittings, the Barrel only has two HP piping connections: one for the pre-treated seawater inlet and one for the concentrate outlet.

The Barrel design results in a significant reduction in the number of HP connections, which means less pressure loss in the system and, therefore, more energy savings.

It allows a significant reduction in electrical consumption of about 0.05 $kWh \cdot m^{-3}$ of fresh water produced, which reflects a large annual economic surplus value.

The Barrel design allows it to be produced locally, close to the site where it will be installed and operated. This reduces delivery time and the carbon footprint.

Made of painted carbon steel, the Barrel is a pressurized vessel designed to withstand the high pressure generated for membrane desalination (60–70 bar). The shell is a conventional ASME vessel, while the interior is coated with a high-quality epoxy paint that was specially developed for seawater applications. Inside, a resin honeycomb structure houses the 8-inch membranes available from any manufacturer. Membrane modules are easily inserted through dedicated maintenance manholes at both ends of the hull. The low pressure permeate is collected outside the shell and sent to the next stage of the process.

Due to its modular design, the Barrel is available in various capacities ranging from 800 to 50,000 $m^{-3} \cdot d^{-1}$. Depending on the quality of the pre-treated water and/or the quality of the water to be produced, it can be arranged in series to operate with multiple passes.

Equipped with sensors connected to databases, expert systems and artificial intelligence, the Barrel provides a permanent status of each membrane through intelligent connectors that are installed in the permeate tube. As a result, the performance of each membrane is continuously monitored and recommendations for optimal operating modes are provided. These devices can communicate wirelessly and provide the monitoring system with various information, such as the local conductivity and temperature of the permeate produced, creating a complete map of the permeate network. The smart connector brings the benefits of digital processing to the Barrel, ensuring the performance of each membrane element. This automatic, remotely accessible monitoring helps operators adopt the most economical strategy for shutting down, rotating or replacing membranes. In addition, the intelligent connector ensures that membrane supplier recommendations are met.

Figure 19.114. *External view of the Barrel. For a color version of this figure, see www.iste.co.uk/gaid/watertreatment4.zip*

This revolutionary innovation is a patented technology developed by SIDEM-Veolia. Its concept, reliability and performance have been proven on an 800 $m^{-3} \cdot d^{-1}$ installation in Saint-Martin in 2017.

Thus, the Barrel encompasses three parameters that become essential for future plants:

– Digitalization: the Barrel allows users to be connected by having continuous access to operating data in order to anticipate and diagnose breakdowns or failures, implement maintenance protocols and have access to the quality of the water produced at any time. Digitization favors anticipation and results in an increase in the life of the equipment, a reduction in the consumption of chemical reagents and in operating costs. The Barrel is a plug and play system, and its innovative design and features result in reduced operating costs and lower capital costs, providing long-term economic performance.

– Safety: this is a parameter that ensures protection insofar as the pressure tubes and membrane modules are inserted in a shell, compared to conventional systems where this equipment is not protected and can, in the event of a mechanical accident, create human and material damage due to the high pressure (60–70 bar) present in the pressure tubes.

– Compactness: the footprint of RO membrane plants is very high and the advantage of the Barrel is that the it reduces the necessary surface (25%) by confining the equipment within a compact block. Installed outdoors, this high-strength painted carbon steel monoblock does not require any building or air conditioning. By reducing the footprint, the amount of equipment is also reduced, resulting in lower capital costs.

Veolia Water Technologies provides the full range of services required to design, deliver, maintain and upgrade water treatment facilities and systems for both industrial and municipal clients.

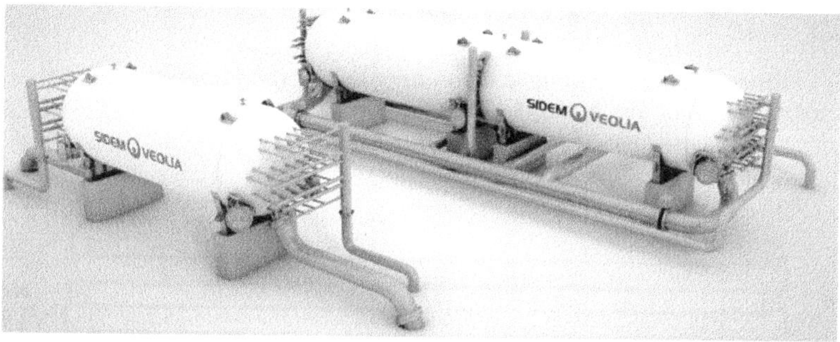

Figure 19.115. *View of the Barrel in a one-block or two-block series*

By optimizing both processes and monitoring, Veolia Water Technologies helps its clients reduce their water footprint while generating significant savings in energy and chemical consumption.

19.14.5. *Hiprode*

Hiprode is an innovative concept of pressure tube implementation and operation, with the integration of a central feed or a feed on both sides of the pressure tube. Hiprode optimizes the performance of the elements in the pressure tubes and benefits from the best performance of the high efficiency membranes put on the market by the manufacturers.

Conventionally, a membrane stage consists of a set of pressure tubes in which seven to eight membrane modules are incorporated. The feed is lateral and the permeate is recovered at the end of the pressure tube (Figure 19.116).

The Hiprode design involves a two-stage distribution of membrane modules (Figure 19.117), with a number N1 of pressure tubes in the first stage and a number N2 (< N1) in the second stage. As a result, the total number of pressure tubes (N1 + N2) can be equal to or less than the number obtained with the conventional single-stage design.

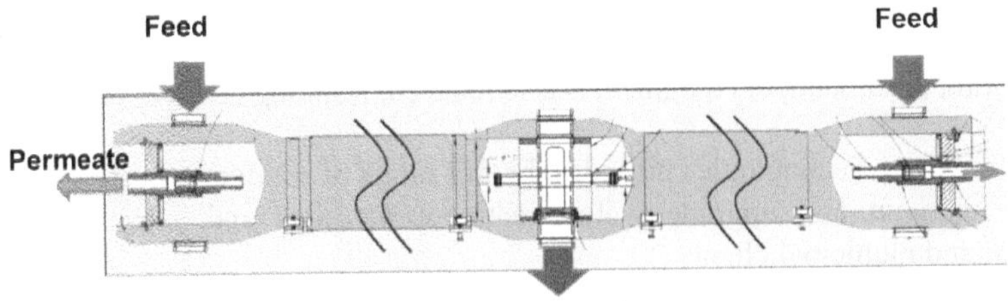

Figure 19.116. *Schematic diagram of the Hiprode operation. For a color version of this figure, see www.iste.co.uk/gaid/watertreatment4.zip*

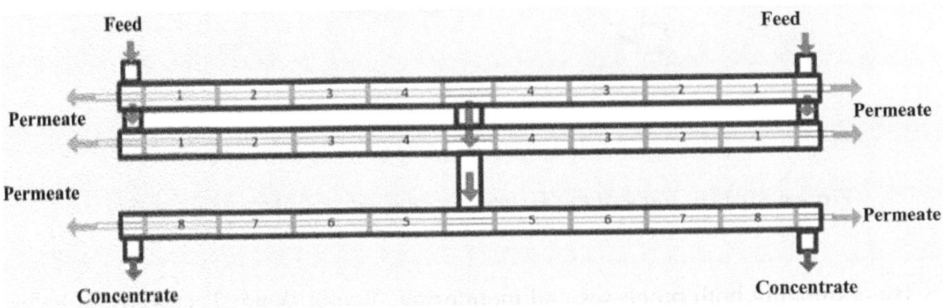

Figure 19.117. *Example of a pressure tube configuration. The numbers represent the position of the module in relation to the feed. For a color version of this figure, see www.iste.co.uk/gaid/watertreatment4.zip*

One of the advantages of this two-stage arrangement is that the first four elements, which represent the highest production rate, are disconnected from the last four elements. This decoupling thus gives greater versatility to the RO membrane process, because the two stages can be designed independently. The distribution of membrane modules can be optimized with more modules at the top, thus increasing the high production rate beyond 50%, depending on the N1/N2 ratio.

The hybrid intermediate membrane design (HID) can be implemented with the Hiprode. This mainly consists of installing different salt flow and rejection elements in the same pressure tube in order to balance the flows along this tube.

For example, there is a two-stage configuration comprising two and one pressure tubes, respectively. The first stage includes two high rejection (HR) membranes and two low-energy (LE) membranes on each side of the tube. The second stage only includes LE membranes (Figure 19.118).

Compared to a conventional configuration, the two-stage Hiprode can save up to 8% of the permeate production. It also reduces investment costs, as fewer pressure tubes, and therefore fewer membrane modules, are required.

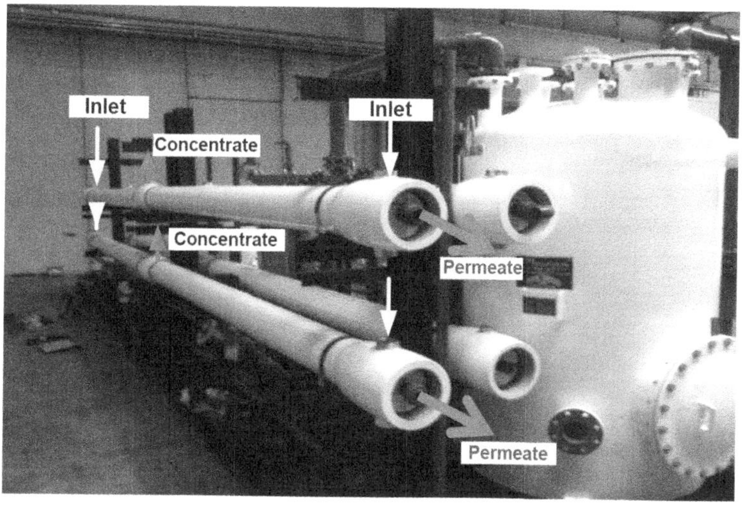

Figure 19.118. *On-site implementation of the Hiprode configuration. For a color version of this figure, see www.iste.co.uk/gaid/watertreatment4.zip*

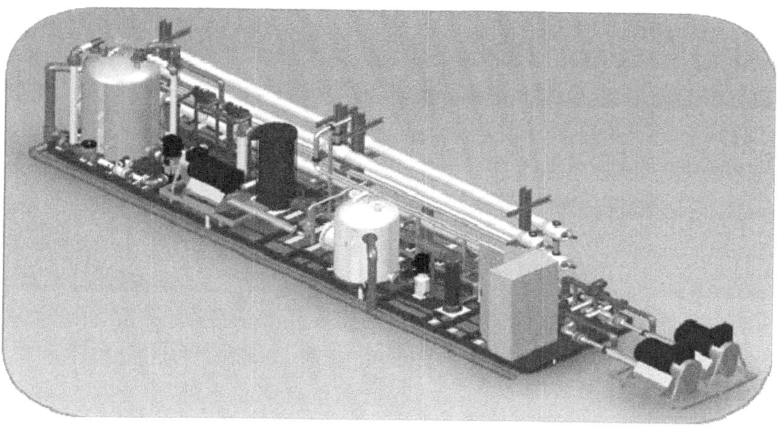

Figure 19.119. *Schematic diagram of a packaged solution in a container. For a color version of this figure, see www.iste.co.uk/gaid/watertreatment4.zip*

19.15. Packaged solutions in desalination

In seawater desalination, packaged solutions involve both conventional systems and innovations like the Barrel.

Conventional systems are integrated in containers, and the pre-treatment consists of pressurized filters with in-line reagent injection. Veolia has already equipped many facilities (hotels, tourist areas, industries, etc.) with these systems.

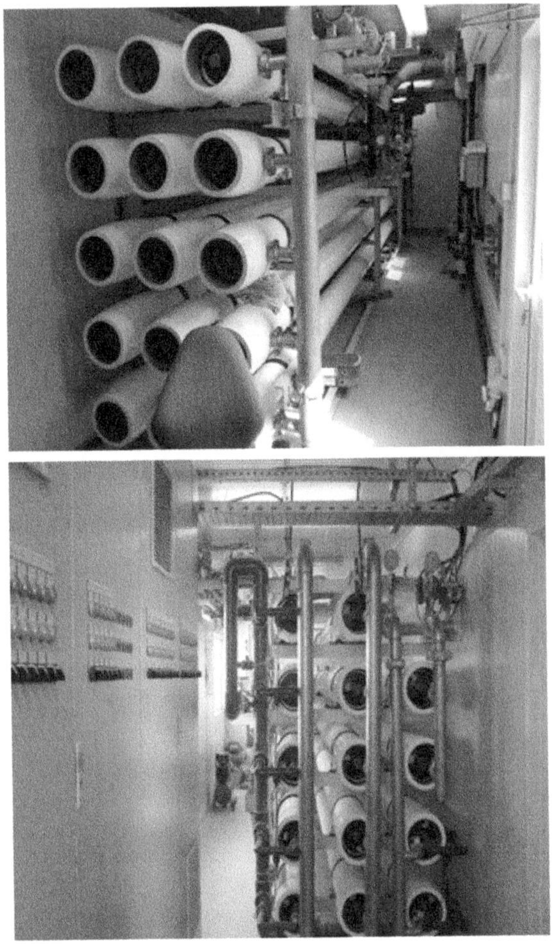

Figure 19.120. *Conventional system integrated in a container. For a color version of this figure, see www.iste.co.uk/gaid/watertreatment4.zip*

Figure 19.121. *Pre-treatment with pressure filters. For a color version of this figure, see www.iste.co.uk/gaid/watertreatment4.zip*

Figure 19.122. *Integration of solar panels as an energy source. For a color version of this figure, see www.iste.co.uk/gaid/watertreatment4.zip*

Veolia has also integrated solar panels into these systems, in order to operate without the need for fossil fuel energy for the HP pumps. They are applicable for brackish water and low flow rates up to 50,000 $L \cdot d^{-1}$ (10 $h \cdot d^{-1}$ period of sunlight).

19.16. Veolia's experience (HP membranes)

Veolia has numerous references with HP membranes throughout the world. Whether they are small or for high-flow plants, they each bring a wealth of experience that enriches Veolia's database. This database helps to create design and operating tools that are specific to the geographical context, but whose methodology and protocols remain common to all plants.

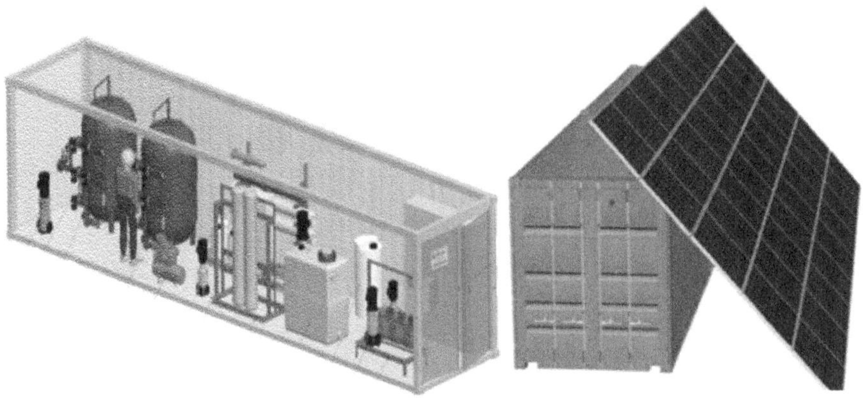

Figure 19.123. *Internal description of a container and a solar panel. For a color version of this figure, see www.iste.co.uk/gaid/watertreatment4.zip*

The following figures present a brief summary of the seawater desalination references built by Veolia in recent years. They include the geographical location, a picture of the plant, the treatment process, as well as the raw and produced water qualities.

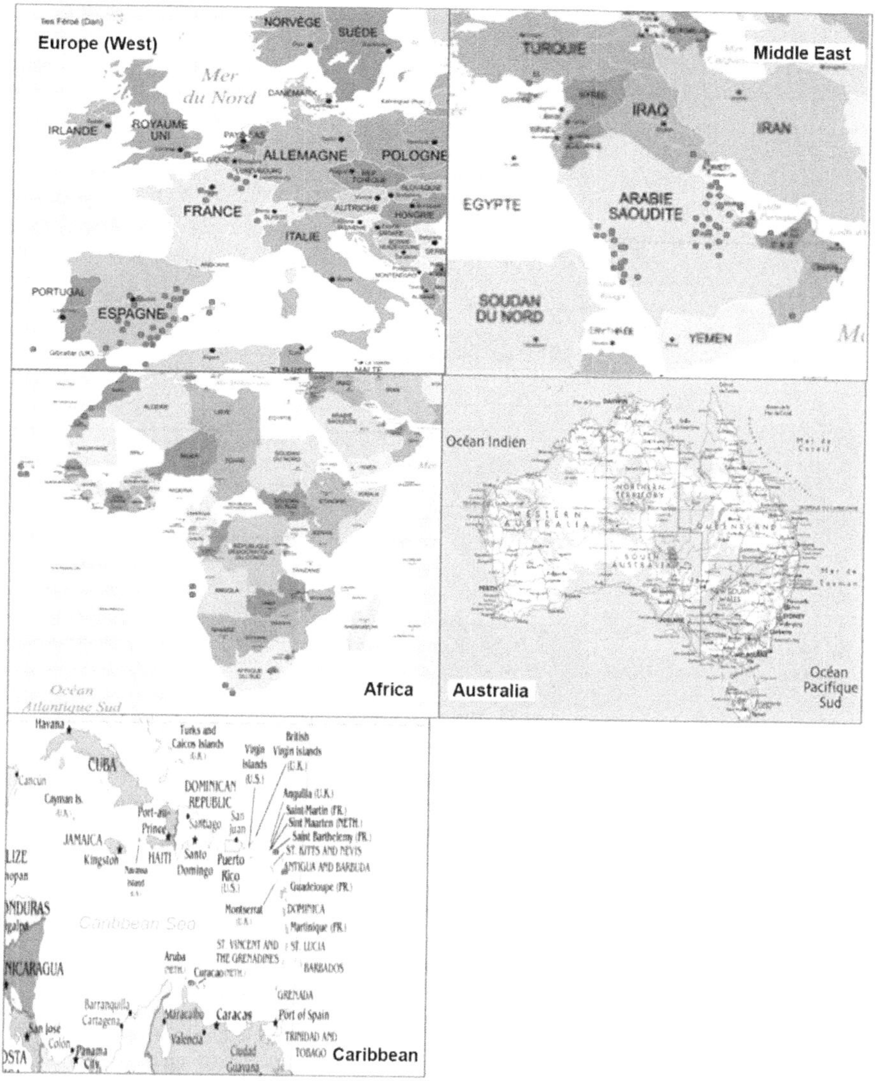

Figure 19.124. *Geographical locations (in red) of high-pressure membrane plants around the world. For a color version of this figure, see www.iste.co.uk/gaid/watertreatment4.zip*

Fujairah (UAE)

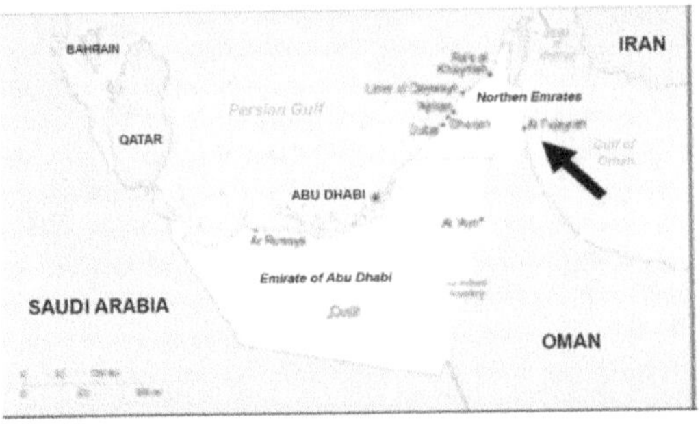

Hybrid treatment system including a distillation-type (MED) channel and a reverse osmosis membrane channel.
Client: Fujairah Asia Power Company
End user: Abu Dhabi Water Energy (ADWEA)
Location: Emirate of Fujairah, UAE
Capacity: 450,000 $m^3 \cdot d^{-1}$ by MED and 136,000 $m^3 \cdot d^{-1}$ by RO.

Raw water quality:

- **Temperature: 22–35°C**
- **pH: 7.8–8.5**
- **Turbidity: 1–14 NTU**
- **TOC: 2–4 mg·L^{-1}**
- **µAlgae: 500-60 10^6 u·L^{-1}**
- **SDI: 15 to >33.3**
- **TDS: 38–40.5 g·L^{-1}**
- **Boron: 5 mg·L^{-1}**

Membrane technology:

- **Offshore pumping,**
- **Spidflow®,**
- **Dual media filtration**
- **Pre-filters**
- **Reverse osmosis**
- **Remineralization (CO_2 – lime water)**
- **Disinfection**

Produced water quality:

- **pH: 7.5–8.5**
- **Turbidity: < 1 NTU**
- **TOC: < 0.5 mg·L^{-1}**
- **µAlgae: Absence**
- **TDS: 80–200 mg·L^{-1}**
- **Boron: 0.5 mg·L^{-1}**
- **LSI: 0 to + 0.3**
- **Cl_2: > 1 mg·L^{-1}**

Figure 19.125. *(A) Geographical site with photo of the plant, treatment process, raw water and produced water qualities. For a color version of this figure, see www.iste.co.uk/gaid/watertreatment4.zip*

Gold Coast (Australia)

Client: JV Queensland and Gold Coast City Council Water secure Queensland
End user: GC Desalination Alliance
Capacity: 125-133,000 $m^3 \cdot d^{-1}$
Location: Gold Coast

Raw water quality:

- Temperature: 17–28 °C
- pH: 7.6–8.2
- Turbidity: < 10 NTU
- TOC: 2 $mg \cdot L^{-1}$
- µAlgae: < 1000 $u \cdot L^{-1}$
- SDI: 15–25
- TDS: 34–38 $g \cdot L^{-1}$
- Boron: 5 $mg \cdot L^{-1}$

Membrane technology:

- Offshore pumping
- Pre-chlorination shocks
- Coagulation-Flocculation
- Dual media filtration
- Pre-filters/Reverse osmosis
- Remineralization (CO_2 – lime water)
- Disinfection

Produced water quality:

- pH: 7.2–8.5
- Turbidity: < 0.5 NTU
- TDS: < 220 $mg \cdot L^{-1}$
- Boron: <1 $mg \cdot L^{-1}$
- CCPP: –5 to –3
- Chlorides: < 50 $mg \cdot L^{-1}$
- Cl_2: 2 $mg \cdot L^{-1}$

Figure 19.125. *(B) Geographical site with photo of the plant, treatment process, raw water and produced water qualities (continued). For a color version of this figure, see www.iste.co.uk/gaid/watertreatment4.zip*

Sydney (Australia)

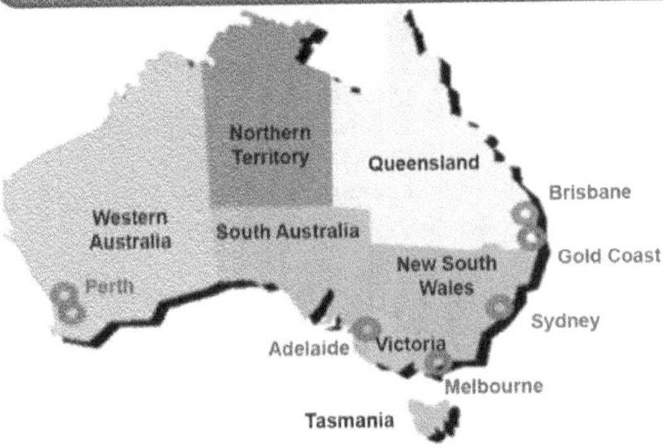

Client: Sydney Water – New South Wales
Capacity: 250,000 $m^3 \cdot d^{-1}$
Location: Sydney

Raw water quality:

- Temperature: 17–28 °C
- pH: 7.6–8.2
- Turbidity: < 10 NTU
- TOC: < 2 $mg \cdot L^{-1}$
- SDI: 15–25
- TDS: 34–38 $g \cdot L^{-1}$
- Boron: 5 $mg \cdot L^{-1}$

Membrane technology:

- Offshore pumping
- Pre-chlorination shocks
- Coagulation-Flocculation
- Dual media filtration
- Pre-filters/Reverse osmosis
- Remineralization (CO_2 – lime water)
- Disinfection

Produced water quality:

- pH: 7.3–8.3
- Turbidity: < 0.3 NTU
- TDS: < 140 $mg \cdot L^{-1}$
- Alkalinity: 35–50 $mg \cdot L^{-1}$
- Boron: < 2 $mg \cdot L^{-1}$
- CCPP: –8 to –1
- Chlorides: < 45 $mg \cdot L^{-1}$
- Cl_2: 1.5 $mg \cdot L^{-1}$

Figure 19.125. *(C) Geographical site with photo of the plant, treatment process, raw water and produced water qualities (continued). For a color version of this figure, see www.iste.co.uk/gaid/watertreatment4.zip*

Oman Sur (Oman)

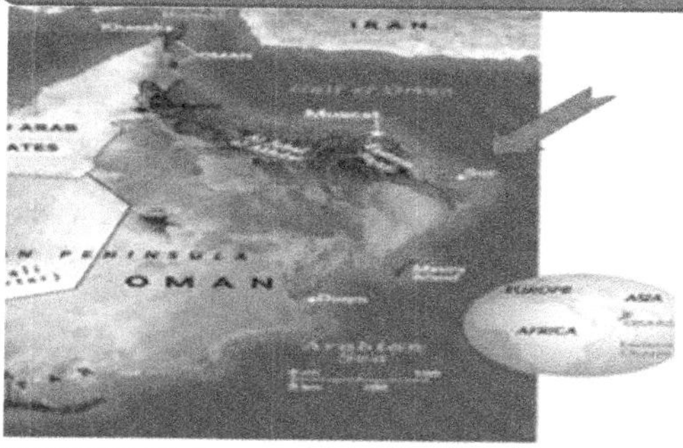

Client: Oman (Ministry of National Economy)
Capacity: 80,200 $m^3 \cdot d^{-1}$
Location: Sur

Beach well quality:

- Temperature: 22–35 °C
- pH: 6.9–8.3
- Turbidity: < 1 NTU
- TOC: < 1 $mg \cdot L^{-1}$
- µAlgae: absence
- SDI: << 3
- TDS: 38–41 $g \cdot L^{-1}$
- Boron: 4.8 $mg \cdot L^{-1}$

Membrane technology:

- Offshore pumping
- Sand filtration
- Pre-filters/Reverse osmosis
- Remineralization (CO_2 – limestone filters)
- Disinfection

Produced water quality:

- pH: 6.5–8.5
- Turbidity: < 2 NTU
- TDS: < 200–500 $mg \cdot L^{-1}$
- Conductivity: < 1000 $mhos.cm^{-1}$
- Boron: 0.5 $mg \cdot L^{-1}$
- LSI: < 2
- Ca: 100–300 $CaCO_3$ $mg \cdot L^{-1}$

Figure 19.125. *(D) Geographical site with photo of the plant, treatment process, raw water and produced water qualities (continued). For a color version of this figure, see www.iste.co.uk/gaid/watertreatment4.zip*

Oman Sur extension (Oman)

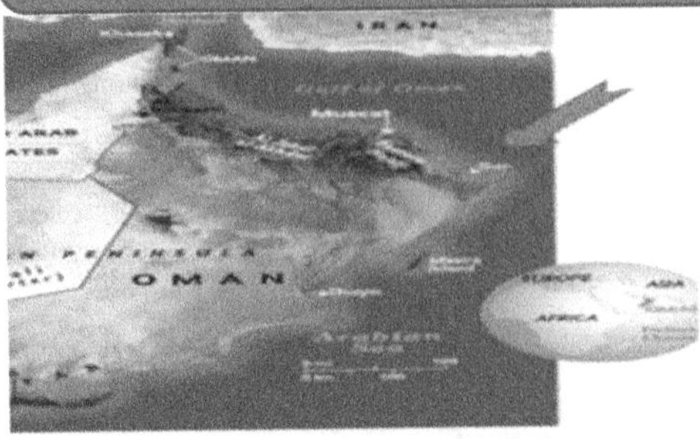

Client: Oman (Ministry of National Economy)
Capacity: 51,600 $m^3 \cdot d^{-1}$
Location: Sur

Raw water quality (surface water):

- Temperature: 22–35 °C
- pH: 6.9–8.3
- Turbidity: < 10 NTU
- TOC: 2–4 $mg \cdot L^{-1}$
- SDI_3: > 15
- TDS: 38–40 $g \cdot L^{-1}$
- Boron: 4.8 $mg \cdot L^{-1}$

Membrane technology:

- Offshore pumping
- Pre-chlorination shocks
- Spidflow
- Dual media filtration
- Pre-filters/Reverse osmosis
- Remineralization (CO_2 – limestone filters)
- Disinfection

Produced water quality:

- pH: 6.5–8.5
- Turbidity: < 2 NTU
- TDS: < 200–500 $mg \cdot L^{-1}$
- Conductivity: < 1000 $mhos.cm^{-1}$
- Boron: 0.5 $mg \cdot L^{-1}$
- LSI: < 2
- Ca: 100–300 $CaCO_3$ $mg \cdot L^{-1}$

Figure 19.125. *(E) Geographical site with photo of the plant, treatment process, raw water and produced water qualities (continued). For a color version of this figure, see www.iste.co.uk/gaid/watertreatment4.zip*

Aruba

Client: Water en Energie Bedrijf Aruba N.V
Capacity: 25,000 $m^3 \cdot d^{-1}$
Location: Aruba

Raw water quality (10 bW):

- **Temperature: 25–30 °C**
- **pH: 7.5–8.3**
- **Turbidity: 0.4–11 NTU**
- **TOC: < 1 $mg \cdot L^{-1}$**
- **SDI: < 5**
- **TDS: 34–38 $g \cdot L^{-1}$**
- **Boron: 5 $mg \cdot L^{-1}$**
- **Fe: 0.7–5 $mg \cdot L^{-1}$**
- **Mn: < 1 $mg \cdot L^{-1}$**

Membrane technology:

- **Pumping**
- **Chemical oxidation**
- **$FeCl_3$ coagulation**
- **Sand filtration**
- **Pre-filters/Reverse osmosis**
- **Remineralization (limestone filters)**
- **Disinfection**

Produced water quality:

- **pH: 8.5–9.5**
- **Turbidity: < 1 NTU**
- **Ca: 7–25 $CaCO_3 \cdot L^{-1}$**
- **Boron: < 0.5 $mg \cdot L^{-1}$**
- **Fe: < 0.3 $mg \cdot L^{-1}$**
- **Mn: < 0.05 $mg \cdot L^{-1}$**

Figure 19.125. *(F) Geographical site with photo of the plant, treatment process, raw water and produced water qualities (continued). For a color version of this figure, see www.iste.co.uk/gaid/watertreatment4.zip*

Sadara (Saudi Arabia)

Client: Aramco-Marafiq
Capacity: 178,650 $m^3 \cdot d^{-1}$
Location: Sadara, Al Jubail

Raw water quality:

- Temperature: 15–39 °C
- pH: 8.1–8.4
- Turbidity: 1–15 NTU
- SS: 5–30 $mg \cdot L^{-1}$
- μAlgae: 2.10^8 $u \cdot L^{-1}$
- SDI: 20 to >33.3
- TDS: 45 $g \cdot L^{-1}$
- TOC: 4–12 $mg \cdot L^{-1}$

Membrane technology:

- Offshore pumping
- Pre-chlorination shocks
- Spidflow
- Ultrafiltration
- Pre-filters/Reverse osmosis
- Remineralization (CO_2 – lime water)

Produced water quality (industrial sector):

- pH: 8–9
- Turbidity: < 1 NTU
- TDS: < 80 $mg \cdot L^{-1}$
- Conductivity: > 70 μS
- LSI: 0.03 to 3
- Ca: 30–50 $CaCO_3$ $mg \cdot L^{-1}$
- Chlorides: < 50 $mg \cdot L^{-1}$
- Sulfates: < 10 $mg \cdot L^{-1}$

Figure 19.125. *(G) Geographical site with photo of the plant, treatment process, raw water and produced water qualities (continued). For a color version of this figure, see www.iste.co.uk/gaid/watertreatment4.zip*

Az-Zour (Kuwait)

Capacity: 136,380 $m^3 \cdot d^{-1}$
Location: Az-Zour

Raw water quality

- **Temperature: 40 °C**
- **pH: 7.9–8.1**
- **Turbidity: < 5 NTU**
- **TOC: < 2 $mg \cdot L^{-1}$**
- **SDI: < 30**
- **TDS: 45 $g \cdot L^{-1}$**
- **Conductivity: 80000 $\mu S \cdot cm^{-1}$**
- **Boron: 5 $mg \cdot L^{-1}$**

Membrane technology:

- **Pumping**
- **Mechanical filtration**
- **Dual media high-speed filtration**
- **Pre-filters/Reverse osmosis**
- **Remineralization (CO_2 – lime water)**

Produced water quality:

- **pH: 7.5–8.5**
- **TDS: < 500 $mg \cdot L^{-1}$**
- **Boron: < 1.2 $mg \cdot L^{-1}$**
- **Ca: 60–80 $CaCO_3$ $mg \cdot L^{-1}$**

Figure 19.125. *(H) Geographical site with photo of the plant, treatment process, raw water and produced water qualities (continued). For a color version of this figure, see www.iste.co.uk/gaid/watertreatment4.zip*

Campo Dalias (Spain)

Client: ACUAMED (Ministry of the Environment)
Capacity: 97,200 $m^3 \cdot d^{-1}$
Location: Campo Dalias

Raw water quality

- **Temperature: 18–22 °C**
- **pH: 7.7**
- **Turbidity: < 15 NTU**
- **SS: 30 $mg \cdot L^{-1}$**
- **Conductivity: 54400 $\mu S \cdot cm^{-1}$**
- **TDS: 36 $g \cdot L^{-1}$**
- **Boron: 4.1 $mg \cdot L^{-1}$**

Membrane technology:

- **Offshore pumping**
- **Pre-chlorination**
- **Coagulation-Flocculation-Settling**
- **Pre-filters/Reverse osmosis**
- **Remineralization (lime milk)**
- **Disinfection**

Produced water quality:

- **pH: 8**
- **Turbidity: < 1 NTU**
- **TDS: < 500 $mg \cdot L^{-1}$**
- **Ca: 50–70 $CaCO_3$ $mg \cdot L^{-1}$**

Figure 19.125. *(l) Geographical site with photo of the plant, treatment process, raw water and produced water qualities (continued). For a color version of this figure, see www.iste.co.uk/gaid/watertreatment4.zip*

Glen Rocky (Gibraltar)

Client: UK Ministry of Defense
Capacity: 1,200 $m^3 \cdot d^{-1}$
Location: Glen Rocky

Raw water quality (Beach wells):

- **Temperature: 25–32 °C**
- **pH: 7.5–8.1**
- **Turbidity: < 3 NTU**
- **TOC: 2–3 $mg \cdot L^{-1}$**
- **SDI: 15–25**
- **TDS: 36–38 $g \cdot L^{-1}$**
- **Boron: 5 $mg \cdot L^{-1}$**

Membrane technology:

- **Pumping**
- **Coagulation**
- **Sand filtration**
- **Pre-filters/Reverse osmosis**
- **Remineralization (CO_2 – lime water)**
- **Disinfection**

Produced water quality:

- **pH: 7.5–8.5**
- **TDS: < 500 $mg \cdot L^{-1}$**
- **Boron: < 2 $mg \cdot L^{-1}$**
- **Ca: 60–80 $CaCO_3$ $mg \cdot L^{-1}$**

Figure 19.125. *(J) Geographical site with photo of the plant, treatment process, raw water and produced water qualities (continued). For a color version of this figure, see www.iste.co.uk/gaid/watertreatment4.zip*

Laayoune (Morocco)

Client: National Office of Drinking Water
Capacity: 26,000 m^3·d^{-1}
Location: Laayoune

Raw water quality (Beach wells):

- Temperature: 21–28 °C
- pH: 6.9–7.3
- Turbidity: 0.2–5 NTU
- TDS: 32–34 g·L^{-1}
- Boron: < 5 mg·L^{-1}
- Total Fe: 0.3–2.7 mg·L^{-1}
- Total Mn: 0.05–0.46 mg·L^{-1}
- SDI: < 3

Membrane technology:

- Offshore pumping
- Pre-chlorination
- Filtration
- Pre-filters/Reverse osmosis
- Remineralization (CO_2 – lime water)
- Disinfection

Produced water quality:

- pH: 6.5–8.5
- Turbidity: < 1 NTU
- TDS: < 500 mg·L^{-1}
- Chlorides: < 200 mg·L^{-1}
- LSI: 0.1 to 0.5
- Boron: < 1 mg·L^{-1}

Figure 19.125. *(K) Geographical site with photo of the plant, treatment process, raw water and produced water qualities (continued). For a color version of this figure, see www.iste.co.uk/gaid/watertreatment4.zip*

Kindasa (Saudi Arabia)

Client: Kindasa Water Services (KWS)
Capacity: 25,500 $m^3 \cdot d^{-1}$
Location: Kindasa

Raw water quality:

- **Temperature 25–34 °C**
- **pH: 8.0**
- **Turbidity: 1–6 NTU**
- **SDI: 20 to 30**
- **TDS: 42.5 $g \cdot L^{-1}$**
- **Boron: < 5 $mg \cdot L^{-1}$**

Membrane technology:

- **Pumping**
- **Mechanical filtration**
- **Coagulation**
- **Dual media high-speed filtration**
- **Ultrafiltration**
- **Pre-filters/Reverse osmosis**
- **Remineralization (lime water)**
- **CaOCl disinfection**

Produced water quality:

- **pH: < 8.75**
- **Turbidity < 1 NTU**
- **TDS: < 250 $mg \cdot L^{-1}$**
- **LSI: 0.05 to 0.1**
- **Chlorides: < 150 $mg \cdot L^{-1}$**

Figure 19.125. *(L) Geographical site with photo of the plant, treatment process, raw water and produced water qualities (continued). For a color version of this figure, see www.iste.co.uk/gaid/watertreatment4.zip*

19.17. References

Adams, L.F. and Ghiorse, W.C. (1987). Characterization of extracellular Mn2+-oxidizing activity and isolation of a Mn^{2+} oxidizing protein from *Leptothrix discophora* SS-1. *J. Bacteriol.*, 169, 1279–1285.

Adham, S. (1997). Control of membrane fouling due to algae, Montgomery Watson Technology Transfer Note. *Applied Research Department*, 12, 1–4.

Al-Leyla, M.A. and Middlebrooks, E.J. (1974). Algae removal by chemical coagulant. *Water Sew. Works*, 121(9), 76–80.

Aleem, F.A.A.E., Al-Sugair, K.A., Alahmad, M.I. (1998). Biofouling problems in membrane processes for water desalination and reuse in Saudi Arabia, Intern.Biodeter. *Biodegradation*, 41, 19–23.

Alessio, L., Campagna, M., Lucchini, R. (2007). From lead to manganese through mercury: Mythology, science, and lessons for prevention. *Am. J. Indust. Med.*, 50(11), 779–787.

Aminot, A. and Kérouel, R. (2004). *Hydrologie des écosystèmes marins : paramètres et analyses*. Éditions Quæ, Ifremer, Brest.

Amy, G. (2008). Fundamentals understanding of organic matter fouling of membranes. *Desalination*, 231, 44–51.

Archer, D.E. and Johnson, K.S. (2000). A model of the iron cycle in the ocean. *Global Biogeochem. Cycles*, 14, 269–279.

Arena, J.T., McCloskey, B., Freeman, B.D., McCucheon, J.R. (2011). Surface modification of thin composite membrane support layers with polydopamine: Enabling use of reverse osmosis membranes in pressure retarded osmosis. *J. Membr. Sci.*, 375, 55–62.

Arnold, S.R., Grubb, T.P., Harvey, P.J. (1995). Recent applications of dissolved air flotation pilot studies and full scale design. *Water Sci. Technol.*, 3(14), 327–340.

Bauman, A.G., Burt, J.A., Feary, D.A., Marquis, E., Usseglio, P. (2010). Tropical harmful algal blooms: An emerging threat to coral reef communities? *Mar. Pollut. Bull.*, 60(11), 2117–2122.

Bernhardt, H. and Clasen, J. (1991). Flocculation of micro-organisms. *J. Water. SRT-Aqua*, 40, 76–87.

Blute, N.K., McGuire, M.J., West, N., Voutchkov, N., Maclaggan, P., Reich, K. (2008). Integration of desalinated seawater into a distribution system: A corrosion pilot study. *Jo. Am. Water Works Assoc.*, 100(9), 117–131.

Boerlage, S.F.E., Kennedy, M., Bonne, P.A.C., Gilbert, G., Schippers, J.C. (1997). Prediction of flux decline in membrane systems due to particulate fouling. *Desalination*, 113, 231–238.

Boerlage, S.F.E., Kennedy, M., Dickson, M.R., El-Hodali, D.E.Y., Schippers, J.C. (2002a). The modified fouling index using ultrafiltration membrane (MFI-UF): Characterization, filtration mechanisms and proposed reference membrane. *J. Membr. Sci.*, 197, 1–21.

Boerlage, S.F.E., Kennedy, M., Tarawneh, Z., Aniye, M.P., Abrogrean, E.M., El-Hodali, D.E.Y., Schippers, J.C. (2002b). The modified fouling index using ultrafiltration membranes (MFI-UF): Characterization, filtration mechanisms and proposed reference membrane. *J. Membr. Sci.*, 197, 1–9.

Boerlage, S.F.E., Kennedy, M., Tarawneh, Z., Faber, R.D., Schippers, J.C. (2004). Development of MFI-UF in constant flux filtration. *Desalination*, 161, 103–113.

Bonnélye, V., Guey, L., Del Castillo, J. (2008). UF/MF as RO pre-treatment: The real benefit. *Desalination*, 222, 59–65.

Bowie, A.R., Achterberg, E.P., Croot, P.L., de Baar, H.J.W., Laan, P., Moffett, J.W., Ussher, S., Worsfold, P.J. (2006). A community-wide intercomparison exercise for the determination of dissolved iron in seawater. *Mar. Chem.*, 98, 81–99.

Brock, T.D. and Clyne, J. (1984). Significance of algal excretory products for growth of epilimnetic bacteria. *Appl. Environ. Microbiol.*, 47(4), 731–734.

Busch, M., Chub, R., Kolbec, U., Mengd, Q., Li, S. (2009). Ultrafiltration pre-treatment to reverse osmosis for seawater desalination three years field experience in the Wangtan Datang power plant. *Desalin. Publ.*, 10, 1–20.

Chen, Y., Dong, B.Z., Gao, N.Y., Fan, J.C. (2007). Effects of coagulation pre-treatment on fouling of an ultrafiltration membrane. *Desalination*, 204, 181–187.

Choi, Y.H., Kweon, J.H., Kim, D.I., Lee, S. (2009). Evaluation of various pre-treatment for particle and inorganic fouling control on performance of SWRO. *Desalination*, 247, 137–147.

Choules, P., Schrotter, J.C., Leparc, J., Gaid, K., Lafon, D. (2007). Reverse osmosis: Improved operation through experience with SWRO plants. In *American Membrane Technology Association's Conference*, Las Vegas.

Choules, P., Ben Boudinar, M., Mack, B. (2009). Membrane (MF&UF) pre-treatment design and operational experience from three seawater RO plants. In *IDA World Congress*, Atlantis, Dubai.

Codemo, G., Awadalla, A.R., Parker, M.J., Banham, J., Rybar, S. (2009). Hamriyah SWRO desa-lination plant: Largest sea water IMS plant. In *IDA World Congress*, Atlantis, Dubai.

Doccko, S. and Han, M. (2004). Fundamental characteristics of bubbles and ramifications for the flotation process. *Water Sci. Technol. Water Supply*, 50(12), 207–214.

Edzwald, J.K. (1995). Principles and applications of dissolved air flotation. *Water Sci. Technol.*, 31(3/4), 1–23.

Edzwald, J.K. (2007a). Fundamentals of dissolved air flotation. *J. New Engl. Water Works Assoc.*, 121, 89–112.

Edzwald, J.K. (2007b). Development of high rate dissolved air flotation for drinking water treatment. *J. Water Supply Res. Technol. Aqua*, 56(6/7), 399–409.

Edzwald, J.K. and Wingler, B.J. (1990). Chemical and physical aspects of dissolved air flotation for the removal of algae. *J. Water Supply Res. Technol. Aqua*, 39(2), 24–35.

Edzwald, J.K., Tobiason, J.E., Amato, T., Maggi, L.J. (1999). Integrating high rate dissolved air flotation technology into plant design. *J. Am. Water Works Assoc.*, 84(3), 92–100.

Fujiwara, N. and Matsuyama, H. (2008). Elimination of biological fouling in seawater reverse osmosis desalination plants. *Desalination*, 227, 295–305.

Futselaar, H., Blankert, B., Spenkelink, F., Rosberg, R. (2009). Ultrafiltration used as pre-treatment for SWRO desalination: Dynamic coagulant control and optimization. In *IDA World Congress*, Atlantis, Dubai.

Gabelich, C.J., Frankin, J.C., Gerringer, F.W., Ishida, K.P., Suffet, I.H. (2005). Enhanced oxidation of polyamide membranes using monochloramine and ferrous iron. *J. Membr. Sci.*, 258, 64–70.

Gaid, K. (2010). Overcoming red tide. *H2O. Middle East*, May, 26–34.

Gaid, A. and Craig, K. (2009). Success story for reverse osmosis technology: Two new plants. In *IDA World Congress*, 7–12 October, Atlantis, Dubai.

Gaid, K. and Rehi, R. (1988). La déminéralisation des eaux saumâtres. Cas de la ville de Abadla. *Techniques & Sciences municipales. L'eau.*, 10, 517–523.

Gaid, K. and Rehi, R. (1992). Conception d'une station de démineralisation par osmose inverse. *Techniques & Sciences municipales. L'eau.*, 12, 623–631.

Gaid, A. and Treal, Y. (2007). Le dessalement des eaux de mer par osmose inverse. *Desalination*, 203, 1–14.

Gaid, K., Gasnier, F., Bourdon, F. (2012a). Spidflow® : la flottation à grande vitesse en eau potable. *L'eau, L'industrie, Les Nuisances*, 342, 95–58.

Gaid, K., Konig, E., Rouxel, S., Ventresque, C. (2012b). Fujairah 2: From pilot to full scale, a success story. *The International Desalination Association (IDA) World Congress*, Dubai.

Gaid, K., Gasnier, F., Girodet, P., Bertrand, S., Becera, L. (2013a). Fujairah 2: Seawater reverse osmosis plant, pre-treatment pilot plant operation results. Report to the consultant FAPCO and the authorities of Municipality.

Gaid, K., Delagarde, S., Muller, G., Leparc, J. (2013b). High-rate filtration for desalination: From pilot plant to full scale plant. *The International Desalination Association (IDA) World Congress*, Tianjin.

Gaid, K., Leparc, J., Bréant, P. (2013c). Current and future technologies for enhanced desalination facilities. *2nd International Symposium on Innovation and Technology in the Phosphate Industry*, Elsevier.

Gaid, K., Ventresque, C., Leparc, J., Pitavy, C., Thubert, J.B. (2015a). Pretreatment of RO desalination, success stories for various water qualities. *International Conference on Desalination Association World Congress*, 30 August–4 September, San Diego, CA.

Gaid, K., Pastorelli, D., Leparc, J. (2015b). A combined flotation-filtration for desalination plant. *International Conference on Desalination Association World Congress*, 30 August–4 September, San Diego, CA.

Gaid, K., Vergnolle, F., Gandre, R., Baaklini, D., Pitt, N. (2016). Optimized one-step seawater pretreatment coupling dissolved air flotation and granular media filtration. *Conference on Desalination for the Environment Clean Water and Energy*, 22–26 April, Rome.

Gaid, K., Vergnolle, F., Gandre, R., Baaklini, D., Pitt, N. (2017). Optimized one-step seawater pretreatment coupling high velocity dissolved air flotation and granular media filtration. *International Desalination Association World Congress, Water Reuse and Desalination*, October 15–20, Sao-Paulo.

Garcia, H. and Gordon, L. (1992). Oxygen solubility in seawater: Better fitting equations. *Limnol. Oceanogr.*, 37(6), 1307–1312.

Guider, R.J. and Roche, J.L. (1994). The role of iron in phytoplankton photosynthesis, and the potential for iron-limitation of primary productivity in the sea. *Photosynthesis Res.*, 39, 275–301.

Haarhoff, J. (2008). Dissolved air flotation and prospects for drinking water treatment. *J. Water Supply Res. Technol. Aqua*, 57(8), 555–567.

Hallegraeff, G.M. (1993). A review of harmful algal blooms and their apparent global increase. *Phycologia*, 32, 79–99.

Henderson, R.K., Parsons, S.A., Jefferson, B. (2008). Surfactants as bubble surface modifiers in the flotation of algae: Dissolved air flotation that utilises a chemically modified bubble surface. *Environ. Sci. Technol.*, 42, 4883–4888.

Her, N., Amy, G., Fos, D., Cho, J., Yoon, Y., Kosenka, P. (2002). Optimization of method for detecting and characterizing NOM by HPLC-size exclusion chromatography with UV and online DOC detection. *Environ. Sci. Technol.*, 36, 1069–1076.

Hermia, J. (1982). Constant pressure blocking filtration laws: Application to power-law non-Newtonian fluids. *Trans. Inst. Chem. Eng.*, 60, 183.

Hong, S. and Elimeleh, M. (1997). Chemical and physical aspects of natural organic matter (NOM) fouling nanofiltration membranes. *J. Membr Sci.*, 132, 159–181.

Hong, K., Lee, S., Choi, S., Yu, Y., Hong, S., Moon, J., Sohn, J., Yang J. (2009). Assessment of various membrane fouling indexes under seawater conditions. *Desalination*, 247, 247–259.

Honga, E.U., Opperman, H., Dietrich, J.A., Venter, W., Tjipangandjara, K.F., Watson, I.C. (2009). Seawater desalination in Namibia: Raw water quality considerations and pre-treatment approach. In *IDA World Congress*, Atlantis, Dubai.

van Hoof, S.C.J.M., Minnery, J.G., Mack, B. (2001). Dead-end ultrafiltration as alternative pretreatment to reverse osmosis in seawater desalination: A case study. *Desalination*, 139, 161–168.

Huntsman, A., Harvey, G.R., Sunda, W.G. (1983). Photoreduction of manganese oxides in seawater and its geochemical and biological implications. *Nature*, 301, 234–236.

Ingham, R.A., Mansour, L., Qadan, T., Al Hindi, M., Attawneh, O. (2009). Pre-treatment design and performance under challenging conditions in the Arabian Gulf. In *IDA World Congress*, Atlantis, Dubai.

Javeed, M.A., Chinu, K., Shon, H.K., Vigneswaran, S. (2009). Effects of pre-treatment on fouling propensity of feed as depicted by the modified fouling index (MFI) and cross-flow sampler modified fouling index (CFS-MFI). *Desalination*, 238, 98–108.

Jerowska, A., Bottino, A., Capanelli, G., Fabbri, C., Migliorini, G. (2009). Ultrafiltration as direct pre-treatment of seawater – A case study. *Desalination*, 245, 723–729.

Johir, A.H., Khorshed, C., Vigneswaran, S., Shon, H.K. (2009). In-line flocculation as pre-treatment to reverse osmosis desalination. *Desalination*, 247, 85–93.

Johnson, K.S., Gordon, R.M., Coale, K.H. (1997). What controls dissolved iron concentrations in the world ocean? *Mar. Chem.*, 57, 181–186.

Junga, C.W. and Son, H.J. (2009). Filtration model: Evaluation of membrane fouling mechanism in various membrane pre-treatment processes. *Desalination Water Treat.*, 2, 195–20.

Khirani, S., Ben Aim, R., Manero, M.H. (2007). Improving the measurement of the modified fouling index using nanofiltration membrane (NF-MFI). *Desalination*, 191, 1–7.

Kim, S.H. and Yoon, J.S. (2005). Optimization of microfiltration for seawater suffering from red tide contamination. *Desalination*, 182, 315–321.

Kim, P.H., Hong, S., Choi, S. (2006). Variation and prediction of membrane fouling index under various feed water characteristics. *J. Membr. Sci.*, 284, 248–254.

Kim, S.H., Chon, K., Kim, S.J., Lee, S., Lee, E., Cho, J. (2009a). Uncertainty in organic matter analysis for seawater reverse osmosis (5SWRO) desalination. *Desalination*, 238, 30–36.

Kim, Y.M., Kim, S.J., Kim, Y.S., Lee, S., Kim, I.S., Kim, J.H. (2009b). Overview of systems engineering approaches for a large-scale seawater desalination plant with a reverse osmosis network. *Desalination*, 238, 312–332.

Knops, F. and Lintelo, R. (2009). Long-term operating experience of seaguard UF as pretreatment to SWRO, in the Mediterranean region. *Desalination and Water Treatment*, 5, 74–79.

Knox-Holmes, B., Hassan, A.M., Williams, E.E., Al-Tisan, I. (1994). Fouling prevention in desalination plants. *Second Gulf Water Conference*, Bahrain.

Krasner, S.W., Croué, J.P., Buffle, J., Perdue, E.M. (1996). Three approaches for characterizing NOM. *J. Am. Water Works Assoc.*, 88, 66–79.

Kremen, S.S. and Tanner, M. (1998). Silt density index (SDI), percent plugging factor (%PF) their relation to actual foulant deposition. *Desalination*, 119, 360–372.

Kruithof, J.C., Schippers, J.C., Kamp, P.C., Folmer, H.C., Hofman, J. (1998). Integrated multiobjective membrane systems for surface water treatment: Pre-treatment of reverse osmosis by conventional treatment and ultrafiltration. *Desalination*, 117, 37–48.

Kuma, K., Nishioka, J., Matsunaga, K. (1996). Controls on iron (III) hydroxide solubility in seawater: The influence of pH and natural organic chelators. *Limnol. Oceanogr.*, 41, 396–407.

Kumar, M., Adham, S., Decarolis, J., Pearce, W. (2007). Comparative seawater RO pre-treatment evaluation using bench and pilot-scale testing. *J. Am. Water Works Assoc.*, 99(9), 168–178.

Kwon, B.N., Park, N., Cho, J. (2005). Effect of algae on fouling and efficiency of UF membranes. *Desalination*, 179, 203–214.

Laîné, J.M., Campos, C., Baudin, I., Janex, M.L. (2003). Understanding membrane fouling: A review of over a decade of research. *Water Sci. Technol.: Water Supply*, 3(5/6), 155–164.

LaZerte, B.D. and Burling, K. (1990). Manganese speciation in dilute waters in the Precambrian shield. *Water Res.*, 24, 1097–1101.

Leparc, J. (2017). Pilot-scale comparison of UF membranes for direct seawater filtration in SWRO desalination. *European Desalination Society Conference*, 4 April, Las Palmas, Canary Islands.

Leparc, J., Gaid, K., Schrotter, J.C., Rapenne, S., Croué, J.P., Lebaron, P., Lafon, D. (2007a). Use of advanced analytical tools for monitoring performance of seawater pretreatment processes. *International Desalination Association World Congress*, October 21–26, Maspalomas, Canary Islands.

Leparc, J., Rapenne, S., Courties, C., Lebaron, P., Croué, J.P., Jacquemet, V., Turner, G. (2007b). Water quality and performance evaluation at seawater reverse osmosis plants through the use of advanced analytical tools. *Desalination*, 203, 243–255.

Liang, H., Gong, W., Li, G. (2008). Performance evaluation of water treatment ultrafiltration pilot plants treating algae-rich reservoir water. *Desalination*, 221, 345–350.

Liu, X. and Millero, F.J. (2002). The solubility of iron in seawater. *Mar. Chem.*, 77, 43–54.

Maa, W., Zhao,Y., Wanga, L. (2006). The pre-treatment with enhanced coagulation and a UF membrane for seawater desalination with reverse osmosis. *Desalination*, 203, 256–259

Matsumoto, K., Aoki, T., Hirao, T., Yoda, K., Morita, S., Nagai, N., Ikuno, N., Uemura, K. (2009). New combined chlorine disinfection for bio-fouling control. In *IDA World Congress*, Atlantis, Dubai.

Millero, F.J. (1998). Solubility of Fe (III) in seawater. *Earth Planet Sci. Lett.*, 154, 3232–329.

Millero, F.J., Sotolongo, S., Izaguirre, M. (1987). The oxidation of Fe (II) in seawater. *Geochim. Cosmochim Acta*, 51, 793–801.

Mitra, S.S., Thomas, A.R., Gang, G.T. (2009). Evaluation and characterization of seawater RO membrane fouling. *Desalination*, 247, 94–107.

Mitrouli, S.T., Yiantsios, S.G., Karabelas, A.J., Mitrakas, M., Føllesdalc, M., Kjolseth, P.A. (2008). Pre-treatment for desalination of seawater from an open intake by dual-media filtration: Pilot testing and comparison of two different media. *Desalination*, 222, 24–37.

Mitrouli, S.T., Yiantsios, S.G., Karabelas, A.J., Kjolseth, P.A. (2009). New granular for dual-media filtration of seawater: Pilot testing. *Sep. Purif. Technol.*, 65, 147–155.

Mosset, A., Bonnélye, V., Petrya, M., Sanz, M.A. (2008). The sensitivity of SDI analysis: From RO feed water to raw water. *Desalination*, 222, 17–23.

Mucci, A. (2004). The behavior of mixed Ca-Mn carbonates in water and seawater: Controls of manganese concentrations in marine porewaters. *Aquat. Geochem.*, 10(1/2), 139–169.

Nakabayashi, S., Kusakabe, M., Kuma, K., Kudo, I. (2001). Vertical distributions of iron (III) hydroxide solubility and dissolved iron in the north-western North Pacific Ocean. *Geophys. Res. Lett.*, 28, 4611–4614.

Passow, U. (2000). Formation of transparent exopolymer particles, TEP, from dissolved precursor material. *Mar. Ecol. Prog. Ser.*, 192, 1–11.

Passow, U. (2002). Transparent exopolymer particles (TEP) in aquatic environments. *Prog. Oceanogr.*, 55, 287–333.

Pearce, G.K. (2007). The case of UF/MF pre-treatment to RO in seawater applications. *Desalination*, 203, 286–295.

Peleka, E.N. and Matis, K.A. (2008). Application of flotation as a pre-treatment process during desalination. *Desalination*, 222, 1–8.

Petrucci, G. and Rosellini, M. (2005). Chlorine dioxide in seawater for fouling control and post-disinfection in potable water works. *Desalination*, 182, 283–291.

Petrusevski, B., Vlaski, A., van Breemen, A.N., Alaerts, G.J. (1995). Optimisation of coagulation conditions for direct filtration of impounded surface water. *J. Water Supply Res. Technol. Aqua*, 44(2), 92–102.

Petrusevski, B., van Breemen, A.N., Alaerts, G.J. (1996). Effect of permanganate treatment and coagulation with dual coagulants on algae removal in direct filtration. *J. Water Supply Res. Technol. Aqua*, 45(5), 316–326.

Poussade, Y., Gaid, K., Vergnolle, F., Pitt, N., Ventresque, C., Vigneron-Larosa, N. (2017a). Impacts of granular media vs membranes filtration on the pre-treatment of SWRO desalination plants. *International Desalination Association World Congress on Water Reuse and Desalination*, October 15–20, Sao-Paulo.

Poussade, Y., Gaid, K., Vergnolle, F., Pitt, N., Ventresque, C., Vigneron-Larosa, N. (2017b). Impact of pretreatment on the operation of a SWRO desalination plant: Return of experience and recommendations. *International Desalination Association World Congress on Water Reuse and Desalination*, October 15–20, Sao-Paulo.

Qin, J.J., Oo, M.H., Kekre, K.A., Knops, F., Miller, P. (2006). Reservoir water treatment using hybrid coagulation-ultrafiltration. *Desalination*, 193, 344–349.

Rue, E.L. and Bruland, K.W. (1995). Complexation of iron (III) by natural organic ligands in the Central North Pacific as determined by a new competitive ligand equilibration/adsorptive cathodic stripping voltametric method. *Mar. Chem.*, 50, 117–138.

Rue, E.L. and Bruland, K.W. (1997). The role of organic complexation on ambient iron, chemistry in the equatorial Pacific Ocean and the response of a mesoscale iron addition, experiment. *Limnol. Oceanogr.*, 42, 901–910.

Schaanning, M., Naes, K., Egeberg, P.K., Bomé, F. (1988). Cycling of manganese in the permanently anoxic Drammensfjord. *Mar. Chem.*, 23, 365–382.

Schippers, J.C. and Verdow, J. (1980). The modified fouling index: A method of determining the fouling characteristics of water. *Desalination*, 32, 137–145.

Serkiz, S.M. and Perdue, E.M. (1990). Isolation of dissolved organic matter from the Suwannee River using reverse osmosis. *Water Res.*, 24, 911–916.

Shawwa, A.R. and Smith, D.W. (2000). Dissolved air flotation model for drinking water treatment. *Can. J. Civil Eng.*, 27, 373–382.

Shin, J.Y. and O'Melia, C.R. (2006). Pre-treatment chemistry for dual media filtration: Model simulations and experimental studies. *Water Sci.Technol.* 53(7), 167–175.

Shon, H.K., Kim, S.H., Vigneswaran, S., Ben Aim, R., Lee, S., Cho, J. (2009). Physicochemical pre-treatment of seawater fouling reduction and membrane characterization. *Desalination*, 238, 10–21.

Smoluchowski, M.V. (1916). Drei Vorträge über Diffusion, Brownsche Bewegung und Koagula-tion von Kolloidteilchen. *Physik. Z.*, 17, 557–585.

Sommariva, C., Al Hindi, M., Fabbri, C. (2009). Palm Jumeirah: The first large SWRO with ultrafiltration plant in the Gulf. In *IDA World Congress*, Atlantis, Dubai.

Stumm, W. and Lee, G.F. (1961). Oxygenation of ferrous iron. *Intl. Eng.Them.*, 53, 143–146.

Tani, H., Nishioka, J., Kuma, K., Takata, H., Yamashita, Y., Tanoue, E., Midorikawa, T. (2003). Iron (III) hydroxide solubility and humic-type fluorescent organic matter in the deep water column of the Okhotsk Sea and the northwestern North Pacific Ocean. *Deep-Sea Res. Part I*, 50, 1063–1078.

Tebo, B.M., Bargar, J.R., Clement, B.G., Dick, G.J., Murray, K.J., Parker, D., Verity, R., Webb, S.M. (2004). Biogenic manganese oxides: Properties and mechanisms of formation. *Annu. Rev. Earth Planet. Sci.*, 32, 287–328.

Tekeuchi, K., Iwahashi, H., Kishi, M., Goto, T., Hirai, M., Al Mohannadi, F.H. (2008). Who plays an important role for bio-fouling in SWRO plant. In *IDA World Congress*, Atlantis, Dubai.

Teng, C.K., Hawlader, M.N.A., Malek, A. (2003). An experiment with different pre-treatment methods. *Desalination*, 156, 51–58.

Teuler, A., Glucina, K., Laine, J.M. (1999). Assessment of UF pre-treatment prior RO membranes for seawater desalination. *Desalination*, 125, 89–96.

Texas Water Development Board (2008). Texas seawater desalination demonstration project, Report, Brownsville Public Utilities Board.

Thornton, D.C.O. (2004). Formation of transparent exopolymeric particles (TEP) from macroalgal detritus. *Mar. Ecol. Prog.*, 282, 1–12.

Ventresque, C., Leparc, J., Lacaze-Elsous, F., Barbé, C., Vergnolle, F., Tréal, Y. (2013). Innovative pressure vessel configuration. *The International Desalination Association (IDA) World Congress*, Tianjin.

Vial, D. and Doussau, G. (2002). The use of microfiltration membranes for seawater pre-treatment prior to reverse osmosis membranes. *Desalination*, 153, 141–147.

Villacorte, L.O., Schurer, R., Kennedy, M.D., Amy, G., Schippers, J.C. (2009). Removal and deposition of transparent exopolymer particles (TEP) in seawater UF-RO system. In *IDA World Congress*, Atlantis, Dubai.

Vlaski, A., van Breemen, A.N., Alaerts, G.J. (1996). The algae problem in the Netherlands from a water treatment perspective. *J. Water SRT Aqua*, 45(4), 184–194.

Vonghia, E., Hagmeyer, G., Fernández, M.G., Sanchez, J. (2009). Escombreras: Large-scale Spanish seawater desalination utilizing innovative membrane pre-treatment technology. In *IDA World Congress*, Dubai.

de Vrind, J.P., Boogerd, F.C., de Vrind-de Jong, E.W. (1986). Manganese reduction by a marine *Bacillus* species. *J. Bacteriol.*, 167(1), 30–34.

Weinberg, E.D. (1989). Cellular regulation of iron assimilation. *Q. Rev. Biol.*, 64, 261–290.

Wells, M.L., Zorkin, N.G., Lewis, A.G. (1983). The role of colloid chemistry in providing a source of iron to phytoplankton. *J. Mar. Res.*, 41, 731–746.

Wittmann, E., Tazi-Pain, A., Chanussot, T., Patterson, A., Niaye, R., Gaid, K., Canto, F. (2002). Clarification of a highly turbid karstic water by microfiltration. *Desalination*, 145, 309–313.

Wittmann, E., Belkhodja, M., Lee, Y., Remize, O.J., Brouwer, H., Schrotter, J.C. (2011). RO membrane biofouling: Problems and solutions. In *World Congress Perth Convention and Exhibition Centre (PCEC)*, Perth.

Wotton, R. (2004). The essential role of exopolymers (EPS) in aquatic systems. In *Oceanography and Marine Biology: An Annual Review*, Gibson, R.N., Atkinson, R.J.A., Gordon, J.D.M. (eds). CRC Press, Boca Raton, FL.

Xia, S., Nan, J., Liu, R., Li, G. (2004). Study of drinking water treatment by ultrafiltration of surface water and its application to China. *Desalination*, 170, 41–47.

Yakubowicz, H. (2010). Large-scale SWRO system supplies Chilean copper gold mine. *World Water*, July–August, 41–42.

Yang, H.J. and Kim, H.S. (2009). Effects of coagulation on MF/UF for removal of particles as a pre-treatment in seawater desalination. *Desalination*, 247, 45–52.

Yao, K.-M. and Habibian, M.T. (1971). Water and waste water filtration: Concepts and applications. *Environ. Sci. Technol.*, 5(11), 1105–1115.

Yoon, J.S., Kim, S.H., Yoon, C.H., Kim, G.T. (2004). Microfiltration as pre-treatment of reverse osmosis for seawater desalination. *J. Korean Soc. Environ. Eng.*, 26(4), 475–480.

Zeng, H., Zhang, J., Ye, C. (2009). Comparison of an ultrafiltration membrane fed with raw seawater, coagulated seawater and cooling tower blowdown. *Desalination*, 244, 199–207.

Index

A, B, C

aeration, 158
algae, 37, 43, 63, 67, 68, 72, 78, 82, 85, 132, 133, 149, 150, 172–174, 206, 207, 212, 222, 226, 228–231, 246–253, 255, 262, 264–266, 268, 271, 273, 276, 280, 281, 286–288, 338–340, 342, 346
aluminum, 161, 177, 247
ammonia, 258, 261, 350
bacteria, 1, 3, 85, 132, 206, 207, 222, 226, 228, 230–232, 246, 254, 255, 262, 286–288
chemical backwash, 50
chlorine, 7, 8, 30, 43, 44, 51, 54, 64, 82, 114, 175, 178, 179, 186, 188, 193, 195, 196, 214, 254–263, 285, 287, 288
Cryptosporidium, 82, 85, 87, 178

D, E, F

decarbonation, 161, 319
disinfection, 159
endocrine disruptors, 176, 179, 234
flocculation, 29, 84, 149, 184, 188, 193, 254, 255, 264, 266, 268, 271, 273, 276, 278, 279, 281, 287, 305, 309, 340, 341, 344, 346
frontal, 9, 19, 21, 75, 128, 208

G, H, I

Giardia, 82, 85, 87, 178
hardness, 52, 127, 146–148, 158, 184, 188, 193, 315, 316, 321, 322, 324, 326, 336
integrity, 16, 19, 35, 36, 56–58, 60, 62, 93, 179
iron, 37, 41, 50, 63, 64, 72, 82, 85, 132, 146, 161, 216, 218–220, 222, 224, 248, 261, 273–275, 279, 281, 289, 316

L, M, O

lime, 158, 189, 263, 322, 335, 336
micropollutants, 64–68, 70, 75, 76, 85, 112, 127, 137, 149–152, 174, 176, 179, 184, 188, 234
mixing, 157, 158, 193, 196, 199
mono media filters, 268, 271

organic matter, 19, 37, 38, 42, 63, 65, 67, 71, 72, 76, 81, 82, 85, 87, 149, 150, 152, 167, 169, 178, 184, 207, 225, 229, 262, 279, 286, 287, 340
oxyhydroxides, 220
ozone, 69, 82, 254

P, R

parasites, 1, 58, 82, 85, 178, 255
permeability, 33, 34, 37, 166
pesticides, 64–68, 70, 75, 76, 85, 112, 137, 149–152, 176, 179, 184, 188, 234
powdered activated carbon, 20, 24, 64, 65, 67, 75, 85
remineralization, 156–159, 177, 189, 315, 319, 321–324, 326–328, 330, 331, 333, 335
resin, 10, 314, 353

S, T, V

sand, 24, 29, 67, 68, 76, 161, 184, 188, 189, 193, 196, 226, 251, 264, 266, 267, 270, 271, 344
SDI (silt density index), 45, 85, 128–131, 133, 149, 168, 183, 206–214, 226, 227, 230, 233, 246, 247, 249–251, 263, 264, 266, 271–273, 276, 278, 281, 282, 288, 337, 340–343, 348, 349
sludge, 65, 82, 276, 305, 343, 344, 346
soda, 44, 52–54, 56, 63, 134, 158, 160, 168, 193, 195, 199, 262, 287, 288, 321, 326, 331, 334
Spidflow®, 82, 252, 277–279, 281, 339–341, 343–345, 347–349
stripping, 186
tangential, 9, 19, 43, 44, 48, 66, 75, 76, 137, 159, 160, 241
TOC, 52, 63, 67, 70, 72, 74, 75, 78, 169, 188, 189, 206, 225–227, 229, 249–253
transmembrane pressure, 34
viruses, 1, 3, 5, 58, 70, 83, 84, 87

Summary of Volume 1

Chapter 1. Introduction

Chapter 2. Physicochemical and Microbiological Composition of Raw Water

2.1. Water resources
2.1.1. Physicochemical parameters
2.1.2. Algae (including cyanobacteria and cyanotoxins)
2.1.3. Tastes and odors
2.1.4. Micropollutants
2.2. Microbiology
2.2.1. Bacteria
2.2.2. Viruses
2.2.3. Parasites (Cryptosporidium and Giardia)
2.3. Quality of water intended for human consumption
2.3.1. Microbiological parameters
2.4. References

Chapter 3. Aeration and Stripping

3.1. Cascade aeration
3.1.1. Characteristics
3.2. Operating principle of a cascade aerator system
3.2.1. Data
3.2.2. Goals
3.2.3. Results

3.3. Aeration by fine bubble diffusers
3.3.1. Air diffusers
3.3.2. Air insufflation by oxytube
3.4. Stripping
3.4.1. Stripping tower design
3.4.2. Description
3.4.3. CO_2 removal
3.4.4. Tetrachlorethylene and trichlorethylene removal
3.5. Synthesis of aeration systems
3.6. References

Chapter 4. Coagulation–flocculation

4.1. Colloidal matter
4.2. Coagulation
4.2.1. Double-layer compression
4.2.2. Adsorption and interparticle bridging (flocculation)
4.3. Flocculation
4.3.1. Perikinetic flocculation
4.3.2. Orthokinetic flocculation
4.3.3. The influence of agitation
4.3.4. G and t
4.4. Coagulants
4.4.1. Metallic coagulants
4.4.2. Synthetic organic coagulants
4.5. Flocculants
4.5.1. Natural organic and synthetic flocculants
4.5.2. Flocculation adjuvants
4.6. Factors affecting coagulation and flocculation
4.6.1. Influence of water temperature
4.6.2. Influence of pH
4.6.3. Coagulation and flocculation times
4.7. How to choose the best coagulant?
4.7.1. How to choose the optimal dose of coagulant?
4.8. Residual aluminum
4.9. Alkalinity consumption
4.9.1. Aluminum and alkalinity
4.9.2. Iron and alkalinity
4.10. Reduction efficiency of some water constituents
4.10.1. Turbidity and SS
4.10.2. Microorganism removal

4.11. Jar tests
4.11.1. The specific case of jar test for Actiflo®
4.12. References

Chapter 5. Settling

5.1. The principles of settling
5.2. Horizontal settlers
5.2.1. Principle
5.2.2. Design
5.2.3. Implementation
5.3. Lamella settlers
5.3.1. Theory and principle
5.3.2. Basic design for lamella settlers
5.3.3. Implementation
5.4. Veolia technologies
5.4.1. Lamella settlers: Multiflo® settler
5.4.2. Ballasted floc settlers
5.5. References

Chapter 6. Flotation

6.1. The scope of DAF
6.2. The main stages of a flotation process
6.2.1. Coagulation
6.2.2. Flocculation
6.2.3. Flotation
6.3. The fundamental mechanisms of flotation
6.3.1. Coagulation
6.3.2. Flocculation
6.3.3. Contact zone
6.3.4. Separation zone
6.4. Design parameters
6.4.1. Coagulation
6.4.2. Flocculation
6.4.3. Flotation zone
6.4.4. Contact time
6.4.5. Temperature
6.4.6. Air saturation tank
6.4.7. Mass balance
6.4.8. Recirculation and injection nozzles

6.5. Operating parameters affecting flotation performance
6.5.1. Choice of coagulant
6.5.2. Rise velocity
6.5.3. Contact time
6.5.4. Bubble volume concentration
6.5.5. Gas solubility
6.5.6. Hydraulic efficiency
6.5.7. Air/water ratio
6.6. Performance and monitoring
6.6.1. Treatment monitoring
6.6.2. Performance of DAF systems in relation to algae removal
6.6.3. Performance against parasites
6.6.4. Performance with the addition of PAC
6.7. Veolia technologies using flotation
6.7.1. Spidflow®: principle
6.7.2. Advantages and limitations of DAF systems
6.7.3. Spidflow® filter
6.7.4. Ozoflot®
6.7.5. Flottazone®
6.7.6. Packaged solutions: Spidflow® Pack
6.8. References

Index

Summary of Volume 2

Chapter 7. Removal of Natural Organic Matter

7.1. Natural organic matter: humic substances
7.2. Methods of quantification and assessment of organic substances in water
 7.2.1. Total organic carbon
 7.2.2. Absorbance of ultraviolet light at 254 nm (UV 254)
 7.2.3. Specific UV absorbance
 7.2.4. Liquid chromatography
 7.2.5. Hydrophobic organic carbon
 7.2.6. Fluorescence
7.3. Conditions for the removal of NOM
7.4. NOM removal techniques
 7.4.1. Coagulation–flocculation
7.5. Adsorption on activated carbon
 7.5.1. Mechanism of NOM removal by activated carbon
 7.5.2. Implementation of activated carbon adsorption for NOM removal
7.6. Ozonation
 7.6.1. Preozonation
 7.6.2. Interozonation
7.7. Biological treatment
 7.7.1. Biological treatment with PAC: operating principle of the Opaline® B process (Veolia)
 7.7.2. Biological treatment combining ozone and GAC
7.8. Treatment of ion exchange resins
 7.8.1. Use of resins for NOM removal: mechanism of NOM removal

7.8.2. Parameters affecting the performance of resins for the removal of NOM
7.8.3. Resin regeneration
7.8.4. The Opalix® process
7.9. NOM removal by high-pressure membranes
7.10. References

Chapter 8. Filtration

8.1. Rapid filters and very high-rate filters (TGV)
8.2. Multimedia filters
8.3. Direct filtration
8.4. Pressurized filters
8.5. Filtration mechanisms
8.5.1. General principle
8.5.2. Mechanisms involved in filtration
8.6. Implementation parameters
8.6.1. Materials
8.6.2. Material height/d_{10} ratio
8.6.3. ES ratio
8.7. Sizing parameters: filtration rate and material height
8.8. Operating parameters
8.8.1. Pressure loss in a clean filter
8.8.2. Pressure loss during clogging
8.8.3. Estimation of the turbidity of filtered water in single-layer filters
8.8.4. Retention capacity
8.8.5. Filter washing conditions: theory and calculations
8.8.6. Cycle time
8.8.7. Main parameters involved in filter sizing
8.9. Veolia filtration technologies: general information
8.9.1. Characteristics of open gravity filters
8.10. Regulation systems
8.10.1. Hydraulic regulators
8.10.2. Control valves
8.11. Recycling and microbiological risks
8.12. Monitoring the operation and performance of filters
8.12.1. Turbidity
8.12.2. Particle count
8.12.3. Problems with the operation of filters
8.13. References

Chapter 9. Adsorption on Activated Carbon

9.1. Activation processes of activated carbon
9.1.1. Chemical activation
9.1.2. Physical activation
9.2. Physicochemical properties of activated carbon
9.2.1. Parameters influencing adsorption
9.3. Transport process in activated carbon: mass transfer
9.3.1. Adsorption mechanisms and isotherms
9.4. The different forms of conditioning of activated carbons
9.4.1. Powdered activated carbon
9.4.2. Micrograin activated carbon (µgrain)
9.4.3. Granular activated carbon
9.5. Adsorption reactors on activated carbon: removal process
9.6. PAC reactors: description of PAC reactors
9.6.1. Mass balance
9.6.2. GAC reactors in adsorption mode: GAC filters
9.6.3. Fluidized bed activated carbon reactors
9.6.4. GAC-µgrain-PAC comparison
9.6.5. Hybrid process with a PAC reactor associated with UF membranes
9.7. Veolia technologies: treatment process with PAC reactors
9.7.1. Implementation in a settler without PAC recirculation
9.7.2. Implementation with PAC recirculation
9.7.3. Opaline® C process: PAC-membrane hybrid
9.8. Micrograin activated carbon reactors
9.8.1. Integration of µgrain activated carbon in the treatment process
9.8.2. The Filtraflo® Carb
9.8.3. Opacarb®FL
9.8.4. Opacarb®MG
9.9. Fixed bed reactors – GAC filters
9.9.1. The GAC filter (Veolia)
9.9.2. GAC filters in series
9.9.3. Implementation modes for GAC filters
9.10. Pressurized GAC filters (Opacarb™ filters)
9.11. References

Index

Summary of Volume 3

Chapter 10. Removal of Micropollutants

10.1. Introduction
10.2. Pesticides
10.3. Pharmaceuticals and industrial waste
10.4. Pesticide removal technologies and emerging MPs
10.4.1. Adsorption onto activated carbon (AC)
10.4.2. Ozonation
10.4.3. Ozone-activated carbon combination
10.4.4. Advanced chemical oxidation
10.4.5. Nanofiltration and reverse osmosis membranes
10.5. Frogbox®: an effective monitoring and control tool
10.6. The evolution of micropollutants in drinking water plants
10.7. References

Chapter 11. Removal of Perfluorinated Compounds

11.1. Physicochemical properties
11.2. Presence in the water
11.3. Drinking water regulations
11.4. Treatments
11.4.1. Coagulation–flocculation–settling (or flotation)
11.4.2. Chemical oxidation
11.4.3. UV oxidation
11.4.4. Activated carbon
11.4.5. High-pressure membranes: nanofiltration and reverse osmosis
11.5. Conclusion
11.6. References

Chapter 12. Biological Removal of Ammonia

12.1. The principle of biological nitrification
12.2. Design parameters
12.2.1. Dissolved oxygen
12.2.2. Filtration rate
12.2.3. NH_4^+ concentration removed as a function of temperature (°C)
12.2.4. Applicable volume load
12.2.5. Contact time
12.2.6. Material height
12.3. Factors limiting oxygen
12.3.1. Mineral carbon
12.3.2. pH
12.3.3. Temperature
12.3.4. Other elements
12.3.5. Biological filter washing
12.4. Implementation
12.4.1. Sand filtration
12.5. Biofilters (Biocarbon® process)
12.6. Water treatment stations
12.6.1. Treatment stations with conventional sand, dual media or GAC filtration
12.6.2. Treatment stations with Biocarbon® filters
12.7. References

Chapter 13. Nitrate Removal

13.1. Biological treatment
13.1.1. Biochemical reactions
13.1.2. Nitrite formation
13.1.3. The bacteriological aspect
13.1.4. Biofilter description (Biodenit® process)
13.1.5. Water treatment stations including biological denitrification
13.1.6. Factors affecting biological denitrification
13.1.7. Design parameters: applied volumic load
13.1.8. Design parameters: minimum contact time (tc min)
13.1.9. Design parameters: height of Biodagen® material (m)
13.1.10. Design parameters: material volume (m^3)
13.1.11. Partial treatment
13.1.12. Sludge production
13.1.13. The reagents
13.1.14. Biological denitrification implementation and exploitation

13.2. Treatment with ion exchange resins
13.2.1. General exchange mechanism (Ecodenit® process)
13.2.2. Ecodenit® process technology
13.2.3. Integration into the water treatment station
13.2.4. Packaged solutions
13.3. Nitrate removal by high pressure membranes
13.4. References

Chapter 14. Removal of Perchlorates

14.1. General aspects
14.2. Main processes for removing perchlorate ions
14.2.1. Ion exchange resins
14.2.2. Nanofiltration membranes
14.3. Conclusions regarding the removal of perchlorates
14.4. References

Chapter 15. Water Softening

15.1. Water hardness
15.2. Alkalinity
15.3. Langelier index (LI or LSI)
15.4. Drinking water hardness goals
15.5. General principles of water softening
15.5.1. The main chemical reactions with lime and soda
15.6. Water softening chemical processes
15.6.1. Limitations of the process and empirical considerations
15.7. Veolia water softening technologies
15.7.1. Clariflocculator
15.7.2. Actiflo® softening
15.7.3. Multiflo® softening
15.7.4. Catalytic water softening
15.8. Saphira® process
15.9. Water softening using high-pressure membranes
15.10. Water softening using ion exchange resins
15.10.1. Resin resistance: use precautions
15.10.2. Ca^{2+} and Mg^{2+} removal
15.10.3. Bicarbonate removal
15.10.4. Resin operating capacity
15.10.5. Operating parameters

15.11. Comparison between the four water softening solutions discussed
15.11.1. Advantages and limitations of the different solutions
15.12. References

Chapter 16. Metal Removal

16.1. Iron and manganese removal: general aspects
16.1.1. Presence of iron and manganese at the resource level
16.1.2. Presence of iron and manganese at the production level
16.1.3. Regulatory aspects of iron and manganese
16.1.4. Iron and manganese treatments
16.2. Arsenic removal
16.2.1. Arsenic chemistry
16.2.2. Arsenic treatments
16.3. Removal of selenium (Se)
16.3.1. The chemistry of selenium
16.3.2. Selenium removal treatments
16.4. Nickel removal
16.4.1. The chemistry of nickel
16.4.2. Nickel removal treatment
16.5. References

Index

Summary of Volume 5

Chapter 20. Calco-carbonic Equilibrium, Correction of Aggressivity and Remineralization

20.1. Characteristics of water leading to calco-carbonic equilibrium
 20.1.1. Chemical equilibria
 20.1.2. Aggressive water
 20.1.3. Scaling water
 20.1.4. Corrosive water
20.2. The equilibrium reactions of water's constituents
 20.2.1. Equilibrium pH
 20.2.2. Langelier equation
20.3. Hallopeau–Dubin diagram
20.4. Indicative criteria to determine the aggressivity or corrosivity of water
 20.4.1. Indicators of aggressivity: concrete pipelines
 20.4.2. Corrosivity indicators
20.5. The calco-carbonic equilibrium of water
 20.5.1. Water quality and regulations
 20.5.2. The correction of aggressivity
 20.5.3. Aggressivity correction treatments
20.6. Remineralization treatments
 20.6.1. Graphic method
 20.6.2. Processes for implementing remineralization: chemical reactions in tanks
20.7. Characteristics of the various reagents used
 20.7.1. Lime
 20.7.2. "Micronized" lime
 20.7.3. Caustic soda

20.7.4. Caustic soda at 50%
20.7.5. Caustic soda flakes or grains
20.7.6. Sodium carbonate
20.7.7. Sodium bicarbonate
20.7.8. Calcium carbonate
20.7.9. Acticalmag limestone
20.7.10. Magno
20.7.11. Calcium sulfate
20.7.12. Calcium chloride
20.7.13. Carbon dioxide
20.7.14. Sulfuric acid (90–98%)
20.7.15. Hydrochloric acid
20.8. References

Chapter 21. Disinfection

21.1. Microorganisms present in the water
21.1.1. Bacteria
21.1.2. Indicator microorganisms or test germs
21.1.3. Viruses
21.1.4. Parasites
21.1.5. Micro-algae
21.2. Quality of potable water
21.2.1. French regulations
21.3. General rules of chemical disinfection
21.3.1. Disinfection mechanisms
21.3.2. The mode of action of chemical disinfectants
21.3.3. Inactivation kinetics
21.3.4. The notion of Ct
21.4. Factors affecting the efficiency of chemical disinfection
21.4.1. Contact time
21.4.2. Turbidity
21.4.3. Presence of oxidizable matter
21.4.4. pH
21.4.5. Injection mode and injection point
21.4.6. Design of the contact tank
21.5. Qualities of a good disinfectant
21.6. Chlorine disinfection
21.6.1. Gaseous chlorine
21.6.2. Hypochlorite
21.6.3. Chlorine application points
21.6.4. Oxidant demand

21.6.5. The implementation of chlorination
21.6.6. Disinfection performances with chlorine
21.6.7. Synthesis of chlorine reactions
21.7. Calcium hypochlorite
21.8. Chlorine dioxide disinfection
21.8.1. Chlorine dioxide preparation
21.8.2. Chlorine dioxide performances
21.8.3. Dechlorination
21.8.4. The advantages of dioxide over chlorine
21.8.5. The special case of the use of chlorine dioxide at a station equipped with ozone
21.8.6. Advantages and drawbacks of using chlorine dioxide
21.9. Chloramination
21.9.1. Principle
21.9.2. Implementation
21.9.3. Performances of monochloramine
21.9.4. Dechloramination
21.9.5. Advantages and drawbacks of chloramination
21.10. Proportion of chlorine in chlorine disinfectants
21.11. Disinfection with ozone
21.11.1. General remarks on ozone
21.11.2. Ozone production
21.11.3. Ozone demand
21.11.4. The implementation of ozonation
21.11.5. Performances of ozone disinfection
21.11.6. De-ozonation
21.12. Criteria for choosing a chemical disinfection technique
21.12.1. Practical implementation of chemical disinfection
21.12.2. Comparative efficiency of the main techniques
21.13. Another chemical disinfectant used: bromine (Br_2)
21.14. Disinfection by ultraviolet radiation
21.14.1. General remarks on UV radiation
21.14.2. Inactivation mechanisms
21.14.3. Lethal dose and inactivation kinetics
21.14.4. Implementation
21.14.5. UV disinfection design parameters
21.14.6. Factors affecting the efficiency of a UV treatment
21.14.7. UV radiation performances
21.14.8. Photoreactivation
21.14.9. Advantages and drawbacks of UV disinfection
21.14.10. Conclusions on UV disinfection
21.15. Comparative criteria between the various chemical disinfectants
21.16. References

Chapter 22. Disinfection By-products

22.1. General aspects
22.2. Reaction by-products
22.3. Formation and evolution of chlorination by-products
22.4. Kinetics and formation mechanisms
22.4.1. Formation kinetics
22.4.2. Mechanisms
22.4.3. Chlorination of HS
22.4.4. Chlorination of carboxylic acids
22.4.5. Factors influencing the formation of DBPs
22.5. Regulations
22.6. Predictive models of CBPs
22.7. Removal of THMs and HAAs
22.7.1. Aeration
22.7.2. Activated carbon
22.7.3. Biofiltration
22.7.4. High-pressure membranes
22.8. The case of nitrosamines and NDMA
22.8.1. Nitrosation mechanism with HOCl
22.9. Oxidation by-products related to chlorine dioxide
22.10. Ozonation by-products
22.11. Recommendations
22.12. References

Chapter 23. Sludge Treatment

23.1. Choosing a treatment chain
23.2. Characteristics of drinking water sludge
23.2.1. The quantity of sludge produced
23.2.2. Sludge concentration estimate at different stages of the chain
23.2.3. Sludge quality: physical and chemical properties
23.3. Handling and storage: shovelable and stackable nature
23.4. Different classes of sludge
23.4.1. Hydroxide sludge
23.4.2. Softening sludge
23.4.3. Metal species sludge treatment
23.4.4. Biological sludge
23.4.5. The case of mixed sludge

23.5. Sludge composition depending on the characteristics of raw water
23.5.1. Surface water sludge
23.5.2. Treatment sludge with coagulants (Fe or Al)
23.5.3. Borehole sludge
23.6. Thickening of drinking water sludge
23.6.1. Function and criteria for choosing a thickener
23.6.2. Thickener design
23.6.3. Implementation of thickeners
23.6.4. Flotation
23.7. Drinking water sludge dewatering
23.7.1. Plate filter
23.7.2. Centrifugation
23.7.3. Belt filters
23.7.4. Filter bags
23.7.5. Drying beds
23.7.6. Sludge lagoon treatment
23.8. Advantages and drawbacks of the different sludge dewatering treatments
23.9. References

Chapter 24. The Treatment Chain: Conception and Design

24.1. The treatment chain
24.2. The definition of a treatment chain
24.3. The stages of a treatment chain
24.4. The renovation of water treatment plants
24.4.1. Adaptation of new goals
24.4.2. The choice of treatment technologies
24.5. References

Chapter 25. The Future of Water

25.1. The major elements of the future of water
25.2. Will there be enough water?

Index

Printed and bound by CPI Group (UK) Ltd, Croydon, CR0 4YY
18/09/2026
14965026-0001